Topics in Applied Physics   Volume 67

Hydrogen in Intermetallic Compounds II

Surface and Dynamic Properties, Applications

Edited by L. Schlapbach

With Contributions by
R. C. Bowman, Jr. N. Gérard R. Hempelmann
I. Jacob M. H. Mintz S. Ono D. C. Richter
G. D. Sandrock L. Schlapbach D. Shaltiel S. Suda
A. Weidinger

With 126 Figures and 25 Tables

Springer-Verlag
Berlin Heidelberg GmbH

Prof. Dr. Louis Schlapbach

Institut de Physique, Université de Fribourg,
Pérolles,
CH-1700 Fribourg, Switzerland

ISBN 978-3-662-31106-6   ISBN 978-3-540-46433-4 (eBook)
DOI 10.1007/978-3-540-46433-4

Typesetting: Thomson Press, New Delhi, India
54/3020-5 4 3 2 1 0 – Printed on acid-free paper

Preface

Interdisciplinary research topics are inevitably associated with both opportunities and risks. These stem, for example, from the different ways of investigating problems and the different expressions used to describe the same phenomenon by scientists from different fields. "Hydrogen in and on Metals" is one such topic. It attracts metallurgists, solid state scientists from both physics and chemistry, mechanical and chemical engineers and energy technology specialists.

Phenomena related to the topic "Hydrogen in Metals" and the physics behind them as they were understood at the end of the 1970s are reviewed in the books *Hydrogen in Metals*, Vol. I and II, edited by G. Alefeld and J. Völkl (Vols. 28 and 29 of Topics in Applied Physics). These books, which deal mainly with hydrogen in elemental metals, are still very valuable and will continue to be important in the coming years.

Since their publication, many new results have been obtained, and have considerably improved both our knowledge and understanding. Much recent work has been stimulated by the extraordinary properties of hydrogen-storing intermetallic compounds and by the constant threat of an energy crisis. The recent progress, which has involved *intermetallic compounds* and alloys rather than elemental metals, has been reviewed in many good articles, either at a technical, purely scientific level, or at a more popular level. Until now, however, the details of this important field have never been collected and presented in the form of a book.

Thus, the aim of the two volumes "Hydrogen in Intermetallic Compounds" is to give a thorough description of the various aspects of the topic in a series of chapters written by specialists in the field, and to review major progress on hydrogen in and on elemental metals. This second volume begins with introductory remarks concerning the latest results, suggestions for future R & D work, safety aspects and experimental techniques and contains a table with the most important physical and chemical properties of hydrogen gas. Then surface properties and activation (Chap. 2), dynamics of hydrogen in long-range diffusion and local modes (Chap. 3), and intrinsic kinetics (Chap. 4) are reviewed. The applications of metal hydrides and the application-oriented material properties are described in Chap. 5. The final chapters, 6 and 7, are dedicated to special experimental techniques.

Volume I concentrated on the preparation and characterization of intermetallics and hydrides, thermodynamic properties, crystal structure, electronic properties, heat of formation models, together with magnetism and supercon-ductivity.

As the interaction of hydrogen with metals and alloys is, and will be, of significant importance for basic research as well as for hydrogen energy tech-nology, fusion, catalysis, getters, electrochemical cells and many other appli-cations, I hope that these two volumes will help many scientists to find the information they are looking for, to spread the fascination which we the authors already share, and to stimulate further work.

Working with hydrogen and its isotopes is sometimes risky, whereby the risk can be of either physical or political origin: Henry Cavendish performed a clever series of experiments on the reaction of acids with metals and in 1766 submitted his results on the evolution of what we call hydrogen gas in the form of three scientific papers on *factitious air* to a scientific journal (Philosophical Transactions). Cavendish was convinced that the evolved factitious air was con-tained in the metals (phlogiston theory). Antoine Laurent Lavoisier, fermier général (tax inspector) and chemist published his revolutionary "Traité élémen-taire de chimie" in 1789. He clearly showed that Cavendish's factitious air originated in the decomposition of water and therefore named it hydrogen. Lavoisier presented his results orally in 1783 at the Royal Academy of Sciences in Paris. He also "went public" and presented the results in a theatre play in which the books on the phlogiston theory were burned. Lavoisier, who had revolutionized chemistry with his modern mind became a victim of the French Revolution: After being condemned to death he asked for a two week stay of execution in order to finish his experiments. "La République n'a pas besoin de Savans, ni de Chymistes" decided the tribunal; thus Lavoisier mounted the scaffold. Recently, outbreaks of cold fusion fever were initiated by a communica-tion in a non-scientific magazine. The idea was revolutionary, but soon showed a lack of reasonable reproducibility. If we were to turn the clock back 200 years, a large part of the scientific community would surely have called for the initiators to mount the scaffold.

I should like to express my thanks to all the authors for their individual contributions and to Drs. H. Lotsch and A. Lahee of Springer-Verlag for co-operation and careful reading of the manuscripts, and to my patient family; continuous support by the Nationaler Energie-Forschungs-Fonds (NEFF) is gratefully acknowledged.

Fribourg, January 1992   Louis Schlapbach

Contents

Contributors

Bowman, Jr., Robert C.
   Aerojet Electronics System Division,
   P.O. Box 296, Azusa, CA 91702, USA

Gérard, Norbert
   Lab. Réactivité des Solides,
   Faculté des Sciences Mirande, Université de Dijon,
   B.P. 138, F-21004 Dijon, France

Hempelmann, Rolf
   Institut für Festkörperforschung,
   Forschungszentrum Jülich GmbH,
   W-5170 Jülich, Fed. Rep. of Germany

Jacob, Isaac
   Department of Nuclear Engineering,
   Ben-Gurion University of the Negev,
   P.O. Box 653, Beer-Sheva, Israel

Mintz, Moshe H.
   Nuclear Research Center-Negev,
   P.O. Box 9001, Beer-Sheva, Israel, and
   Department of Nuclear Engineering,
   Ben-Gurion University of the Negev,
   P.O. Box 653, Beer-Sheva, Israel

Ono, Shuichiro
   National Chemical Laboratory for Industry,
   Tsukuba Research Center,
   1-1 Higashi, Ibaraki 305, Japan

Richter, Dieter
   Institut für Festkörperforschung,
   Forschungszentrum Jülich GmbH,
   W-5170 Jülich, Fed. Rep. of Germany

Sandrock, Gary
    SunaTech Inc.
    113 Kraft Place, Ringwood, NJ 07456, USA

Schlapbach, Louis
    Institut de Physique,
    Université de Fribourg,
    Pérolles, CH-1700 Fribourg, Switzerland

Shaltiel, David
    The Raccah Institute of Physics,
    The Hebrew University Jerusalem, Jerusalem, Israel

Suda, Seijirau
    Chemical Engineering Department,
    Kogakuin University,
    2665-1, Nakano-machi, Hachioji-shi,
    Tokyo 192, Japan

Weidinger, Alois
    Bereich Schwerionenphysik,
    Hahn-Meitner-Institut Berlin GmbH,
    Glienicker Str. 100, W-1000 Berlin 39, Fed. Rep. of Germany

1. Introduction

Louis Schlapbach

With 2 Figures and 1 Table

Volume 1 of "Hydrogen in Intermetallic Compounds" begins with a simple introductory description of the phenomena related to the formation of a metal hydride from molecular hydrogen gas and a host metal matrix. It contains thereafter detailed treatments of thermodynamic properties, crystal structure, electronic properties, heat of formation models, magnetism and superconductivity as well as the preparation and characterization of metal hydrides. In the effects and phenomena described there, one of the most typical aspects of hydrogen in metals was completely neglected: the high mobility of hydrogen and of its isotopes (D, T, μ^+).

A high jump rate (typically 10^{12} jumps per second) together with a low activation energy for hydrogen bulk diffusion (typically $100\,meV$) are the origin of long range diffusion, local modes and metal–metal hydride phase transitions, which—at a given temperature—occur orders of magnitude faster than those of other interstitially dissolved elements such as O, N and C [Ref. 1.1, Fig. 12.1]. At surfaces these differences are much less pronounced; the mobility of hydrogen is still higher than that of O, N and C, but its reaction rate is comparable to that of the other elements.

This volume gives a comprehensive review of the surface properties, dynamics of bulk hydrogen, and intrinsic kinetics, which are explicitly related to the hydrogen mobility. Further chapters are devoted to applications and to shorter descriptions of some special experimental techniques. In order to underline the interdisciplinary character of the field, a more chemical approach is given in Chap. 4 on intrinsic kinetics, and metallurgical and chemical engineering aspects are emphasized in Chap. 5 on applications.

1.1 About the Topics of This Volume

The *surface-related phenomena* refer to well-defined systems such as a few hydrogen molecules and their dissociation, surface diffusion and adsorption on a single crystal surface [1.2] as well as catalytic reaction [1.3] and the process of surface activation and poisoning for hydrogen uptake [1.4], permeation [1.5] and embrittlement [1.6]. The most experimental and theoretical studies on the interaction of hydrogen with metal surfaces today involve elemental d-transition metal surfaces; some concern transition metal alloy surfaces; a few deal with noble metal surfaces, but hardly any concern the reactive simple metals like

Ca, Mg and Al and the important lanthanides and actinides. Surfaces of binary alloys have an additional interesting parameter as compared to those of elemental metals, the surface composition and its variation upon sample treatment. Studies of the surface properties of hydrides, which of course differ in many respects from those of the host metal, are also very scarce. Recent progress in the understanding of surface phenomena of hydrogen-metal systems is a good example of a stimulating interdisciplinary collaboration: surface scientists learnt that adsorbed hydrogen might diffuse underneath the top layer of metal atoms, that subsurface sites can be filled and surface hydrides form, and "bulk hydriders" accepted that surface science explains phenomena like activation and poisoning of hydride forming alloys and the inhibition of embrittlement.

The *dynamics of bulk hydrogen* are studied primarily to understand the usually rapid long distance diffusion and local modes at various temperatures. Experimental methods establishing a nonequilibrium distribution and measuring the time to reach equilibrium (e.g. by Gorsky effect) and methods like quasielastic neutron scattering and nuclear magnetic resonance, which allow measurements at equilibrium hydrogen distribution, are most commonly used (Chap. 3). Perturbed angular correlation (Chap. 6) is developing into a powerful method, particularly for low-temperature studies of very dilute systems. The analysis of the results provides detailed information on hydrogen sites, site-dependent activation energies, trapping, blocking and tunneling as well as on the character of the hydrogen potential. It has become evident in recent years that the dynamics of hydrogen in metals clearly shows quantum mechanical features (Sect. 1.2). Coherent and incoherent tunneling has been observed far below $10\,\mathrm{K}$ [1.7, 8]. Ordered and disordered binary alloys again offer the unique possibility of looking at a more complex variety of sites and site distributions. The Hall-type voltage induced by hydrogen diffusion [1.9] and electromigration and thermotransport of hydrogen [1.10] are further interesting diffusion-related phenomena.

The overall *kinetics* of hydrogen absorption and desorption is determined [1.11] by the slowest of the series of steps: transport of molecular hydrogen to the surface, dissociation and surface migration, transition from surface to subsurface and bulk sites, diffusion (across host metal or hydride phase) and nucleation and growth of the hydride phase. Chapter 4 is devoted to a more chemical description of the intrinsic kinetics. It is clearly seen that the effect of superposition of several steps makes unequivocal interpretations difficult and renders comparisons of results of different authors (always measured under different conditions) virtually impossible. Two very different nucleation and growth mechanisms are elaborated: In saturated $LaNi_5H_x$ solid solution nucleation and growth begin at many sites distributed throughout the bulk and further diffusion occurs across the saturated intermetallic compound. In Mg, however, nucleation begins near the surface and the hydride phase grows from the surface into the bulk. Diffusion occurs across the hydride phase.

The field of potential *applications of metal hydrides* has developed enormously over the last ten years and many prototype units have been built

(Chap. 5). Their commercial use, however, is still restricted to a few cases. The enthusiam of the "post-energy-crisis" years for a near future "hydrogen economy", which was a strong driving force for R & D in the use of hydrides to store the fuel hydrogen, deflated together with the oil price. Nevertheless the pioneering Daimler–Benz program to develop metal-hydride storage units and incorporate them into cars and the very successful test period of a fleet of five vans and five personal cars over a distance of more than 200 000 km proved the technical feasibility, reliability, safety and low pollution of hydrogen-fuelled cars with metal hydride storage tanks [1.12]. Growing environmental risks will force continued development of cleaner energy distribution and transformation systems. Reviews of national activities can be found in [1.13].

The use of metal hydrides today is concentrated on closed-system thermal machines and electrochemical cells. The variation of pressure and temperature and the heat of reaction upon the absorption and desorption of hydrogen by one or several hydrogen storage alloys is used to build compressors, heat pumps refrigerators and actuators. A metal-hydride-filled device, which converts a temperature rise into a pressure rise, is widely (and economically) used as a very reliable fire detector in aircrafts.

The potential of metal hydrides in these various applications had strong feedback on the studies of hydride properties, particularly the thermal conductivity of metal hydrides and hydride beds, hysteresis, degradation, and poisoning effects. Unfortunately, the feedback on the development of new materials was smaller. Apart from the progress in vanadium-based ternary solid solution alloys, which show very interesting thermal and kinetic properties [1.14], materials engineering by substitutions was generally the approach for finding alloys for specific applications.

Many *experimental techniques* were described in Vol. I in the chapters on preparation, crystal structure, electronic and magnetic properties, surfaces, dynamics and kinetics. Additional experimental methods are described in the last two chapters of this volume. They are methods that are relatively new in the context of hydrogen in and on metals, namely:

— perturbed angular correlation (Chap. 6), which measures the electric field gradient and the internal magnetic field at the site of a local probe, and yields information on trapping and tunneling as well as on phase diagrams.
— time-of-flight analysis of direct recoils (Chap. 7), which is sensitive to composition, chemistry and structure of the surface and detects surface hydrogen directly
— nuclear resonant scattering of gamma rays, which probes bonding strengths (Chap. 7)
— thermal desorption spectroscopy of bulk and surface hydrogen (Chap. 7).

Some important phenomena that are not treated in detail in these volumes are the following:

Hydrogen embrittlement still is a significant technical problem. Its understanding involves the description of decohesion and hydrogen-defect interaction in the bulk of single phase grains as well as at the grain boundaries, nucleation and growth of hydride phases and various surface phenomena. Decohesion and inhibition are mentioned in Chap. 5 of Vol. I and Chap. 2 of the present volume. Further references can be found in [1.6, 15–19].

Effects of *tritium* and of the naturally related *helium in metals* were reviewed in [1.20–22].

A few results from studies of *hydrogen in liquid alloys* can be found in [1.16, 18].

1.2 Major Recent Developments and Outlook

The exact calculation of the total energy of a hydrogen–metal system from the positions and potentials of hydrogen and host metal atoms—an idealistic fundamental approach to the hydrogen-metal interaction—is generally an impossible task. Reasonable simplifications, small and large scale computations, and much experimental effort have brought us significantly closer to a coherent description of the hydrogen–metal interaction.

Solid metallic hydrogen: Condensed H_2 forms an electrically insulating molecular solid at normal pressure below 14 K. On the other hand condensed H with appropriately small H–H distance is theoretically expected to be the simplest metal, one which could even show superconductivity. The transformation of solid molecular H_2 into a metal is a fundamental problem in condensed matter and astrophysics [1.23–27]. Recent calculations [1.27], indicate that the insulator–metal transition may occur in the 250–400 GPa range. A superconducting transition temperature of 230 ± 85 K was estimated using ab initio calculations of the electron–phonon coupling in a distorted hexagonal high pressure phase of hydrogen. Direct optical observation [1.26] of solid hydrogen in the 250 GPa range at 77 K shows opaque hydrogen consistent with a band-overlap mechanism of metallization.

Metallization of hydrogen dissolved in a solid solution metal phase or in a metal hydride, i.e. the formation of a metallic hydrogen sublattice, is another approach to obtaining metallic hydrogen and thus possibly superconductivity. *Fukai* comprehensively details the range of composition of metal–hydrogen systems from pure metals to metal hydrides and finally pure hydrogen [1.28]. A universal compression behaviour of hydrogen in a metallic environment was established. Two-dimensional metallic hydrogen has been reported recently based on the interpretation of proton NMR measurements on hydrogen-potassium-graphite intercalation compounds [1.29].

Order-disorder effects, strain fields, amorphization: Amorphization of bulk samples by hydrogen absorption has been demonstrated for several intermetallic compounds including Zr_3Rh, $SmNi_2$ and $GdCo_2$ [1.30]. Recently amorphization of the more stable cubic Laves phase (C14) compound $GdFe_2$ was achieved

[1.31]. Strain in the structurally most perfect superlattice Nb/Ta can be tuned by dissolution of hydrogen. This effect has been used to study the physics of a lattice gas in the presence of a periodic perturbation potential [1.32]. A specific heat anomaly related to a premelting effect was observed during vitrification of ErFe$_2$ hydride [1.33].

Hydrogen as a probe: By analyzing the diffusional properties of hydrogen, strain fields of dislocations and vacancies and grain boundaries can be probed [1.34]. The number of Zr$_4$, Zr$_3$Ni, Zr$_2$Ni$_2$ tetrahedral sites filled by hydrogen in amorphous Zr–Ni alloys has been determined from the diffusion coefficient and concentration dependence of the chemical potential [1.35]. The usefulness of hydrogen as a local probe of the chemical and geometrical structure, surface tension and subsurface sites on nanocrystals has been demonstrated. It was shown that pressure–composition isotherms allow far more accurate measurements of the volume derivative of the chemical potential than does X-ray diffraction [1.36].

Dynamics: Enormous progress has resulted from recent experimental and theoretical studies of hydrogen dynamics, performed mostly on elemental host metals. Diffusion properties are found to vary considerably with temperature. At the lowest temperatures coherent tunneling dominates. Bloch states, bandwidth and electron–hole pair excitations determine the motion [1.37]. Diffusion constants much larger than those expected from the extrapolation of 100–300 K values are the result. Tunneling has been studied in detail on the model system Nb(OH)$_x$ by neutron spectroscopy from 0.5–10 K [1.7, 38] and in Ta by perturbed angular correlation [1.39]; see also Chap. 6. The concentration and temperature dependences of the tunnel splitting of the vibrational ground state of hydrogen have been analyzed. Tunneling at low temperatures has been observed not only for hydrogen in bcc metals, but also for chemisorbed hydrogen [1.37] and for the positive muon in fcc metals [1.37, 40]. Quantum tunneling has also been detected for HD molecules in molecular solid hydrogen below 1 K [1.41].

As the temperature is increased, incoherent tunneling, assisted by thermally activated phonons (small polarons), becomes possible [1.42]. The lifetime of the Bloch states decreases. The motion becomes completely diffusive, with a diffusion constant $D = W^2\tau a^2$ when the coherence time τ is shorter than the average time between jumps (renormalized bandwidth W, lattice constant a) [1.43]. At even higher temperatures the frequently observed classical "over the barrier jump" diffusion dominates and at still higher temperature the motion of hydrogen is liquid-like. The hydrogen motion is strongly affected by lattice defects at all temperatures.

The earlier reported stress-induced superdiffusion of hydrogen in V single crystals [1.44], which was tentatively described as hydrogen tunneling at room temperature, could not be confirmed. A reinvestigation by means of electromigration [1.45] and neutron spectroscopy [1.46] revealed no stress-induced diffusional changes.

Measurements of the local vibrational properties of hydrogen in solid solution and hydride phases by means of neutron vibrational spectroscopy yield new information on the strength and assymmetry of the hydrogen potential in these systems [1.47] and permit e.g. the determination of the anharmonicity parameter of the potential. Spectacularly well-resolved hydrogen vibrational spectra were measured by *Hempelmann* et al. [1.48] on single crystalline β-V_2H. The direction of the fundamental excitations were determined, and overtones of these vibrations were observed up to the fourteenth order (Fig. 1.1). The H potential derived up to 1.4 eV is shown in Fig. 1.2. Sample quality and experimental resolution have attained a level which will allow detailed diffusion studies

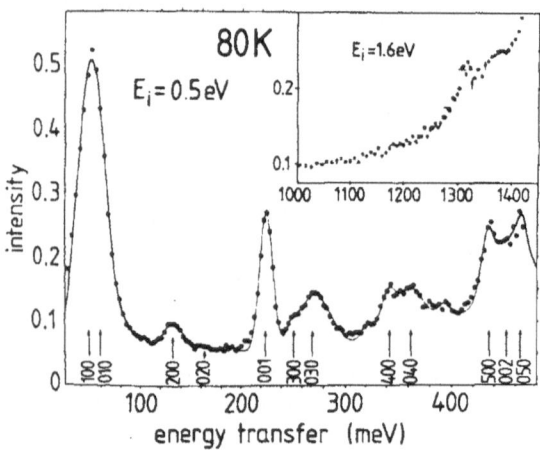

Fig. 1.1. Neutron spectra from single crystalline $\beta - V_2H$ at different incident neutron energies E_i. The indices *lmn* are the assignments of the different transitions [1.48]

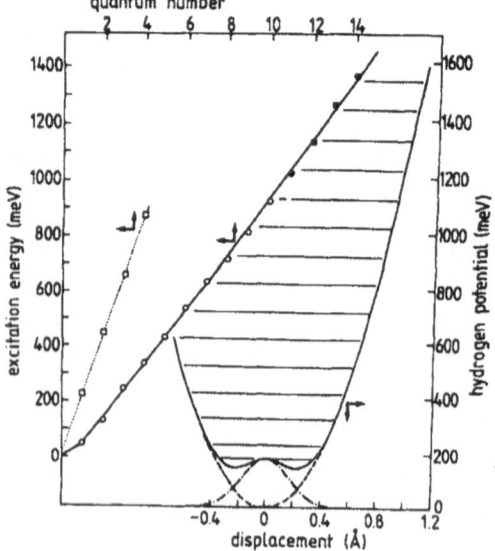

Fig. 1.2. Left-hand side: Hydrogen excitation energies in the stiff (001) direction (squares) and in the soft (100) direction of $\beta - V_2H$ [multidomain crystal, open circles; single-domain crystal, filled circle]. Right-hand side: Resulting H potential in the soft (100) direction and the corresponding energy levels [1.48]

over order–disorder transitions. For ZrH_2, harmonic hydrogen wavefunctions were found for the first, second and third excited states [1.49].

Rare-earth hcp metals such as Y, Sc, Er, Tm and Lu are preferentially studied materials because of their ability to retain hydrogen in solid solution up to relatively high concentrations. The c-axis vibrations of hydrogen in Y show an unusual splitting caused by hydrogen pairs dynamically coupled across metal atoms [1.50, 51]. Furthermore the rare earths allow hydrogen diffusion studies up to high temperature due to the strong hydrogen–metal bonding. The spin lattice relaxation time of ScH_2 shows a first minimum at 500 K, a maximum at 800 K and again a minimum at 1100 K, possibly related to H–H repulsion at the saddle point and to long-lived fluctuations in the local environment of the Sc nuclei [1.52].

Some experimental results on hydrogen and tritium diffusion and solubility in Zr, Hf, Mn, and Ti are summarized in [1.18]; recent theoretical approaches (small polaron theory, path integral method) are given in [1.53].

A new and relatively simple model, which incorporates a site energy distribution, site blocking, and effective interstitial–interstitial interactions describes quantitatively the variation of the diffusion coefficient of hydrogen (and other interstitials) in disordered Nb-V alloys as a function of alloy composition and interstitial concentration [1.54]. From pressure–composition isotherms and diffusion data information on site energies, site density and saddle point energies has been derived. Unscreened proton behaviour was found for H electromigration in Nb-V alloys [1.55].

Surfaces: Some features that emerged from recent experimental and theoretical studies of well-defined surfaces are [1.56, 57]: for open surfaces of some transition metals the dissociation probability of an incident H_2 molecule of thermal energy is essentially unity and, within a wide range, it is almost independent of incident energy and substrate temperature. For other substrates this probability is low for low incident energy and increases somewhat at higher energy. The former behaviour, which is observed on transition metals, displays little or no entrance channel barrier for dissociative adsorption and generally has large binding energies. The latter behaviour corresponds to high entrance channel barriers and small or negative binding energies and is observed for some noble-metal and simple-metal surfaces. Some interesting exceptions to this general behaviour are known, but not well investigated: H_2 on Mg has an extremely low dissociation probability though the H–Mg bond is quite strong. Careful studies of the reaction of H_2 with well-defined Mg surfaces up to high incident energies are needed, firstly to provide an understanding of the phenomena, and secondly to aid the development of new light weight hydrides.

In order to study the hydrogen uptake of Mg in the absence of the dissociation limit, *Krozer* and *Kasemo* [1.58] modified the surface of Mg films by the deposition of a thin Pd layer. The results indicate the formation of a hydride at the Pd–Mg interface which further limits uptake due to the slow hydrogen diffusion across the hydride.

As a consequence of the activation energy barrier for dissociative adsorption, H_2 molecules desorbing from hot surfaces (no tunneling) can have a mean translational energy far in excess of the surface temperature. Systematic investigations confirmed the important role of surface defects and surface impurities in dissociative chemisorption. The bound state energy levels for H_2 and D_2 physisorption on Cu(100) were evaluated at 80 K by the analysis of diffractive adsorption scattering resonances [1.57].

Hydrogen-sensitive Pd–MOS structures [1.59, 60] might become important tools for the study of such surface and interfacial problems and as sensors.

A low temperature surface transition from semiconducting to metallic behaviour was recently found on rare-earth hydrides [1.61]. Octahedral-like surface hydrogen was assumed to diffuse into the bulk leading to a metallic surface dihydride. Quantum phenomena could well be observable in this two-dimensional electron gas.

Experimental techniques: Microtechniques and microprocessors enter in many forms into the experimental equipment used to study hydrogen–metal systems. A very small high pressure diamond anvil cell was developed for the synthesis of metal hydrides and for in situ resistivity measurements at pressures of 100 kbar and down to liquid He temperature [1.62]. Automated equipment for the analysis of thermochemical properties and hysteresis has been described by *Dantzer* and *Marcelet* [1.63]. The characterization of metal–hydrogen systems by a thermodynamic evaluation of phase diagrams and electrochemical measurements was proposed by *Lüdeckl* et al. [1.64].

Hydrogen in permanent magnets and superconductors metallization: In recent years there have been enormous advances in the performance of rare-earth–transition-metal permanent magnets. The maximum energy product $(BH)_{max}$ increases from $SmCo_5$-type magnets through $SmCo_{17}$-type to those based on $Nd_2Fe_{14}B$. They all experience a considerable effect on the magnetic properties upon hydrogen absorption (cf. Chap. 7, Vol. I). The technologically important effect is hydrogen decrepitation. The hydrogen-decrepitated powder can be used directly in magnet production [1.61]; see also Sect. 5.3.13.

The non-stoichiometric oxide superconductors of the YBaCuO type also absorb some hydrogen. The maximum hydrogen content reached so far was $H_{6.8}YBa_2Cu_3O_7$ [1.66]. Whereas annealing at high hydrogen pressure often destroys superconductivity, a progressive small increase of T_c was observed in some superconductors upon hydrogen absorption [1.66, 67]. Upon hydrogen implantation into superconducting YBaCuO films a rise of T_c was observed possibly induced by strain effects between film and film substate [1.68].

The new oxide superconductors rekindled interest in the old question of high-temperature superconductivity in metallic hydrogen. Instead of the compression of hydrogen by very high external pressure, the hydrogen density theoretically needed for superconductivity could be reached in metal hydrides. Band-structure calculations have been performed for $LiBeH_3$ [1.69, 70] and MgH_2 [1.71], but unfortunately these hydrides turned out to be nonmetallic as pure compounds.

New materials, improvements of material properties and preparation: The effort invested in analyzing the fundamental properties of hydrogen–metal systems and in R & D on the application of metal hydrides has been concentrated to a large extent on transition and rare-earth metals and their compounds. In order to reach a breakthrough in the applications, new lightweight hydrides of alloys of cheap and abundant components like Mg, Al, and Ca have to be developed. The H–Mg bond in MgH_2, for example, needs to be weakened, possibly by a transition to metallic behaviour. Some promising work has begun, including the preparation of Na_2PdH_2, K_2PdH_4, and Li_2PdH_2 [1.72], amorphous $Mg_2PdC_yH_{1-1.4}$ [1.73], and metallic $Ca_3Pd_2H_7$ [1.74]. The development and properties of Mg_2FeH_6 are described in Chap. 4 of Vol. I.

The precipitation of hydrides from organometallic solutions is developing into a novel and interesting method for the preparation of highly reactive compounds [1.73, 75].

The development of disintegration-resistant $LaNi_5$-Ni eutectic alloys prepared by unidirectional solidification has also been reported [1.76].

Applications and application-oriented properties: Hydrides of binary and ternary intermetallics and alloys are thermodynamically unstable and tend to disproportionate into a stable binary hydride and further compounds. Only recently was it shown that reproportionation occurs even at moderate temperature (100–200°C). By the appropriate choice of the cycling parameters, particularly by allowing reproportionation between two subsequent cycles, major disproportionation can be suppressed [1.77].

Much progress has been made in the area of gas purification systems: By flowing impure noble gas over reactive hydride-forming intermetallics of high surface area gas purification down to a 100 ppb impurity level can be attained [1.78] (cf. Chap. 5). The application of metal hydrides in thermal machines is described in detail in Chap. 5.

A major commercial breakthrough is expected for rechargeable batteries with metal hydride electrodes (Chap. 5 and [1.79, 80]).

The 1989 cold fusion fever raised the question of whether two or more deuterium nuclei (d) at the surface, in the near surface region or in the bulk of a metal deuteride could come closer than 0.35 Å to raise the d–d or possibly d–p fusion rate. Theoretical arguments about a suppression of the repulsive Coulomb barrier or an enhancement of the fusion rate were discussed [1.81] and the possible occurrence of fusion products like heat, radiation and particles was checked by a variety of more-or-less sophisticated experimental setups. Experimentalists reporting the observation of cold fusion did not succeed with a reasonable reproducibility. Many experimentalists reported the detection of particles or radiation; only very few reported the observation of heat, and very many observed no particles, radiation or heat. For a summary see [1.82, 83]. A weak phenomenon, possibly new, possibly hot fusion in strong electric fields over short distances, does seem to exist, although it is probably too weak to be of interest in the energy business. It was shown that, after electrochemical charging in ordinary D_2O, the electrode surfaces are covered by monolayers of heavy metal impurities (Chap. 2).

1.3 Physical and Chemical Properties
of Hydrogen; Safety Aspects

Some physical and chemical properties of atomic, gaseous, liquid and solid hydrogen are given in Table 1.1. More engineering design data can be found in [1.88].

Hydrogen is a nontoxic but highly inflammable gas. Compared to other inflammable gases (e.g. methane, propane) its range of inflammability in air is much broader (4–74.5%), but it also vapourizes much more easily. Handling of hydrogen in the form of metal hydrides enhances safety enormously (Chap. 5). Experiments with gaseous and liquid hydrogen should be performed in rooms with good ventilation. Simple and efficient safety arrangements consist in a hydrogen detector on the ceiling and an open window.

For the construction of hydrogen reactor vessels and other equipment, the following materials selection criteria should be considered:

Table 1.1. Physical and chemical properties of atomic, gaseous, liquid and solid hydrogen [1.84–90]

Atomic hydrogen	
Atomic weight (natural)	1.0079
Atomic weight (H^1)	1.00782
Proton rest mass	$1.672623 \cdot 10^{-27}$ kg
	$= 1836.1527\, m_e$
Proton spin	1/2
Proton magnetic moment	$2.7928474\, \mu_N$
	$= 1.4106071 \cdot 10^{-26}$ JT^{-1}
Natural abundance H^1	99.985%
Ionization potential	15.427 eV
Gaseous hydrogen (H_2)	
Para-hydrogen	nuclear spins ↑↓
Ortho-hydrogen	nuclear spins ↑↑
Natural abundance, ratio para:ortho at	
300 K, 1 bar	25:75
below 20 K	100:0
Density (STP)	$0.090\, \mathrm{kg\,m^{-3}}$
Specific heat c_p (273 K)	$14.15\, \mathrm{Jg^{-1}K^{-1}}$
Ratio of specific heats c_p/c_v (273 K)	1.41
Viscosity (273 K)	$0.84 \cdot 10^{-5}\, \mathrm{N\,s\,m^{-2}}$
Thermal conductivity NTP	0.194 W/mK
Inflammability limits in air, 20°C, 1 bar	4–74.5%
Minimum autoignition temperature, 1 bar	570°C
Diffusion coefficient in air (STP)	$0.61\, \mathrm{cm^2/s}$
Heat of combustion (lower, H_2O vapour)	120 000 kJ/kg
	33.33 kWh/kg
(upper, H_2O liquid)	141890 kJ/kg
	39.41 kWh/kg
Solid–liquid–gaseous hydrogen (H_2)	
Boiling point (1 atm)	20.38 K
Melting point	13.96 K
Triple point	13.95 K, $0.0^{-}2$ bar
Critical point	12.98 bar; 33.24 K

Room temperature: most elastomers are compatible with hydrogen [1.84]. Hydrogen is noncorrosive and can therefore be employed with all commonly used non-reactive metals at low pressure. However, at high pressure hydrogen causes embrittlement, particularly of cold-worked ferritic steels.

High temperature: to avoid embrittlement and carbon grain-boundary segregation, austenitic stainless steels of the quality AISI 316 and 304, and above 300 °C 316 L and 304 L, have to be used. Alumina and silicon carbide and nitride are suitable refractory materials.

Low temperature: para-hydrogen, the stable low-temperature form, requires the use of metals which retain good ductility at low temperatures. Austenitic stainless steels of the quality AISI 304 and 304 L or aluminium and aluminium alloys (series 5000) are recommended. Teflon$^{®}$ and Kel-F$^{®}$ also behave satisfactorily.

Safety aspects of hydrogen technology are reviewed in [1.90].

The thermodynamic properties of hydrogen between 100 K and 1000 K at pressures up to 1 Mbar, as derived from an equation of state, are described in [1.91]. The theory of the properties of solid H_2, HD, and D_2 has been reviewed by *Kranendonk* [1.92]. The equation of state of solid molecular H_2 and D_2 at 5 K up to 37 GPa has been measured by *van Straaten* and *Silvera* [1.93].

References

1.1   J. Völkl, G. Alefeld: In *Hydrogen in Metals I*, Topics Appl. Phys. Vol 28, ed. by G. Alefeld, J. Völkl (Springer, Berlin, Heidelberg 1970) p. 321
1.2   K. Christmann: Surf. Science Rep. 9, 1 (1988)
1.3   see e.g. J. H. Sinfelt: Catalysis by Metals, J. Phys. Chem. 90, 4711 (1986); *Bimetallic Catalysts* (Wiley, New York, 1983) p. 86; A.A. Elattar, W.E. Wallace: in *The Rare Earth in Modern Science and Technology*, ed. by G.J. McCarthy, J.J. Rhyne, H.B. Silber (Plenum, New York 1980) p. 533
1.4   L. Schlapbach: In *Hydrogen in Disordered and Amorphous Solids*, ed. by G. Bambakidis, R.C. Bowman, NATO ASI B 136, 397 (1986); B.J. Berkowitz, J.J. Burton, C.R. Helms, R.S. Polizotti: Scripta Metall. 10, 871 (1976)
1.5   T.N. Kompaniets, A.A. Kurdyumov: Prog. Surf. Science 17, 75 (1984)
1.6   see e.g. M.F. Ashby, J.P. Hirth (eds.): *Perspectives in Hydrogen in Metals* (Pergamon, Oxford 1986); H.K. Birnbaum: J. Less-Common Met. 104, 31 (1984), H.J. Flitt, J.O'M. Bockris: Int. J. Hydrogen Energy 6, 119 (1981)
1.7   A. Magerl, A.J. Dianoux, H. Wipf, K. Neumair, I.S. Anderson: Phys. Rev. Lett. 56, 159 (1986); H. Wipf, K. Neumair: Phys. Rev. Lett. 52, 1308 (1984); H. Wipf, K. Neumair, A. Magerl: Z. Phys. Chem. N.F. 145, 237 (1984)
1.8   Y. Fukai: Japan J. Appl. Phys. 23, L596 (1984)
1.9   A.H. Verbruggen, R. Griessen, J.H. Rector: Phys. Rev. Lett. 52, 1625 (1984)
1.10 A.H. Verbruggen, R. Griessen, D.G.de Groot: J. Phys. F 16, 557 (1986); H. Wipf, B. Herth, A. Bauberger: Z. Phys. Chem. NF 143, 205 (1985)
1.11 M.H. Mintz, J. Bloch: Prog. Solid State Chem. 16, 163 (1985)
1.12 J. Töpler, K. Feucht: Proc. Int. Symp. on Metal-Hydrogen Systems, Stuttgart 1988, Z. Phys. Chem. NF 164, 1451 (1989)
1.13 T.N. Veziroglu, A.N. Protsenko (eds.): *Hydrogen Energy Progress* (Pergamon, New York 1989) Chap. A, Vol. 1
1.14 G.G. Libowitz, A.J. Maeland: J. Less-Common Met. 131, 275 (1987)
1.15 G.M. Pressouyre: Acta Metall. 28, 895 (1980), Metall. Trans. 14A, 2189 (1983); A.J. Markworth: Mater. Lett. 2, 333 (1984)

12   *Louis Schlapbach*

1.16 Int. Conf. on "Effect of Hydrogen on Behaviour of Materials" organized and published regularly by the Metallurgical Society AIME, USA
1.17 H.J. Cialone, J.H. Holbrook: Metall. Trans. **16A**, 115 (1985); J. Richter, P. Deimel: "Hydrogen Embrittlement of Pipeline Steel", IEA-Report (1984), unpublished. Available from Staatl. Materialprüfungsanstalt Stuttgart, FRG
1.18 W. Dahl: Wasserstoff in Metallen, Ergebnisse eines Schwerpunktprogramms 1977–1985, Deutsche Forschungsgemeinschaft 1986; D. Behrens: Wasserstofftechnologie, DECHEMA, Frankfurt, 1986
1.19 T. McMullen, M.J. Stott, E. Zaremba: Phys. Rev. B **35**, 1076 (1987)
1.20 R. Lässer: *Tritium and Helium-3 in Metals*, Springer Ser. Mater. Sci. Vol 9, (Springer, Berlin, Heidelberg 1989); Z. Phys. Chem. NF **143**, 23 (1985)
1.21 T. Schober: In *Hydrogen in Disordered and Amorphous Solids*, ed. by G. Bambakidis, R.C. Bowman, NATO ASI B **136**, 377 (1986); K.S. Forcey, D.K. Ross, L.G. Earwaker: Z. Phys. Chem. NF **143**, 213 (1985)
1.22 A.A. Lucas: Physica **127B**, 225 (1984)
1.23 E. Wigner, H.B. Huntington: J. Chem. Phys. **3**, 764 (1935)
1.24 N.W. Ashcroft: Phys. Rev. Lett. **21**, 1748 (1968); Nature **340**, 345 (1989)
1.25 T. Schneider, E. Stoll: Physica **55**, 702 (1971)
1.26 H.K. Mao, R.J. Hemley: Science **244**, 1462 (1989); Phys. Rev. Lett. **63**, 1393 (1989)
1.27 T.W. Barbee III, A. Garcia, M.L. Cohen, J.L. Martins: Phys. Rev. Lett. **62**, 1150 (1989); Nature **340**, 369 (1989)
1.28 Y. Fukai: Bull. Am. Phys. Soc. **35**, 580 (1990); J. Less-Common Met. (1991) in press
1.29 S. Miyajima, M. Kabasawa, T. Chiba, T. Enoki, Y. Maruyama, H. Inokuchi: Phys. Rev. Lett. **64**, 319 (1990); S. Mizuno, K. Nakano: Phys. Rev. B **40**, 577 (1989)
1.30 K. Samwer: Phys. Rep. **161**, 1 (1988); J. Less-Common Met. **140**, 25 (1988)
1.31 K. Aoki, M. Nagano, A. Yanagitani, T. Matsumoto: J. Appl. Phys. **62**, 3314 (1987); A. Yanagitani, K. Aoki, T. Matsumoto: Sci. Report Tôhoku University A **34**, 195 (1989)
1.32 H. Zabel, P.F. Micelli: Z. Phys. Chem. NF **163**, 3 (1989); P.F. Miceli, H. Zabel: Z. Phys. B **74**, 457 (1989)
1.33 H.J. Fecht, Z. Fu, W.L. Johnson: Phys. Rev. Lett. **64**, 1753 (1990)
1.34 R. Kirchheim, X.Y. Huang: Z. Phys. Chem. NF **164**, 821 (1989)
1.35 F. Jaggy, W. Kieninger, R. Kirchheim: Z. Phys. Chem. NF **163**, 431 (1989)
1.36 E. Salomons, R. Griessen, D.G. de Groot, A. Magerl: Europhys. Lett. **5**, 449 (1988); R. Feenstra, R. Brower, R. Griessen: Europhys. Lett. **7**, 425 (1988)
1.37 K.W. Jacobsen, J.K. Nørskov, M.J. Puska: Phys. Rev. B **35**, 7423 (1987)
1.38 H. Wipf, K. Neumair, A. Magerl: Z. Phys. Chem. NF **145**, 237 (1985)
1.39. A. Weidinger, R. Peichl: Phys. Rev. Lett. **54**, 1683 (1985) see also M. Weiser, S. Kalbitzer: Z. Phys. Chem. NF **143**, 183 (1985)
1.40 K. Yamada: Prog. Theor. Phys. **72**, 195 (1984); J. Kondo: Physica **125B**, 279 (1984); A. Schenk: *Muon Spin Rotation Spectroscopy* (Hilger, Bristol 1985)
1.41. C.M. Edwards, N.S. Sullivan, D. Zhou: Physica Scripta T **19**, 458 (1987)
1.42 Y. Fukai, H. Sugimoto: Adv. Phys. **34**, 263 (1985)
1.43 P. Hedegård: Phys. Rev. B **35**, 6127 (1987)
1.44 T. Suzuki, H. Namazue, S. Koike, H. Hayakawa: Phys. Rev. Lett. **51**, 798 (1983)
1.45 R.C. Brower, H. Douwes, R. Griessen, E. Walker: Phys. Rev. Lett. **58**, 255 (1987)
1.46 D. Steinbinder, H. Wipf, G. Kearley, A. Magerl: J. Phys. C **20**, L321 (1987)
1.47 A. Magerl, J.J. Rush, J.M. Rowe: Phys. Rev. B **33**, 2093 (1986); S. Ikeda, N. Wanatabe: J. Phys. Soc. Japan: **56**, 565 (1987)
1.48 R. Hempelmann, D. Richter, D.L. Price: Phys. Rev. Lett. **58**, 1016 (1987)
1.49 S. Ikeda, M. Furusaka, T. Fukunaga, A.D. Taylor: Submitted to J. Phys. C
1.50 I.S. Anderson, N.F. Berk, J.J. Rush, T.J. Udovic: Phys. Rev. B **37**, 4358 (1988); J.-W. Han, C.-T. Chang, D.R. Torgeson, E.F. Seymour, R.G. Barnes: Phys. Rev. B **36**, 615 (1987)
1.51 F. Liu, M. Challa, S.N. Khanna, P. Jena: Phys. Rev. Lett. **63**, 1396 (1989)
1.52 R.G. Barnes, M. Jerosch-Herold, J. Shinar, F. Borsa, D.R. Torgeson, D.T. Peterson: Phys. Rev. B **35**, 890 (1987); P.M. Richards: Phys. Rev. B **36**, 7417 (1987)
1.53 H.R. Schober, A.M. Stoneham: Phys. Rev. Lett. **60**, 2307 (1988); M.J. Gillan: Phys. Rev. Lett. **58**, 563 (1987)
1.54 R.C. Brower, E. Salomons, R. Griessen: Phys. Rev. B **38**, 10217 (1988)
1.55 R.C. Brower, R. Griessen: Phys. Rev. Lett. **62**, 1760 (1989)
1.56 J. Harris: Appl. Phys. A **47**, 63 (1988); K.D. Rendulic: Appl. Phys. A **47**, 55 (1988)

1.57 S. Andersson, L. Wilzén, M. Persson: Phys. Rev. B **38**, 2967 (1988)
1.58 A. Krozer, B. Kasemo: J. Vac. Sci. Technol. **A5**, 1003 (1987); Phys. Rev. Lett., submitted
1.59 H.M. Dannetun, L.G. Petersson, D. Söderberg, I. Lundtröm: Appl. Surf. Science **17**, 259 (1984)
1.60 T. Greber, L. Schlapbach: Z. Phys. Chem. NF **164**, 1213 (1989)
1.61 L. Schlapbach, J.P. Burger, P. Thiry, J. Bonnet, Y. Petroff: Phys. Rev. Lett. **57**, 2219 (1986); Surf. Science **189/190**, 747 (1987)
1.62 H. Hemmes, A. Driessen, J. Kos, F.A. Mul, R. Griessen, J. Caro, S. Radelaar: Rev. Sci. Instrum. **60**, 474 (1989)
1.63 P. Dantzer, F. Marcelet: J. Phys. E **18**, 536 (1985)
1.64 C.M. Lüdeckl, G. Deublein, R.A. Huggins: Int. J. Hydrogen Energy **12**, 81 (1987)
1.65 I.R. Harris, P.J. McGuiness, D.G.R. Jones, J.S. Abell: Phys. Scripta **T19**, 435 (1987), P.J. McGuiness, I.R. Harris, U.D. Scholz, H. Nagel: Z. Phys. Chem. NF **163**, 687 (1989)
1.66 J.R. Johnson, M. Suenaga, P. Thompson, J.J. Reilly: Z. Phys. Chem. NF **163**, 721 (1989)
1.67 J.P. Burger: Private communication
1.68 G. Wang, G. Pang, C. Luo, S. Yang, Y. Li, Z. Ji, Z. Sun: Phys. Lett. **130A**, 405 (1988)
1.69 A.W. Overhauser, Phys. Rev. B **35**, 411 (1987); M. Gupta: Z. Phys. Chem. NF **163**, 517 (1989); M.R. Press, B.K. Rao, P. Jena: Phys. Rev. B **38**, 2380 (1988); R. Yu, P. Lam: Phys. Rev. B **38**, 3576 (1988); M. Gupta, A. Percheron-Guegan: J. Phys. F **17**, L201 (1987)
1.70 M. Seel, A.B. Kunz, S. Hill: Phys. Rev. B **39**, 7949 (1989)
1.71 R. Yu, P. Lam: Phys. Rev. B **37**, 8730 (1988)
1.72 D. Noreus, K.W. Törnroos, A. Börje, T. Szabo; W. Bronger, H. Spittank, G. Auffermann, P. Müller: J. Less-Common Met. **139**, 233 (1988); K. Kadir, M. Kritikos, D. Noreus, A.F. Andresen: J. Less-Common Met. (1991) in press
1.73 B. Bogdanovic, S.C. Huckett, B. Spliethoff, U. Wilczok: Z. Phys. Chem. NF **163**, 337 (1989)
1.74 J.P. Burger, L. Schlapbach, I. Vedel, U. Maier: Z. Phys. Chem. NF **163**, 569 (1989) and references therein
1.75 H. Imamura, Y. Murata, S. Tuchiya: J. Less-Common Met. **123**, 59 (1986); **123**, L1 (1986); H. Imamura, T. Nobunaga, S. Tsuchiya: J. Less-Common Met. **106**, 229 (1985)
1.76 T. Ogawa, K. Ohnishi, T. Misawa: J. Less-Common Met. **138**, 143 (1988)
1.77 G. Sandrock: Z. Phys. Chem. NF **164**, 1285 (1989)
1.78 O. Bernauer: Int. J. Hydrogen Energy **13**, 181 (1988); O. Bernauer: Z. Phys. Chem. NF **164**, 1381 (1989)
1.79 T. Sakai, H. Ishikawa, K. Oguro, C. Iwakura: Prog. Batteries and Solar Cells, **6**, 221 (1987); J. Electrochemical Soc. **134**, 558 (1987)
1.80 Y. Matsumara, L. Sugiura, H. Uchida: Z. Phys. Chem. NF **164**, 1545 (1989)
1.81 S.E. Koonin, M. Nauenberg: Nature **339**, 690 (1989); A.J. Leggett, G. Baym: Nature **340**, 45 (1989)
1.82 D.E. Williams, D.J.S. Findlay, D.H. Craston, M.R. Sené, M. Bailey, S. Croft, B.W. Hooton, C.P. Jones, A.R.J. Kucernak, J.A. Mason, R.I. Taylor: Nature **342**, 375 (1989)
1.83 F. Close: New Scientist **129**, 46 (1991); J. Bockris: New Scientist **129**, 50 (1991); E. Storms: Fusion Technology **20**, 433 (1991)
1.84 L'Air Liquide, Encyclopédie des Gaz (Elsevier, Amsterdam 1976)
1.85 A. Seeger: In *Hydrogen in Metals I*, ed. by G. Alefeld, J. Völkl, Topics Appl. Phys. **28** (Springer, Berlin, Heidelberg 1978) p. 351
1.86 R.C. Weast, M.J. Astle: Handbook on Chemistry and Physics (CRC Press, Boca Raton, Florida 1979)
1.87 R.D. Harrison: *Datenbuch Chemie, Physik* (Vieweg, Braunschweig 1982)
1.88 R.D. McCarty, J. Hord, H.M. Roder: Selected Properties of Hydrogen, National Bureau of Standards, Boulder, CO, NBS Monograph **168** (1981)
1.89 E.R. Cohen, B.N. Taylor: Physics Today **41**, August 69 (1988)
1.90 C.J. Winter, J. Nitsch: *Wasserstoff als Energieträger* (Springer, Berlin, Heidelberg 1986)
1.91 H. Hemmes, A. Driessen, R. Griessen: Physica **139B**, **140B**, 116 (1986)
1.92 J.V. Kranendonk: *Solid Hydrogen* (Plenum, New York 1983)
1.93 J.v. Straaten, I.F. Silvera: Phys. Rev. B **37**, 1989 (1988)

2. Surface Properties and Activation

Louis Schlapbach

With 43 Figures

The first step in the formation of metal hydrides and solid solutions from molecular hydrogen occurs on the surface of the host metal. This chapter contains a description of the properties of the surface and of the interaction of that surface with hydrogen. The interaction can consist e.g. of the sticking of hydrogen with a reaction probability of 0.99 on the Pd(111) surface, physisorption, dissociation, chemisorption and solution on surface, near-surface and bulk sites, or of the retarded but nevertheless fast reaction and disintegration of oxide-covered LaNi$_5$, or in the other extreme case, the complete absence of any reaction of H$_2$ with previously air exposed and thus oxide-covered FeTi.

The chapter begins with a description of effects and phenomena occurring on clean surfaces (structural, thermodynamic, electronic and magnetic properties, dynamics) and on precovered surfaces (activation, poisoning) and of experimental methods. Then results on the hydrogen adsorption on clean surfaces of transition, rare earth, simple and noble metals are summarized and finally the activation and hydrogen adsorption characteristics of intermetallic compounds and crystalline and amorphous alloys are reviewed.

The results reviewed here are of great relevance not only for the formation of metal hydrides from the gas phase or electrochemically, but also for the activation of getters and catalysts and for the prevention of hydrogen embrittlement.

The few available results on deuterium at metal surfaces are also included.

2.1 Introduction

The particular importance of surface effects in hydrogen adsorption and absorption by metals, for getters, permanent magnets, in catalytic reactions, battery electrode reaction, H embrittlement and plasma–wall interaction in fusion stems from two facts: The first relates to surface itself. The sharp discontinuity of matter with electric charges and potentials of electrons and atom cores at the surface together with the loss of periodicity in the direction orthogonal to the surface leads to

— structural properties
— electronic and magnetic properties
— a chemical composition of the uppermost layer and of near-surface layers and
— dynamical properties of surface and near-surface atoms

which are quite different from those of the bulk. It belongs to the general goals of solid state and surface sciences to describe these phenomena, which, of course, determine the reactivity of the surface.

The second important factor is the interaction of H with that surface. Gaseous hydrogen (mainly H_2, but also H) and its isotopes (D, T) and H^+ (protons) in an electrolyte interface with a surface in the process of adsorption and may or may not penetrate it. The most common reaction proceeds from the physisorption of gaseous H_2 via dissociative adsorption of H and subsurface H to H dissolved in the bulk.

The surface is considered to comprise not only the top atomic layer of the substrate, but also the first few layers which differ from the bulk. In the case of a clean single crystalline substrate the top three or possibly four monolayers consitute the surface. On oxidized and contaminated substrates, e.g. multiphase alloys, however, the surface or surface layers might be as much as 10 nm thick. Usually these covered surfaces are not reactive towards hydrogen or much less so; they have to be activated. The activation process is of particular importance for H absorption by intermetallic compounds, in catalysis and for getters. A chapter on surface effects should, of course, describe the phenomena observable in a clean well-defined simple H—metal system as well as the activation of more complex H—metal systems, e.g. H_2 on an oxidized surface.

General references to surface science are given in [2.1–18]. H adsorption on single crystalline substrates of elemental d-transition metals was studied very extensively as a prototype reaction and because of the dominant role of H in heterogeneous catalysis [2.18–20]. The results are well-documented in many excellent reviews focussing on experimental [2.18, 20–23] and theoretical work [2.24–27].

Relatively few investigations and reviews [2.28–34] deal with H sorption on other metallic substrates, mainly because these other substrates were considered as being less important in catalysis or because their preparation and characterization in a reasonably clean form was too difficult: Surface properties and surface reactions of and on rare earth metals have been reviewed by *Netzer* et al. [2.31, 32]. A few results on actinides have been published [2.33, 34]; probably many more exist in classified form. Significant progress has been made recently in studies of H sorption on Cu [2.35, 36]. The dissociative adsorption of H_2 on Cu is a prototype reaction with a high activation barrier. In view of the need for light-weight hydrides, Mg, Al, and Ca are promising starting materials because of their abundance, low density, low price and high H content in their hydrides. But very few surface studies have yet been performed on these elements and their compounds [2.29, 36–39].

Studies of the reaction of H with surfaces of binary alloys began at the end of the 1970s with the growing interest in bimetallic catalysts [2.28] and with the first successful explanation of the astonishingly high reactivity of some intermetallics towards H by a surface segregation model [2.29, 40, 41]. Surface studies of bimetallic catalysts were performed mostly on clean, well-defined surfaces. In contrast, oxidized and contaminated surfaces and their modification upon activation treatments were the subject of most surface studies on hydride forming intermetallics. It was shown that the investigation of these precovered surfaces led not only to an understanding of the activation process, but is of great relevance for the control of H embrittlement [2.42–44] and H-first wall interaction in fusion [2.45] as well as in understanding the getter effect [2.46, 47], the catalytic properties of intermetallics [2.29, 48–50] and to some extent even rare-earth-type permanent magnets [2.51]. Studies of surface effects on H storage electrodes for reversible batteries have recently started (Sect. 2.5).

A rather simple reaction, the catalytic formation of water ($H_2 + \frac{1}{2}O_2 \rightarrow H_2O$) occurs on intermetallics and on transition metals, and with a lower reaction rate also on transition metal oxides and carbides. This reaction is becoming more important in catalytic heaters and in safety devices to eliminate hydrogen or oxygen from explosive mixtures. Dissociation of O_2 and H_2 are important steps prior to water formation [2.52].

This chapter is organized as follows: after this introduction the physical effects related to the topic "H at the surface of clean and of precovered metals" and chemisorption models are described in Sect. 2.2. Section 2.3 gives an overview of experimental methods. In Sects. 2.4 and 2.5, results are presented for elemental metals and alloy surfaces, respectively. Particular surface properties of hydrides are also included.

For consistency with the title of this book, Sect. 2.4 gives only an incomplete review and highlights on H at the surfaces of elemental metals. Section 2.4, however, is more comprehensive. The reader will be astonished by the differing levels of knowledge described in Sects. 2.4 and 2.5: Whereas the former summarizes fundamental studies on structural, electronic and vibrational properties of H on well-defined single crystal surfaces, the latter concerns mostly the activation process; indeed, very few studies of the fundamental H adsorption properties of intermetallic compounds have been made so far.

2.2 A Survey of Effects, Phenomena and Models

The adsorption of H is conveniently described in terms of simplified one-dimensional potential energy curves for an H_2 molecule and for two H atoms on a clean metal surface (Fig. 2.1). Far from the surface the two curves are separated by the heat of dissociation $E_D = 218 \, \text{kJ/mol}$ H (4.746 eV, [2.30]). The flat minimum in the "$H_2 + M$"-curve corresponds to physisorbed H_2 (heat of physisorption $E_P \approx 10 \, \text{kJ/mol}$ H), and the deep minimum in the "$2H + M$"-curve describes chemisorbed, dissociated H (heat of chemisorption $\approx 50 \, \text{kJ/mol}$ H). If the two curves intersect above the zero energy level, the chemisorption requires an activation energy E_A, which slows down the kinetics of dissociative adsorption and recombinative desorption.

The adsorbed single H atoms may have a high surface mobility. They interact with each other at sufficiently high coverage and form surface phases. Many surface properties such as heat of chemisorption and sticking probability become coverage dependent. In further steps the chemisorbed H atoms penetrate the surface and are dissolved exothermically or endothermically in the bulk, where a hydride phase may nucleate and grow. There are two general pathways

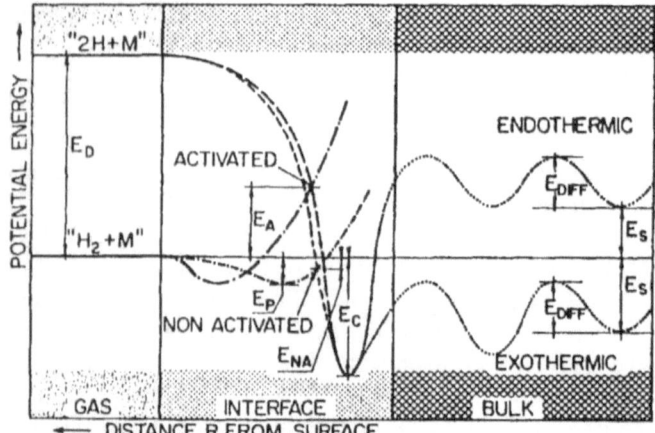

Fig. 2.1. Potential energy curves for activated and non-activated chemisorption of hydrogen on a clean metal surface and exothermic or endothermic solution in the bulk. A more pronounced minimum just below the surface allows for subsurface hydrogen (one-dimensional Lennard–Jones potential [2.53])

for H penetration into the bulk: H may either equilibrate in the chemisorption state and then penetrate into the bulk, or it may pass directly into the bulk. Isotope experiments recently showed that direct penetration occurs on Pd at low temperature (Sect. 2.2.8).

The H adsorption and absorption process is reversible, the equilibrium being determined by pressure and temperature. In the desorption process two H atoms recombine to an H_2 molecule to regain the heat of dissociation in order to overcome the heat of chemisorption (associative desorption). The desorption of atomic H has also been observed recently [2.54].

The H–metal bonding is of electronic nature both in the bulk and at the surface. An understanding of the surface and of the adsorption of H on that surface requires a description of the structural properties, thermodynamic properties (heat of adsorption, equilibrium composition, phases), electronic and magnetic properties (nature of the chemical bond) as well as dynamic and kinetic properties (sticking, vibration, rotation, diffusion), all of which are, of course, interrelated.

Experimental and theoretical studies of H adsorption are mostly performed on idealized surfaces, i.e. single crystals. Adsorption on real surfaces is a more complex phenomenon, mainly because of the physical and chemical nonuniformity of the surfaces. Surface defects (steps, kinks, grain boundaries) and the presence of impurity atoms strongly affect the adsorption [2.55]. Striking examples are Al and Nb: the sticking probability for H_2 on Al single crystal surfaces is almost zero ($\approx 10^{-4}$) but Al clusters are very reactive to H_2 [2.38, 56, 57]. The particular Nb clusters Nb_8, Nb_{10} and Nb_{16} exhibit significantly reduced reactivity towards H_2 compared to other Nb clusters. Surfaces of alloys and intermetallic

compounds show additional phenomena related to the fact that the chemical
composition at the surface may differ from that in the bulk.

2.2.1 Surface Structure

In order to minimize the free energy of a crystal the equilibrium position of
surface atoms can be different from that given by the lattice periodicity of the
bulk. Very often the interatomic spacing between the top atomic layers differs

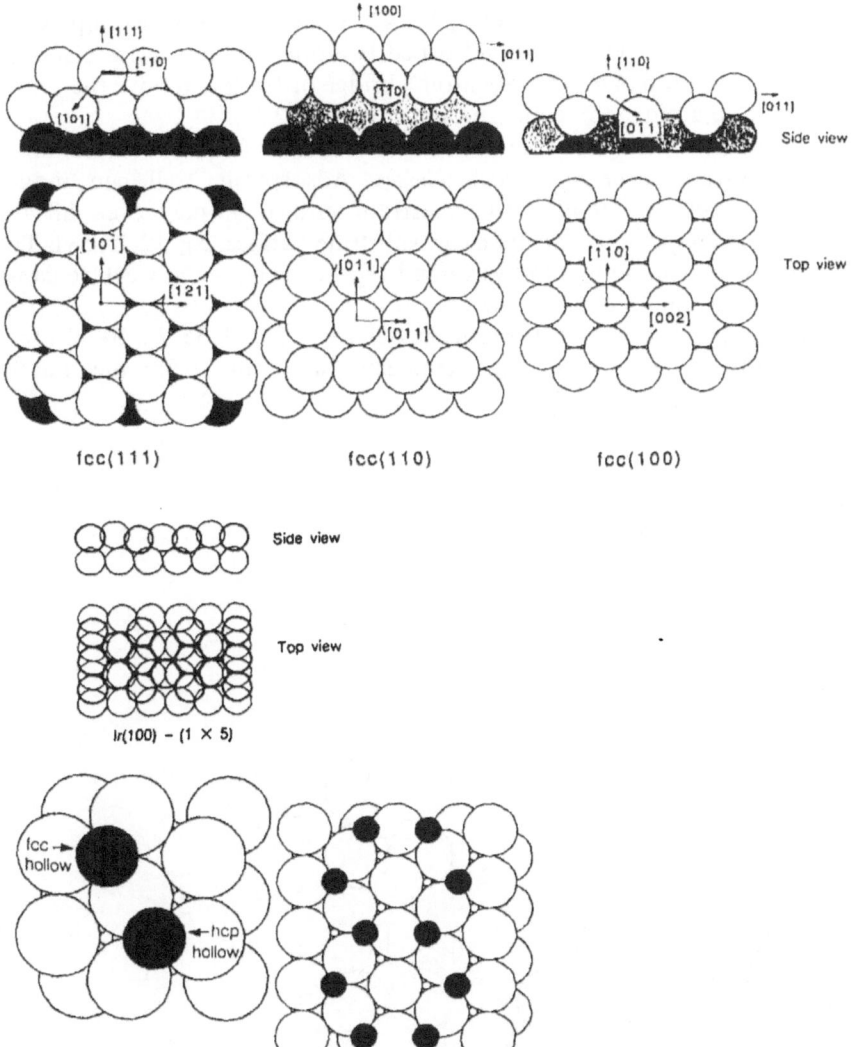

Fig. 2.2. Some surface planes and
surface crystallographic notation [2.3]

from that deeper in the bulk, without noticeable change in the lateral symmetry. This effect is called *surface relaxation*. Often it amounts to a 5–10% contraction between the first and second layer and a smaller, but measurable expansion between the second and third, and third and fourth atomic planes. In few cases clean metal surfaces are *reconstructed* [2.58, 59] (e.g. the (100) surface of Ir, Pt, Au) i.e. the lateral symmetry of the surface differs from that of the bulk. Usually additional diffraction spots indicate a lowering of the symmetry. Generally the crystallographically open surfaces [fcc (110), bcc (111) and (211), hcp (1010)] are more susceptible to relaxation and reconstruction than the close-packed surfaces [fcc (111), bcc (110)] [2.58]. The notation used to describe surface reconstructions is elucidated in Refs. [2.2–4]. Some illustrative examples are shown in Fig. 2.2. The positions of adsorbed H atoms on the substrate surface are called atop (on top of a substrate atom), bridge or twofold (above the centre of two substrate atoms), threefold and fourfold (Fig. 2.3a).

Upon increasing the coverage adsorbed H atoms form *disordered* or at lower temperatures *ordered surface phases*. Adsorption itself can induce relaxation or reconstruction of the substrate surface or may even lift the relaxation or reconstruction of the clean substrate surface, e.g. [2.21, 58]. The van der Waals type bonding of physisorbed H_2, however, is too weak to cause noticeable displacement of substrate atoms.

There is experimental and theoretical evidence [2.20, 21, 35, 36, 60–62] that chemisorbed H does not necessarily occupy sites on top of the first metal atom

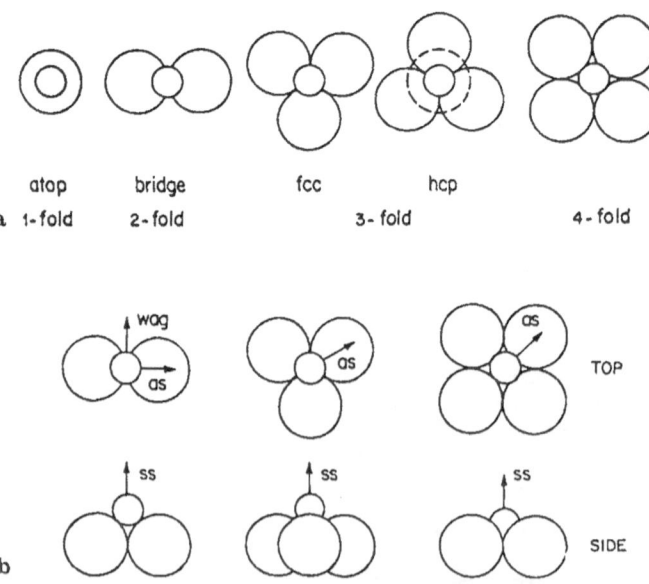

Fig. 2.3. (a) High symmetry sites of low index surfaces [2.53]. (b) Vibrational modes of high symmetry sites: symmetric stretch (ss), asymmetric stretch (as), and wag [2.25]

layer, but also sites between and below top layer metal atoms. *Subsurface* H is generally accompanied by a strong surface reconstruction (*surface hydride* formation) and can be considered as an intermediate stage between adsorbed H and bulk hydride formation. The H–H (and D–D) equilibrium distances are comparable to those between bulk interstitial sites, i.e. $\geqq 2$ Å. The distance between planes of surface H and subsurface H (e.g. on Pd (111) [2.63]) can be shorter. However, it is unlikely that a surface H (or D) sits on top of a subsurface H (or D) in thermal equilibrium.

On alloy substrates the reconstruction of the clean surface includes not only a geometrical rearrangement of the surface atoms, but also a redistribution of atoms of the alloy components. It results in an alloy composition at the surface different from that of the bulk and is known as *surface segregation*.

The important experimental methods for surface structural analysis are the most widely used LEED and the very powerful He diffraction, but also STM, surface EXAFS, photoelectron spectroscopy and photoelectron diffraction, and neutron scattering (Sect. 2.3). Results of structure analysis of H layers on transition metals are summarized in [Ref. 2.21, Tables 5, 10–12]. Absorption sites, the formation of ordered adsorbate phases and order-disorder transitions can be calculated in the mean field or embedded atom approach [2.24, 30, 43, 64, 65]. Theoretical descriptions of surface relaxation and reconstructions can be found in [2.5, 36, 63, 65, 66].

2.2.2 Thermodynamic Properties, Surface Segregation

Although the H–metal bond is of electronic nature, a thermodynamic description of some adsorption properties is adequate. The relevant thermodynamics, with emphasis on surface problems, can be found e.g. in [2.1, 67].

A weak van der Waals attraction is the origin of *physisorption* of molecular H_2 [2.68, 69]. Accordingly heats of physisorption are very small (≈ 5 kJ/mol H), and experimental physisorption studies have to be performed at temperatures $\leqq 20$ K. The growing interest in physisorption has been stimulated by the successful use of molecular beam scattering as a probe of surface structure.

It is well established that stable physisorbed H_2 can exist on the noble metal surfaces Cu [2.70, 71], and Ag [2.72, 73]. It has also been studied on Au (111) [2.74]. Low energy molecular beam studies show sharp diffraction patterns due to physisorbed H_2 and have allowed the determination of the interaction potential. The H_2–Cu physisorption well, for example, was determined to be 23 meV deep [2.75]. Studies of H_2 physisorption on simple metals were initiated both theoretically [2.69, 76] and experimentally [2.37]. Adsorption of molecular H_2 at temperatures well above 20 K might occur on special sites of lower coordination, e.g. on edge sites of stepped Ni (100) [2.77].

Chemisorption of atomic hydrogen occurs through electronic bonding states formed by H 1s and substrate metal states. The heat of chemisorption is of the order of 30–60 kJ/mol H and varies smoothly with the number of d-electrons within the 3d-transition metal series [2.24]. Tabulated values of the heat of H chemisorption on most elemental metals studied so far can be found in [2.18, 20, 21, 78, 79]; calculations are reviewed in [2.24, 80].

The transition from physisorbed H_2 to chemisorbed 2H involves dissociation. Dissociative chemisorption turns out to be one of the most important surface-related steps in the formation of hydrides of intermetallic compounds [2.40]. Dissociation and chemisorption are called activated if the H_2 molecule approaching the surface has to overcome an activation barrier (Fig. 2.1). Activation barriers of up to 400 meV were found on s-electron metals; d-electrons tend to reduce the barrier [2.27]. No spontaneous dissociation of molecular H_2 is observed on s-electron metals such as Cu (20–300 K); atomic H, however, readily chemisorbs even above room temperature [2.71, 81].

On irregular surfaces (Fig. 2.4) with high defect concentrations (e.g. catalysts, clusters), H is preferentially adsorbed at sites of high coordination and near steps [2.20, 21]. This leads to an adsorption energy which may be initially several kJ/mol H higher and shows that H absorption can be used as a probe for surface defects.

Salomons et al. [2.82] recently showed how to use H as a local probe of the structure of small H-absorbing metallic particles. They determined the energy of subsurface and bulk sites from pressure–composition isotherms.

If a new surface is created by the fracture of a metal (in absolute vacuum) the chemical potential of a surface atom μ^s is different from that of a bulk atom (μ^b). Thermodynamic equilibrium requires $\mu^s = \mu^b$ so that surface relaxation and reconstruction are induced. Alloy surfaces have an additional degree of freedom, the composition. Thermodynamic equilibrium requires that the chemical potential of the components A and B be uniform right up to the surface,

$$\mu_A^s = \mu_A^b, \quad \mu_B^s = \mu_B^b, \tag{2.1}$$

and induce a surface composition which generally differs from that of the bulk, the phenomenon of surface segregation.

Surface segregation of binary alloys is often described [2.1, 18, 83] in the form

$$\frac{c_A^s}{c_B^s} = \frac{c_A^b}{c_B^b} \exp(\Delta H/kT) \tag{2.2}$$

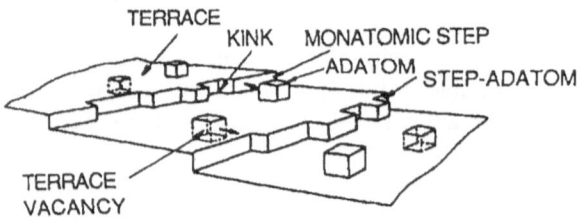

Fig. 2.4. Model of a heterogeneous solid surface, depicting different surface sites. These sites are distinguishable by their number of nearest neighbours [2.18]

where c stands for concentration. The heat (enthalpy) of segregation ΔH results from the differences in the chemical potentials

$$\Delta H = (\mu_A^s - \mu_A^b) \pm (\mu_B^s - \mu_B^b). \tag{2.3}$$

In the absence of chemisorption ΔH can be approximated [2.83] by the difference in surface enthalpy of the constituents A and B.

The above argumentation assumes that surface atoms A can be replaced by surface atoms B within the same phase. This is correct for solid solution alloys and within the homogeneity range (at the surface) of intermetallic compounds. If segregation, i.e. enrichment of one of the components, exceeds that range, one or both components precipitate in a new phase or in new phases, a transition which has as yet never been analyzed in detail.

Adsorbates that interact strongly with the substrate (e.g. oxygen, sulphur) may induce an additional surface segregation which is generally much stronger than the segregation expected or observed in (absolute) vacuum. Different bond strengths between the adsorbate and the different components act as driving forces. Thus differences of the heats of adsorption of the adsorbate with the different components have to be added to the heat of segregation in (2.2).

Surface segregation induced by the adsorption or absorption of H is reported to be weak or even absent in most cases studied so far (Sect. 2.5). However, oxygen-induced segregation effects are strong and are very important in the activation and poisoning of intermetallics for hydrogen absorption, in bimetallic catalysts, getters, and in surface magnetism [2.28, 29, 46, 47, 51, 84, 85].

Cu–Ni and Au–Ni are the best investigated solid solutions [2.85–90]. Strong Cu enrichment was observed on the topmost surface layer of Ni–Cu alloys containing up to 84% Ni. On the Ni-rich side of the alloy surface segregation of Ni extending over the top surface layer and near surface layers was observed. H chemisorption was found experimentally and theoretically [2.90] to reduce the Cu segregation.

Two special forms of surface segregation have been observed in binary and ternary hydrides and in H solid solution phases: First, the hydrogen-to-metal ratio in the surface or near-surface region differs from that in the bulk. The fact that the distance between the first and second metal atom planes can be adjusted more easily than between planes in the bulk, generally leads to an enhanced hydrogen content in this subsurface region. Nb and Pd, for example, are known to exhibit strongly enhanced near-surface solubility for H [2.91–93]. In rare earth hydrides REH_{2+x} the opposite occurs: At low temperatures some surface hydrogen seems to diffuse into the bulk leading to a H-depleted REH_2 surface [2.94]. The depth concentration profile or near-surface H is strongly affected by the presence of impurities like oxygen [2.92]. Second, the presence of hydrogen in an intermetallic compound or alloy (hydride or solid solution) may induce a redistribution of the surface metal atoms and affect the surface composition. The effect is very weak or absent in all hydrides of intermetallics studied so far by surface sensitive techniques, with the exception of a few Cu and Ca alloys. Strong modifications of the alloy composition upon hydrogen absorption were noticed on Pd–Cu [2.95] and recently on Ca_3Pd_2 [2.96], although one has to keep in mind that it is not always an easy task to distinguish clearly between artefacts induced by e.g. oxygen and other impurities.

Numerous theoretical studies have predicted the surface composition of ordered and disordered alloys from bulk thermodynamic properties and from microscopic electronic theories [2.65, 67, 83, 86, 88–90, 97, 98]. They describe quite successfully the surface concentration ratio as a function of the bulk concentration ratio at a fixed temperature, which is mostly room temperature or zero temperature. However, they do not describe the temperature dependence of segregation. Experimentally strong variation of the surface concentration ratio and even cross-overs were observed in chemisorption-induced segregation as a function of temperature.

Surface segregation is normally studied experimentally by the surface sensitive techniques of photoelectron spectroscopy (ESCA, XPS), Auger electron spectroscopy, and atom probe.

2.2.3 Electronic Properties, Chemisorption Models

The electronic properties of elemental metals, ordered and disordered alloys, and of metal hydrides, and the experimental and theoretical methods to study them are described in Chap. 5 of Vol. I. Here we raise the question of how the electronic states at and near the surface of clean metals and alloys differ from those in the bulk and how they change upon the adsorption and absorption of hydrogen and of contaminants. There are two common approaches for calculating the electronic structure in the bulk and at the surface [22, 99, 100]:

i) relatively crude, quick and easy approaches such as the nearly free electron model or the tight binding model; they are invaluable for identifying gross features and trends;

ii) precise solutions of an appropriate one-electron Schrödinger equation which includes correlation terms, e.g. by the density functional method [2.14, 101] in which the ground state energy of the many body problem can be written as a unique functional of the ground state charge density. Many-body effects have to be treated explicitly in order to correctly describe collective excitations, such as surface plasmons.

At the surface, the semi-infinite lattice gives rise to boundary conditions different from those in the bulk; the periodicity perpendicular to the surface is lost and surface atoms have a lower coordination number. A modified local electronic structure results and one may observe surface states, surface magnetism or a surface valence that is different from the bulk.

In the *jellium* model, which describes sp-metals quite well, the discrete ion cores are replaced by a uniform positive background charge that fills the half-space up to the surface and vanishes outside the surface. The resulting ground state electron density is translationally invariant parallel to the surface. However, the variation of the electron density perpendicular to the surface reveals two characteristic features (Fig. 2.5):

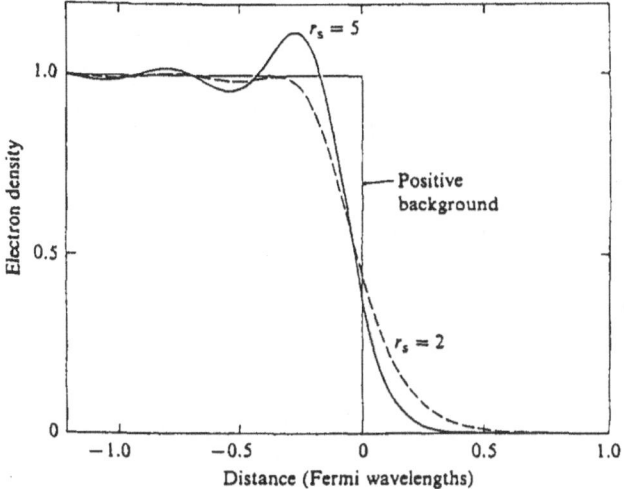

Fig. 2.5. Electron density profile at a jellium surface for two choices of the background density, r_s [2.1]

— the electron density only goes to zero far away from the edge of the positive background, i.e., the electrons spill out into the vacuum region and create an electrostatic surface dipole layer which affects the work function
— the electron density oscillates in the near-surface region.

The tight binding (LCAO) method provides a useful starting point for understanding transition metal surfaces, where the d electrons dominate the bonding.

In the *effective medium* approximation [2.24] and *embedded atom* method [2.26, 43] one calculates the change in energy occurring when a free atom is immersed or embedded in an otherwise uniform electron gas. The immersion energy is calculated as a function of the charge density of the effective medium. In the simplest version the repulsive potential is proportional to the electron density of the substrate.

Band theory shows that the surface boundary conditions lead to the existence of surface states, i.e. wavefunctions which are localized at the surface. In addition to sp-like and d-like occupied surface states, a third class of (empty) surface states, arising from the long-range image potential experienced by an electron in front of a metal surface, were discovered by inverse photoemission [2.102].

Self-consistent calculations of electronic energy bands and charge densities are routinely carried out using a slab geometry. The electronic properties of transition metals are dominated by fairly localized d-states. Surface d-states display essentially atomic character and their energy remains near the centre of the bulk band. Illustrative results for Ni(100) obtained by self-consistent local orbital calculations for a slab of nine atomic layers are shown in Fig. 2.6a (total density of states) and Fig. 2.6b (charge density) [2.103]. Total energy calculations and their significance for surface properties have been reviewed e.g. by *Ihm*

Ni(100)

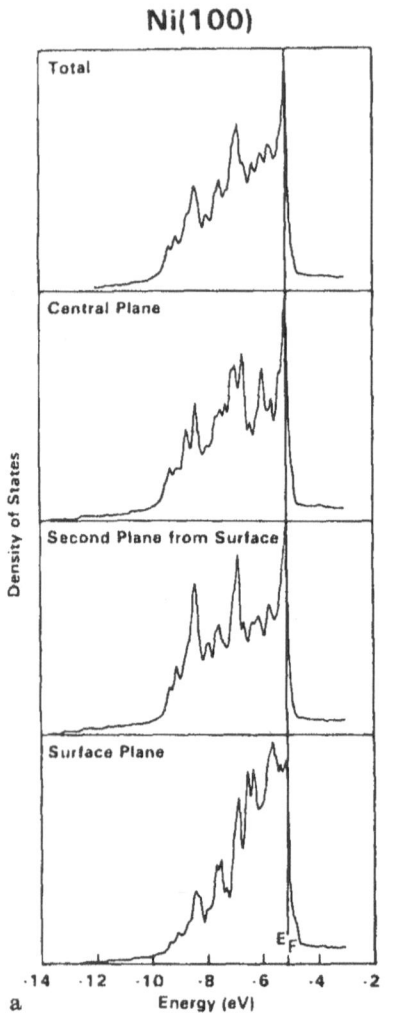

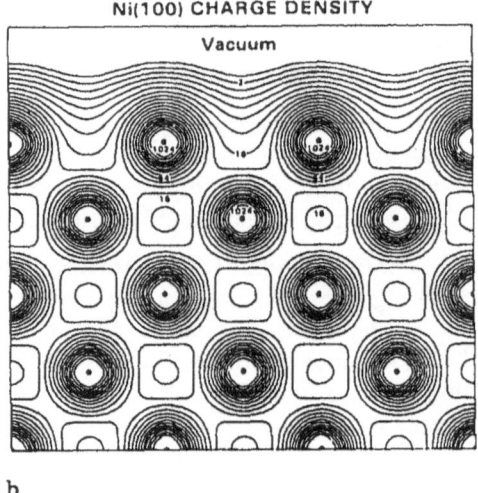

Fig. 2.6. (a) Ni(100) total density of states and planar density of states for central plane, second plane from the surface, and the surface plane [2.103]. (b) Ni(100) charge density plotted in a (110) plane perpendicular to the nine-plane slab and passing through the centre of the atoms [2.103]

[2.104] and applied to H on Ni(100) by *Umrigar* and *Wilkins* [2.105]. A spin fluctuation theory for finite temperature band calculations of bulk, surfaces, interfaces and superlattices has recently been developed [2.106].

The modification of the local density of states at the surface also affects the local spin density and is thus the origin of surface magnetism. Recent progress in the self-consistent spin-polarized energy band calculations and spin-polarized photo- and Auger-electron emission and in spin-polarized low energy electron diffraction have enabled detailed analysis of magnetism at surfaces, interfaces and superstructures [2.107–111] and of two-dimensional second-order magnetic phase transitions.

A surface-enhanced magnetic order as well as a magnetic surface reconstruction have been observed on Gd(0001) [2.112]. The magnetic hysteresis loop of the Fe(100) surface, recently measured by means of the spin polarization of secondary electrons (Fig. 2.7), shows a softer behaviour within the outer-most 5 Å due to reversed domain nucleation [2.113].

Rare earth elements have the free atom electron configuration [Xe] $4f^n6s^2$. In the condensed phase a (usually non-bonding) $4f$ electron is transferred to the $5d$ band and the elemental solids are mostly trivalent in the bulk but some remain divalent at the surface. This leads to interesting transitions in surface valence and surface magnetism [2.31, 32, 114–116].

Experimentally observed *core level shifts* are a straightforward illustration of the variation of the valence electron properties at the surface. The change in the amount of charge in the valence orbitals of the surface atoms produces a

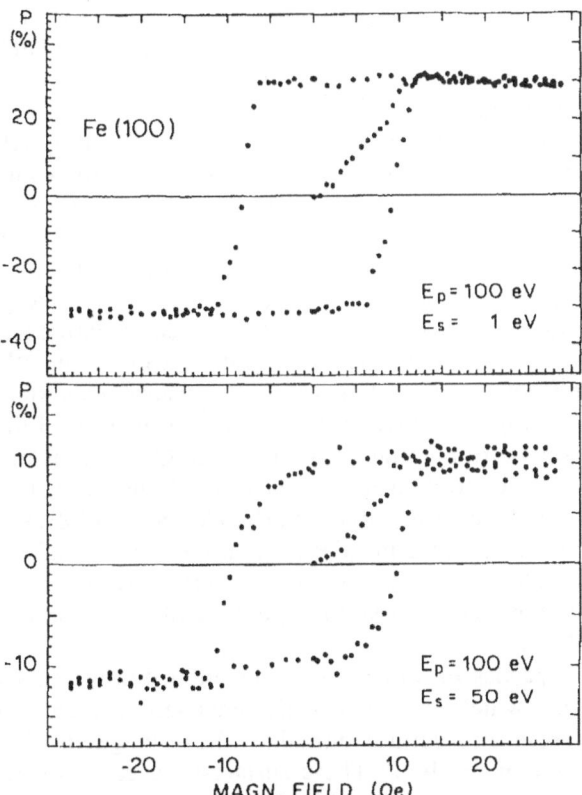

Fig. 2.7. Magnetic hysteresis loops $P(H)$ at the surface of Fe(100) recorded with secondary-electron spin polarization at 300 K using electrons of primary energy $E_P = 100$ eV [2.113].
Top: secondary electron energy $E_S = 1$ eV corresponding to a probing depth of ≈ 50 Å, bulk hysteresis loop.
Bottom: $E_S = 50$ eV, probing depth ≈ 5 Å, surface hysteresis loop

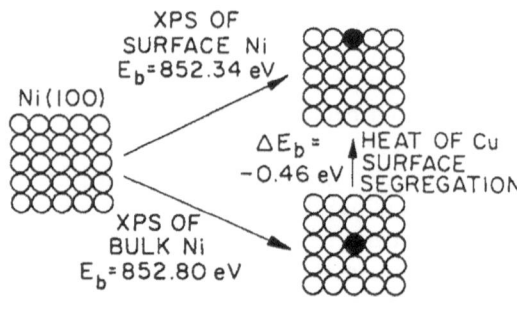

Fig. 2.8. An illustration of the single initial state and the two possible final states which, together with the equivalent-core approximation, relate the surface core-level shift to the heat of surface segregation of Cu in Ni. Open circles represent Ni atoms, and closed circles the equivalent-core version of Cu atoms [2.123]

different electrostatic potential at the surface site and hence shifts the energy of deep core levels [2.117–121].

The electronic structure of an alloy surface is complicated by the fact that surface segregation may lead to a wide range of surface composition. Fully ordered and fully disordered random distributions as well as clustering of the components on a two-dimensional lattice have to be considered. It was recently shown [2.122, 123] that the heat of surface segregation of a binary alloy AB can be evaluated from measured surface core level shifts of the elements A and B, if A and B are adjacent elements in the periodic table, so that a "Z + 1 approximation" is valid (Fig. 2.8).

The electronic structure of ordered surfaces of alloys, e.g. of intermetallic compounds, can be accurately calculated using the techniques of band theory. Disordered or random alloy surfaces are harder to treat, but great progress has been made in the last ten years. The Bloch theory cannot be applied directly because of the absence of long range order. By replacing the disordered system by an ordered one of suitably chosen effective atoms, average properties of the alloy can be obtained. The coherent potential approximation (CPA) turns out to be the most appropriate scheme for choosing the effective atoms [2.99, 124, 125]. Bulk band structures of many disordered alloys, e.g. Cu–Ni [2.87, 89, 97] have been approximated successfully by using CPA; quantitative calculations of electron states at disordered alloy surfaces, however, are very scarce [2.1, 99]. Once they are developed fully they will be easy to apply to artificial structures such as monolayers and multilayers.

The magnetic surface properties of alloys are again often dominated by surface segregation phenomena which may lead to ferromagnetic surfaces on paramagnetic bulk or vice versa and surface domains whose magnetization direction is antiparallel to that of the bulk. These surface-induced magnetic structures are particularly strong in $3d$–$4f$ alloys and affect important magnetic properties such as coercivity [2.40, 51, 126–128].

A coherent picture of the electronic aspects of hydrogen adsorption is now developing, based on a series of experimental findings, an improved understanding of the adsorbate-induced electronic structure and calculations of potential energy surfaces of adsorbates [2.24–27, 30, 36, 39, 65, 129, 130].

The isolated H_2 molecule possesses an occupied $1\sigma_g$ bonding level well below the bottom of most metal bands and a $1\sigma_u$ antibonding level above E_F. At large distances from the metal surface the electronic structure of H_2 is little affected by the presence of the surface. The physisorption well is determined by the dynamic polarization properties of H_2 (van der Waals attraction) and the steep rise in energy due to Pauli repulsion as the separation is reduced. H_2 acts as neutral but polarizable adsorbate. A physisorbed state of that nature is expected on all simple and noble metal surfaces. The corresponding potential energy curves were calculated for the simple metals Al, Mg, Li, Na, and K and for the noble metals Cu, Ag, and Au [2.68–70, 76, 129, 130]. They compare well with the few available experimental results [2.70, 131].

The 4.7 eV binding energy of H_2 is comparable to the cohesive energy of metals. The bond length of 1.4 a.u. is much smaller than a typical metal lattice constant. Thus dissociative chemisorption requires hard work to increase the proton–proton (deuteron–deuteron) distance. A change of the orbital structure with sharing of electrons between H and the substrate and the formation of new electronic configurations are needed. H acts as a reactive adsorbate. For simple and noble metals the antibonding $1\sigma_u$ level of H_2 mixes with metal (M) levels of the same symmetry forming H–M bonding and antibonding levels. The H–M bonding and H–M antibonding levels move downwards as H_2 approaches the metal surface. If the H–M bond wins the competition dissociative chemisorption occurs.

On transition metal surfaces an additional effect comes into play as a result of the unfilled d-band. The d-orbitals are quite strongly localized and thus overlap little with the H_2 orbitals. However, when H_2 orbitals approach the surface, substrate s-electrons can be transferred to the d-band, avoiding the penetration of the H_2 $1\sigma_g$ orbital and thus weakening the Pauli repulsion. The strong influence of d-holes is illustrated in Fig. 2.9 [2.129] where potential energy surfaces for Cu_2H_2 (no empty d-states) and Ni_2H_2 clusters are shown. The Cu_2H_2 potential energy curves display strong Pauli repulsion and the potential energy rises steadily to a value of 1.3 eV. In Ni_2H_2, however, the gradual filling of d-holes lowers the barrier to ≈ 0.1 eV. In the chemisorption region strong M–H bonds are formed without a complete break of the H–H and M–M bonds. The total energy falls -1.25 eV relative to that of isolated Ni_2 and H_2.

Due to the weakening of the Pauli repulsion the activation barrier may be reduced practically to zero and the physisorbed state may cease to exist, allowing direct entry of H_2 into the chemisorption region.

The mechanisms summarized above [2.27, 129] correctly describe the observed fact that most transition metals adsorb H_2 dissociatively whereas most simple and noble metals do not. They do not, of course, account for detailed variations of the H adsorption properties, e.g. within transition metals series, but have eliminated earlier controversy about the role of s and d electrons in the H–M bonding.

The potential energy curves, activation energy and chemisorption energy cannot be calculated accurately, but satisfactory approximations are available for clean surfaces of simple and noble metals and for clean transition metal surfaces.

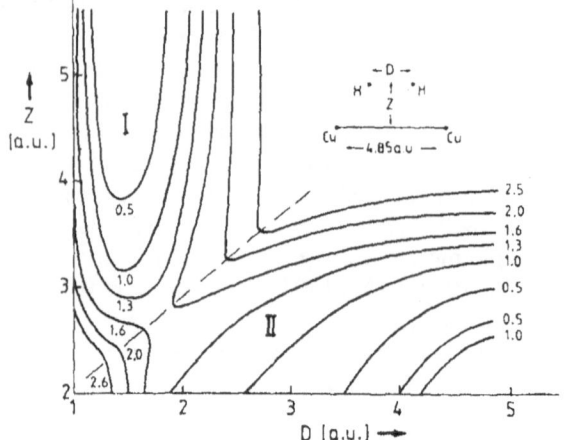

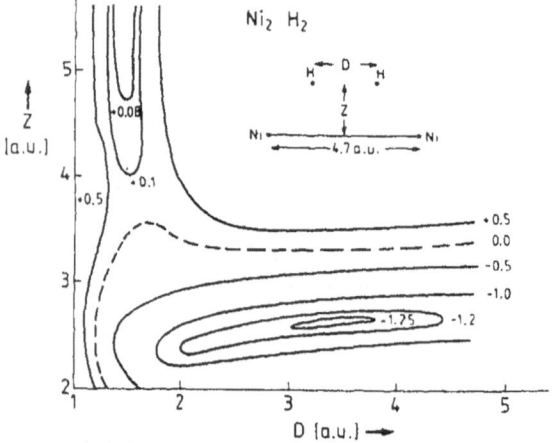

Fig. 2.9. Energy contours for Cu_2H_2 (top) and Ni_2H_2 (bottom) as a function of the H–H separation D and the normal distance Z between H_2 and Cu_2 or Ni_2. Energies in eV relative to the value at $Z = \infty$, $D = 1.4$ [2.129].
Top: the dashed line marks the "seam" where the electronic configuration switches.
Bottom: the dashed line shows the zero-energy contour. The electronic configuration changes continuously for $Z \gtrsim 3.5$, with fractional occupations, but is constant, with integral occupations, throughout the chemisorption region. Note that the H–H separation can vary by as much as 2.5 a.u. with essentially no change in the energy. The diagram refers to surface dissociation and not to gas-phase Ni_2H_2

The methods and models mentioned in the preceding section on clean surfaces are also used to treat chemisorption. Accuracies of ≈ 0.1 eV are reached. For reviews the reader is referred to [2.22, 24, 30, 36, 43, 68–70, 76, 80, 132–136]. In full analogy to the H–metal bond in the bulk, low lying bonding levels are formed by hybridization.

For chemisorption non-transition metal substrates are again treated in a jellium model starting with single atom chemisorption, e.g. an Anderson impurity model (Anderson–Newns model of chemisorption, cf. [2.30, 65] and references therein) and introducing the adatom interaction in a further step. Transition metal substrates are treated by tight binding [2.137], effective medium [2.24, 80], and embedded atom approximations [2.26, 43, 64]. Reasonably good accuracy is achieved even with pseudopotential and augmented plane wave total energy calculations [2.25, 62]. Whereas the simpler effective medium calculations correctly reproduce trends in H chemisorption energy across the series of transition metals (Fig. 2.10) [2.24, 80, 138, 139], they only allow general statements about local phenomena like surface reconstruction and subsurface H [2.36], which are accurately described by first principal calculations [2.25, 62]. For H on Ni(100) total energy calculations yield much stronger H binding energy than effective medium theory [2.105].

The reversible migration of H from sites on top of the surface via subsurface sites to bulk sites is of growing interest. This migration is dominated by the electron density in the effective medium theory. Upon relaxation of the metal atoms the electron density decreases. Because relaxation in the top metal layer is energetically favourable as compared to deeper layers, a subsurface H site can become more stable than a bulk site.

The most striking valence band feature is the appearance of a H-induced structure below the bottom of the substrate d-band, in analogy to the H-induced band in the bulk density of states [2.140]. The positions of these bands relative to E_F for H on transition metals are given in [2.30].

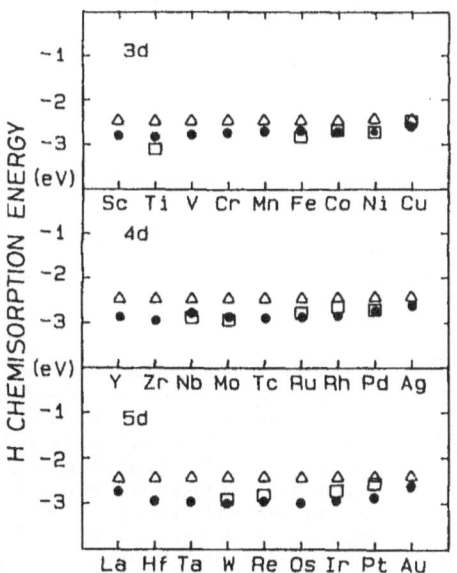

Fig. 2.10. Comparison between calculated and experimental chemisorption energies for hydrogen on the transition metals. ●: calculated values for the (100) surfaces; □: compilation of experimental data; △: contribution to the chemisorption energy from the binding energy of H in a homogeneous electron gas [2.24]

The adsorption of H induces core level shifts which are experimentally rather easily measurable. They can be interpreted (as a ground state effect) in terms of charge redistribution, or can be related via a Born/Haber cycle to thermodynamic parameters [2.117–123, 141]. Evaluation of changes of the work function yields similar information on the charge redistribution. Measured work function changes for H saturation coverage on transition metal surfaces are of the order of + 0.1 to + 0.9 eV [exception: − 85 meV on Fe(110)] and are tabulated in [2.20, 21, 30].

The reciprocal influence of surface magnetism and H adsorption is particularly interesting and also controversial [2.142]. H adsorption has been shown to change the magnetic properties of the surface (magnetic moment and ordering temperature) of e.g. Ni and Gd [2.29, 143, 144]. In turn, the surface magnetism seems to influence the H adsorption kinetics, although the old "magneto-catalytic" effect probably does not exist [2.145].

Very little is known about the surface electronic structure of bulk metal hydrides. A semiconductor-metal transition has been observed on the surfaces of light and heavy rare earth trihydrides at low temperature [2.94].

The perturbation of surface electronic structure of clean surfaces by the adsorption of well-known catalytic poisons such as S has been calculated self-consistently for S on Rh(001) [2.146]. The results (Fig. 2.11) reveal a substantial reduction of the local density of metal states at E_F. In terms of effective medium theory [2.24, 133] coadsorbed atoms can change the interaction energy of H with the metal surface so that the activation energy rises (for a poison) or decreases (for a promoter).

The most valuable techniques for probing surface electronic structure are photoemission and inverse photoemission of the occupied and empty part of the density of states and photoemission from deep core levels (Sect. 2.3).

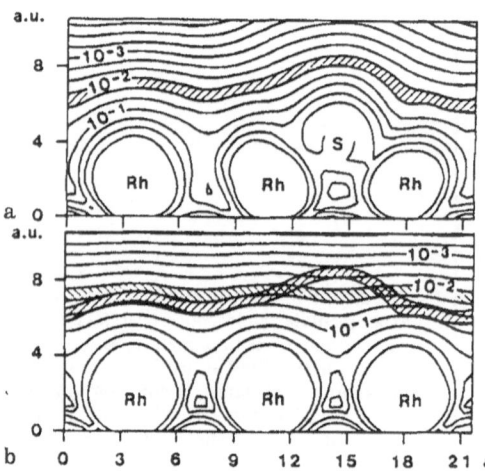

Fig. 2.11. Fermi-level local of states (LDOS), in $(eV \times a_B^3)^{-1}$, for two-layer Rh(001) films (**a**) with and (**b**) without a $S(3 \times 1)$ adlayer. Regions of equal LDOS in the two plots are hatched. Hatched region from the upper plot transcribed to the lower plot for comparison. Notice that at a given height above the Rh with no S neighbour, the LDOS is reduced by a factor of 2 [2.146]

2.2.4 Dynamics of Hydrogen at Surfaces: Sticking, Vibration, Migration

When H_2 (or H) strikes a surface it may either be trapped or scattered back into the gas phase. A trapped H_2 molecule, however, will not stick, dissociate, chemisorb and diffuse into the bulk instantaneously. The translational (kinetic) and adsorption energy first have to be dissipated and rotational and vibrational modes may be created or destroyed. During that time the H atoms are not in thermodynamic equilibrium with the substrate and its surface. The excitation and annihilation of substrate phonons and electron–hole pairs are the major mechanisms in the process of relaxation to equilibrium. The adsorbed H atoms vibrate around their equilibrium positions and migrate (diffuse) along the surface before being dissolved in the bulk or being recombined with another H atom to desorb as H_2.

Experimental and theoretical efforts to understand the surface dynamics thus concern

— the sticking probability and its dependence on temperature, crystal plane and precoverage [2.147]
— absent and additional vibrational modes of adsorbed H_2 and H and frequency shifts of substrate phonons [2.148]
— surface migration, transition to subsurface and bulk sites
— angular and velocity distributions of desorbing H_2
— the effect of the surface steps on the overall kinetics of H uptake by metals ([2.33] and Chap. 4).

Dynamical processes are much faster than overall kinetics due to the occurrence of Arrhenius factors $\exp(\Delta E/kT)$ in kinetics with potential energy well depths ΔE usually being large compared to the temperature T. In other words the residence time or lifetime of the vibrational states is large ($\approx 10^4$) compared to a typical vibrational period [2.149].

The terminology for vibrational modes of adsorbed H on top of simple surfaces is shown in Fig. 2.3b together with the geometry of the adsorption sites [2.25]. The modes follow selection rules and not all modes necessarily produce a signal in a given experimental setup [2.150].

The vibrational properties of H adsorbed on single crystal surface of transition metals have been thoroughly studied and have given very valuable information on the nature of the chemisorptive bond; for reviews see [2.8–10, 21, 25]. In recent studies the equilibrium adsorption geometry and force constants of the vibrational modes of H on low-index faces of transition metals were evaluated using local density functional total energy calculations. In particular, H at the twofold sites on W(001), threefold sites on Ru(0001) and fourfold sites on Rh(001) was analyzed and compared with the experimental results [2.151]. The H potential in the vicinity of the fourfold hollow site on Rh(001) contains important anharmonic components.

Only a few results on the vibrational properties of H or H_2 on nontransition metals and on alloys are available to date. They concern mainly H_2 on Cu and Au (Sect. 2.4), on Cu–Ni alloys, and a first study of H on CoTi (Sect. 2.5).

In recent years a great deal of research has been devoted to elucidate the relative importance of the various mechanisms of transferring energy between adsorbate and substrate [2.152] in the adsorption and desorption process. The substrate phonons and electrons serve as a heat bath. Multi-phonon processes and electron–hole pair excitations are considered to be important [2.153–156].

At low temperatures the sticking of H_2, D_2 and HD on Cu(100) was found to occur not via direct phonon excitation but via resonant processes involving quasi-bound states at the metal surface and rotational excitations of the molecule [2.157–159]. Radiative transitions were not observed in the relaxation of excited states at metal surfaces. The lifetime of excited H near a jellium metal surface was shown recently to be orders of magnitude longer than earlier estimates had predicted [2.160]. Furthermore experimental evidence was given that transitional and vibrational temperatures of molecules leaving the surface are much higher than expected from the surface temperature [2.161, 162]. Where high entrance channel barriers to dissociative chemisorption occur (simple and noble metals), H_2 necessarily desorbs with energy in excess of that of the surface temperature.

In comparison, very little work has been undertaken to analyze possible effects of strong electric or magnetic fields on the H adsorption properties and particularly on the surface dynamics. Such an analysis could be very valuable for comparing the gaseous H_2/metal and electrolyte/metal interfaces. In view of the rather long lifetime of excited H near a jellium surface, the evidence of high translational and vibrational temperatures of desorbing species and the ongoing discussion on the possible existence of cold fusion, such analysis and comparisons should be done. Furthermore, local magnetic fields at the surface are crucial for the ortho–para conversion of adsorbed H_2 and D_2. The sources of the magnetic fields are the electronic and nuclear spins and the orbital motion of the electron [2.163]. The magnetization of a Ni(110) sample has been reported to influence H desorption kinetics [2.164]. Dilatometric and voltammetric measurements in magnetic fields up to 10T showed that absorption by well-annealed Pd electrodes increases up to 1T and saturates thereafter [2.165].

Electron–hole pairs are excited when a substantial charge rearrangement occurs. The excited pairs ultimately decay by phonon excitation and heat up the lattice. It was shown that the probability for excitation of an electron–hole pair is proportional to the square of the density of electron states at E_F [2.153]. This could explain why H sticks far better on transition metals than on noble and simple metals. In another picture (Sect. 2.2.3) empty d-states allow an s–d charge transfer on the substrate surface and thereby reduce the energy barrier for chemisorption.

Only a few results are available on the surface migration/diffusion of H [2.105, 166]. Lateral transport, reordering and site filling between successive thermal desorption pulses are studied to measure diffusion [2.167, 168]. Around room temperature surface diffusion is thermally activated. At low temperature, however, quantum mechanical tunnelling becomes important.

The filling of subsurface sites and more tightly bound surface H adsorption sites results not only in the occurrence of vibrational modes which are inconsis-

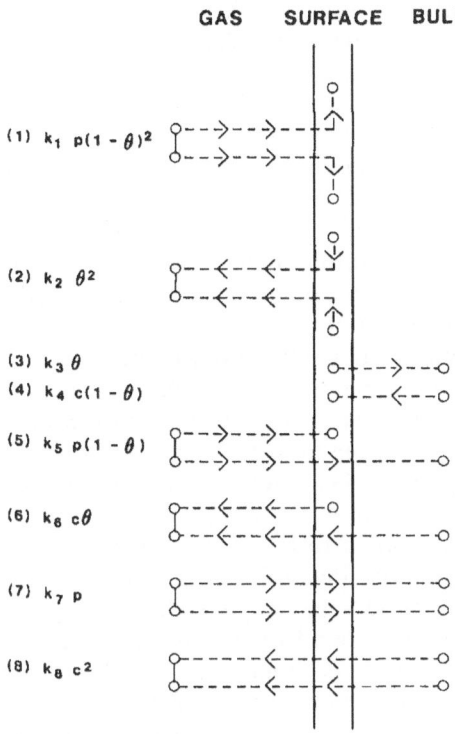

GAS   SURFACE   BULK

(1) $k_1\, p(1-\theta)^2$

(2) $k_2\, \theta^2$

(3) $k_3\, \theta$

(4) $k_4\, c(1-\theta)$

(5) $k_5\, p(1-\theta)$

(6) $k_6\, c\theta$

(7) $k_7\, p$

(8) $k_8\, c^2$

Fig. 2.12. Schematic of rate processes. Expressions for the rates give explicit dependences on surface coverage θ, bulk concentration c (for $c \ll 1$) and pressure p. The k are constants dependent on energies and (except for k_3 and k_4) on gas and dissociation/recombination kinetics. (1) molecule dissociates and both atoms stick on surface. (2) Two atoms from surface recombine and leave the surface [reverse of (1)]. (3) Atom from surface enters bulk. (4) Atom from bulk enters surface [reverse of (3)]. (5) Molecule dissociates. One atom sticks on surface, the other immediately goes into bulk. (6) Atom from bulk recombines with one from surface and both escape [reverse of (5)]. (7) Molecule dissociates and both atoms immediately go into bulk. (8) Two atoms jump directly from bulk, recombine and escape without "pausing" at the surface [reverse of (7)] [2.171]

tent with either surface or bulk H modes, but also significantly affects the overall uptake kinetics, which has been a matter of some controversy [2.60, 169, 170].

The rates of the various steps are shown schematically in Fig. 2.12 [2.171] with rate constants k and coverage (θ) dependence. The mechanism of the transition from adsorbed H to dissolved H has not been studied in detail [2.170]. Surface hydride formation is found to slow down the kinetics [2.172].

The greatest body of data on H dynamics at surfaces comes from electron energy loss spectroscopy (vibrations), infrared absorption and Raman spectroscopy, and from the inelastic scattering of thermal neutrons.

2.1.5 Chemisorption on Metal Oxides

The surfaces of metal oxides and their H_2 chemisorption characteristics have been far less studied than the surfaces of elemental metals and semiconductors [2.173–175]. Cation surface states are formed on ideal oxide surfaces at about 2 eV below the bottom of the conduction band. The charge of the surface ions is found to be reduced compared with that of the bulk ions and this leads to an enhanced covalency at the surface. The reduction amounts to less than 10% for oxides of simple metals as MgO and to 20–30% for transition metal oxides.

Cluster and slab calculations reveal that special surface state bands with metallic character can be formed on polar surfaces by charge compensation effects. To what extent the metallic band accounts for special catalytic activity is not yet known [2.175].

A lattice defect on the oxide surface usually introduces deep localized states around it. Figure 2.13 [2.175] illustrates the differences in the calculated density of states of the TiO_2-rutile(110) surface without (Ti_4O_{16}) and with the oxygen vacancy (Ti_4O_{15}). Such defects play a crucial role as active centres for surface reactions of otherwise inert surfaces. Chemisorption on defect sites is generally much stronger than on normal sites. The photo-catalytic activity of the rutile surface is greatly enhanced by the introduction of oxygen vacancies.

Studies of the adsorption of H_2 and H on single crystal oxides of TiO_2, SrTiO, WO_3, ZnO and NiO paint a consistent picture of the behaviour of oxide surfaces towards hydrogen: most perfect or nearly perfect oxide surfaces are essentially inert to H_2. However, atomic H has been shown to adsorb on stoichiometric oxides, e.g. on WO_3(001) and ZnO [2.173]. NiO(100) surfaces cleaned by ion bombardment and annealing—and thus probably slightly reduced [2.176]—become further reduced when exposed to H_2 at 400–600 K.

The defect-free surface of TiO_2(110) is inert. Below 370 K adsorption of H_2 was observed on TiO_2(110) surfaces having O-vacancy point defects. H_2 was postulated to dissociate at the defect site and bond to the adjacent partially coordinated cations resulting in Ti^{4+}–H–hydride bonds [2.177]. Results of earlier studies [2.178] of H_2 adsorption on Ar^+-bombarded surfaces were interpreted in terms of surface OH formed by dissociated H_2 with H bonding to surface oxygen ions.

Though H_2 does not dissociate on perfect TiO_2, the diffusion of atomic H across single crystalline TiO_2 is fast: the crystal structure of rutile TiO_2 is characterized by large open channels parallel to the c-axis which allow rapid diffusion of light interstitials [2.179]. For H the diffusion is described by $D = 1.8 \cdot 10^{-3} \exp(-0.59\,eV/kT)\,cm^2/s$ and amounts to about $2 \cdot 10^{-13}\,cm^2/s$ at 300 K [2.180]. Lattice defects sharply inhibit H diffusion. Accordingly TiO_2 coating prevents effectively H-embrittlement [2.181].

H_2 and H adsorption on simple metal oxides has been studied even less than that of transition metal oxides. H_2 adsorption onto various defects on the MgO(100) surface has been treated theoretically using defect lattice techniques,

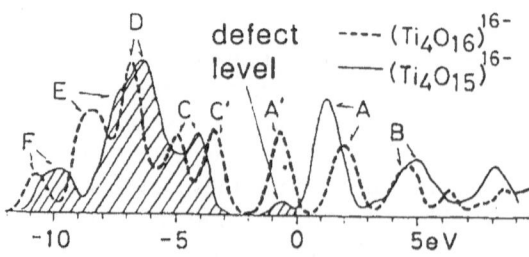

Fig. 2.13. The density of states of the TiO–rutile (110) surface without (dashed curve) and with (full line) an oxygen vacancy. Correspondence of the peaks is indicated. A defect level is formed below $0 = E_F$ [2.175]

including the relaxation of the lattice around the defects. Surface defects (F_s and V_s centres and self-trapped holes) were all found to activate dissociative chemisorption, resulting in the formation of OH^- radicals [2.182, 183]. This is in general agreement with the observed catalytic activity of MgO after the creation of F-centres by X-rays [2.184].

2.2.6 Effect of Thin Film Coating and Precoverage on H_2 Dissociation and H Adsorption and Absorption

The ability to deposit monolayers of metal atoms on single crystal metallic substrates and the formation of multilayer structures initiated a surge of activity in surface and interface science, which has also had an impact on H sorption and permeation studies. The modification of structure, electronic structure and magnetism at the interface and in the overlayer are widely studied properties [2.111, 185–188]. Strongly enhanced magnetic moments were found for some transition metal overlayers as were induced magnetic moments at the interface of otherwise nonmagnetic elements. Epitaxial constraints limit lattice expansion due to H absorption in Nb-Ta superlattices and depress superconductivity [2.189]. It was noticed [2.190] that the deposition of more than two monolayers of Pd or Pt dramatically enhances the H uptake rate of Nb or Ta substrates. It is the variation of the surface atomic and electronic structure and the weakening of subsurface bonding, and not just protection from surface oxidation, that are at the root of the enhanced uptake ([2.170] and Sect. 2.5) (Fig. 2.14).

It is well known from catalysis that electropositive (e.g. Na, Cs, K) and electronegative (e.g. S, O, C, Cl) adatoms decrease or increase the reaction rate and thus poison or promote the reaction, respectively [2.133, 191, 192].

Alkali-metal influenced adsorption on transition metals was reviewed recently by *Bonzel* [2.191]. Coadsorption of alkali metals and H_2 or D_2 on Al(100) revealed that the sticking coefficient and dissociation rate are extremely weak ($\geq 10^{-4}$) at all alkali coverages [2.193]. Upon exposing alkali-covered metal substrates to a beam of atomic H or D, alkali hydride formation was observed.

The effect of oxygen precoverage of transition metals on the H_2 chemisorption increases with the degree of oxide formation and was investigated in relation to the inhibition of H sorption and embrittlement and strong metal support interaction in catalysis. Monolayer amounts of oxygen reduce the H_2 adsorption and desorption on single and polycrystalline transition metals by orders of magnitude [2.29, 40, 41, 194, 195]. On polycrystalline metals O precoverage seems to deteriorate somewhat less than on monocrystals: Oxidation of polycrystalline Ti caused a decreased D_2 absorption rate; complete inhibition of the reaction occurred at oxide thicknesses exceeding 20 Å [2.196]. *Ko* and *Gorte* observed a complete suppression of H_2 adsorption even for TiO [2.197]. At elevated temperatures some substrate metals dissolve surface oxygen, a mechanism which is particularly important in the activation of getter alloys [2.47, 198]; cf. Sects. 2.1.7 and 2.5. From detailed volumetric studies *Fromm* and coworkers [2.199] conclude

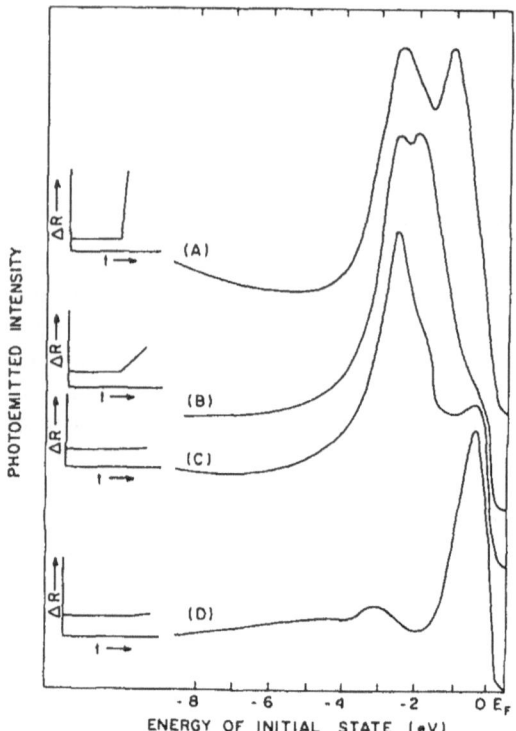

Fig. 2.14. Normal-photoemission spectra at $h\omega = 21.2\,eV$: Curve A, a Pd(111) overlayer on Nb(110); curve B, a Pd overlayer on Nb(110) at the Pd(111) to Pd*(110) transition; curve C, a Pd*(110) overlayer on Nb(110); and curve D, Nb(110). The insets show schematically the hydrogen-uptake curves where the change in the resistance (ΔR) of the Nb foil is plotted against time. Hydrogen uptake is measured by the change in the resistance of the Nb foil with hydrogen bulk concentration [2.190]

that thin suboxide layers, formed during the initial stages of oxidation or by partial oxide reduction in H_2, do not impede the H_2 reaction drastically. A decrease of the H_2 reaction probability of typically one order of magnitude was observed for a Ti substrate film covered with 3 nm thick metal overlayers after precoverage of the overlayers with 10 monolayers of oxygen (Fig. 2.15). The results indicate that H_2 dissociation is strongly impeded if the oxygen precoverage exceeds a critical value which is given by the maximum amount of oxygen that can be adsorbed with sticking probability of one in the initial stage of oxidation. H_2 on TiO_2 dissociates in neither the rutile nor anatase structure modifications [2.41]. TiO_2 is used, as mentioned earlier, as an effective coating to prevent embrittlement [2.181]. Various authors have reported that transition metal overlayers on the oxide layer restore the original uptake rate of the substrate [2.196, 199, 200]. This clearly indicates that oxidation reduces the dissociation rate and is in agreement with the observed rather fast diffusion of atomic H across TiO channels [2.179, 180] mentioned earlier.

Golczewski et al. [2.201] observed slow desorption of H from polycrystalline α-phase $VH_{0.015}$ through an oxide layer freshly formed at air. The type of oxide formed, however, was not characterized. It may well be different from that formed in the absence of H. The permeation slows down with time either by healing

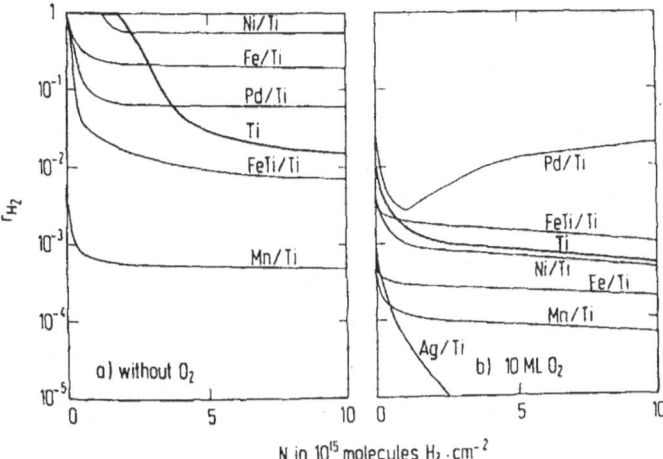

Fig. 2.15. H$_2$ reaction probability of titanium film covered with 3 nm thick metal overlayers (Pd, 1 nm). (a) Without O$_2$ precoverage and (b) overlayers precovered with 10 ML O$_2$ [2.199]

inhomogeneities in the oxide layer or by further oxidation. *Richards* et al. [2.202] compared deuterium desorption from clean and fully oxidized iron. They conclude that D$_2$ recombination most likely takes place from sites of the same energy in samples with both clean and oxidized surfaces. As it is unlikely that chemisorption sites on the bare oxide surface have the same energy as such sites on the bare metal surface, they suggested that recombination occurs at the metal–oxide interface and that the oxide forms a barrier to D$_2$ desorption from shallow precursor states. Diffusion of molecular D$_2$ across the oxide seems rather unlikely, apart from in channels along grain boundaries. An experimental verification is desirable.

Petersson et al. [2.203] showed that (at 473 K) the effect of oxygen coverage of Pd on the dissociation process is much more severe than for other molecules. The measurements showed a trend where H$_2$ dissociation was blocked completely, C$_2$H$_4$ dissociation was blocked only above a certain large oxygen coverage, C$_2$H$_4$ dissociation was not influenced and NH$_3$ dissociation was even promoted.

The adsorption of sulphur on clean transition metal surfaces completely inhibits H$_2$ dissociation and H adsorption (from gas phase and cathodic in aqueous acid medium) [2.55, 204–206]. A major effect seems to be the blocking of H adsorption sites, possibly even the blocking of several H sites by one S atom. *Marcus* and *Protopopoff* [2.205] conclude from S and H adsorption experiments on Pt(110) and Pt(111) that one S atom blocks 8 ± 1 and 12 ± 3 H adsorption sites, respectively. Self-consistent linearized APW calculations of the electronic structure perturbations induced by S on the Rh(001) surface reveal a substantial reduction of the density of states at the Fermi level even at non-adjacent sites (Fig. 2.11) [2.146]. Thus blocking is explained by local electronic structure arguments. Also in the analysis of *MacLaren* [2.206] enhancement and decrease of the local density of states determine the sensitivity of different surfaces of Ni and Rh to poisoning by S. Surface poisoning by adsorption of sulphur species is known to inhibit H desorption from Pd hydride [2.207].

Poisoning of the surface reaction $H_2 \rightleftarrows 2H$ in the absorption and desorption process by SO_2 surface treatment has been observed on rare earth intermetallic compounds [2.40, 208, 209]. Permeation of H through V was also reported to slow down by an order of magnitude upon S segregation to the surface [2.210].

The effect of precovering metal surfaces by species other than O and S on the H_2 and H adsorption is not well known or understood. Rare gas (He, Ar) admixtures to H_2 have been reported to slow down the H uptake rate [2.211] which is, at least to some extent, a blanketing effect (Chap. 5). Intermetallic compounds in particular often contain significant amounts of carbon on the surface. Its effect on H uptake has not yet been studied. Dissociative chemisorption in the threefold hollow site of the TiC(111) surface with a rather high sticking probability of ≈ 0.5 has been reported [2.212].

In an effective medium approach to poisoning and promotion by *Norskov* et al. [2.133] evidence is given for the importance of empty (antibonding) states near the Fermi level. Competition between the closed-shell kinetic energy repulsion and the attraction due to the gradual filling of the antibonding level determines the height of the molecular adsorption barrier, the depth of the molecular wall and the height of the dissociation barrier in the reaction path. Possible ways of altering the balance between the two competing terms are the addition of a partially empty d-band around the Fermi level (transition metal atoms) or changes of the work function of the surface.

The close relation between inhibitors of H embrittlement and H_2 dissociation poisons was noticed early by *Berkowitz* et al. [2.42]. The source of hydrogen determines whether e.g. surface S, which slows down the $H_2 \rightleftarrows 2H$ reaction in both directions, inhibits or promotes embrittlement. For dissolved H in the host metal, e.g. during electrochemical treatment, S prevents recombinative desorption and thus promotes embrittlement. In contrast, S prevents dissociation and solution of gaseous H_2 e.g. in a H_2 pipeline tube, and thus inhibits H embrittlement under these circumstances.

After electrochemical charging of Pd in D_2O and H_2O surface contamination by heavy metals like Hg, Pd and Zn in monolayer amounts has been observed [2.213]. The heavy metals are present in H_2O and D_2O as impurities; after electrodeposition on Pd or other electrodes they raise the overpotential.

2.2.7 Activation of Intermetallic Compounds and Alloys as H Absorbers, Getters, and Catalysts

Air-exposed elemental metals and alloys are covered by a 0.5–10 nm thick layer which contains, in addition to the bulk constituents, the elements C, O, H, and N and possibly further elements, mixed together in ill-defined composition and morphology. This surface layer forms naturally on the originally clean and reactive metal surface. In some cases it passivates the metal surface completely (e.g.

Al_2O_3 on Al). The passivation is highly desirable in some ways, e.g. to avoid corrosion of stainless steel or H embrittlement, but it is very annoying in all cases where a reactive surface is needed: i.e. for catalysts, getters, and hydride-forming elemental metals and alloys. An activation process is needed to make the surface reactive again. The reactive or activated surface becomes poisoned by the adsorption of gases like SO_2, CO and H_2S. Reactivation is usually feasible.

The term activation has been in use for a long time to describe the process applied to prepare catalysts or to make catalysts and getters reactive after their exposure to air, e.g. during preparation and storage. Hydrogenation catalysts consist of transition metals finely dispersed on a high surface area oxide support. They facilitate the dissociation of H_2. The typical activation procedure consists of heating the supported catalyst to temperatures between 100 and 400 °C in the reducing atmosphere of H_2 gas flow. Getter alloys chemically pump the residual gas of evacuated and sealed volumes, e.g. X-ray and TV tubes and vacuum insulated dewars, by adsorption and absorption of the residual gas. Activation consists of heating up to 400 or 500 °C in vacuum. The activation procedure for catalysts and getters should be performed in situ, i.e. in the position of final use, or if this is not possible subsequent handling must be under vacuum or noble gas atmosphere. The reactivity of hydride forming metals and alloys is also reduced by the surface layer formed upon exposure to air. The rather noble Pd, for example, does not react at all with H_2 at room temperature even at high H_2 pressure.

In analogy to catalysts and getters the term activation is also used for hydride forming materials. As most hydride forming intermetallics decompose into powder upon the first H adsorption (cf. Chap. 2 of Vol. I) and thereby produce freshly fractured clean surface, a more precise definition of activation is needed. In agreement with *Sandrock* [2.214] and our earlier work [2.29, 40, 41] and in view of technical applications of hydrides we distinguish between a first stage or surface activation which makes the previously air-exposed or poisoned surface reactive, and a second stage or bulk activation, during which the bulk hydride is formed and the material may disintegrate into powder. Evidently the bulk activation is related to bulk properties such as H diffusion, hydride nucleation and growth and to mechanical properties of the host metal such as fracture toughness and microstructure. That part of the activation is not considered further in this chapter; the reader is referred to Chap. 2 of Vol. I and to Chaps. 4 and 5 of this volume.

What are the mechanisms involved in the surface activation? For gaseous H the crucial step is the dissociation of molecular H_2 in the H absorption process and the recombination of atomic H to H_2 molecules in the desorption process. As we know from preceding sections, dissociation and recombination proceed fast and have a negligible activation barrier on most clean surfaces of transition metals. Thus the contaminating surface layers have to be removed or made transparent to H_2 in the activation process. Ways of achieving this include [2.29, 40, 41, 215]:

a) mechanical or chemical (etching) surface cleaning at room temperature,
b) diffusion of H_2 across cracks in the contaminating surface layer during incubation time, and mechanical separation of the contaminating layer by hydride formation in the underlying host metal,
c) chemical reduction of surface contaminants,
d) evaporation/volatilization of surface contaminants,
e) solution of surface contaminants in the bulk,
f) formation of specific H-absorbing near-surface compounds from the host metal and the contaminants,
g) surface segregation with selective oxidation and the formation of nonoxidized surface species accessible to H_2.

Whereas steps (a)–(f) are in principle feasible on all kind of host metals, step (g) is possible on multicomponent host metals only, i.e. on crystalline and amorphous alloys and intermetallic compounds.

In the process of activation of multicomponent systems — usually bimetallic alloys or layered structures containing some oxygen — at various temperatures and gas (H_2, O_2, air) pressure, the following surface and near-surface oxidation, segregation and diffusion mechanisms have often been observed:

i) Alloys $A_x B_{1-x}$ between very reactive, rather electropositive elements a and rather inert electronegative elements B show, around room temperature, a strong selective oxidation of A, which, for sufficiently strong oxidation, leads to a decomposition into an oxide or suboxide AO_{n-y} and clusters or precipitates of B. A is a strongly exothermic oxide- and hydride-forming element. The concentration ratio A:B is strongly enhanced on top of the surface if the mobility of the A and B atoms — i.e. the temperature—is high enough that the equilibrium distribution can be reached. At elevated temperature A oxidizes completely, B remains metallic and the thickness of the decomposed layer grows. At typical laboratory pressures (10^{-13}–1 bar O_2, 10^{-13}–10^2 bar H_2) oxygen has a much stronger effect than hydrogen. The partial pressures of O_2 and H_2 affect the A:B ratio considerably.

ii) Alloys $A_x B_{1-x}$ of electropositive A and electronegative B, but with only small differences in the enthalpies of oxide or hydride formation between A and B, show, at around room temperature, oxidation of both components and sometimes the formation of ternary A–B–O or quaternary A–B–O–H compounds. The concentration ratio A:B on top of the surface is close to that of the bulk. At elevated temperatures (≈ 450–700 K) diffusion of A, B and O becomes the most important factor in reaching the equilibrium distribution. The composition depends strongly on temperature and partial pressure of O_2 and H_2 as well as on the bulk concentration ratio A:B. Strong oxidation of one of the components A or B and reduction of the other to the metallic state as well as dissolution of all oxygen in the near surface region and in the bulk together with the restitution of a metallic alloy surface can be observed as extreme cases. Due to the temperature dependence of enthalpies of formation, surface composition and oxidation may be reversed as compared to those around room temperature. This reversed segregation is observed after initial oxidation (mechanism (i)) is completed.

iii) Thin films of B (usually a rather electronegative transition metal) deposited on the stoichiometric oxide AO_n remain stable around room temperature and oxidize slightly at the outer surface. At intermediate temperatures (≈ 450–700 K) in H_2 atmosphere nucleation of B particles occurs over the entire oxide support.

iv) At high temperature ($\geqq 700$ K) and low oxygen partial pressure (vacuum, H_2 or rare gas atmosphere) one observes wetting and spreading of the metallic precipitates (B) over the oxide (AO_{n-y}), interdiffusion and the formation of near surface ternary A–B–O (rarely quarternary A–B–O–H) compounds.

Very generally speaking mechanism (i) dominates activation for hydride formation of intermetallic compounds around room temperature. Most hydride-forming intermetallic compounds absorb and release H_2 extremely rapidly (more rapidly than their constituents under the same conditions) even after having been exposed to air. That "truly remarkable feature of intermetallics which sets them apart from the metallic elements as hydrogen hosts" [2.216] was explained successfully [2.40, 217] by the surface segregation mechanism (i) (Fig. 2.16). The ongoing selective oxidation and surface segregation of A (La) prevents the formation of passivating oxide layers. H_2 molecules penetrate the decomposed surface layer and reach dissociation sites on the B (Ni) precipitates or A_xB_{1-x} (LaNi$_5$) surface.

Besides the activation of intermetallic compounds for H sorption, mechanism (i) is of importance for oxidation and corrosion phenomena at the solid–gas [2.218] and solid–liquid interfaces, particularly electrode–electrolyte [2.40, 219], for the formation of hydrogenation catalysts from rare earth–d-transition metal intermetallic compounds [2.48, 220] and for the oxidation and deterioration of rare earth–transition metal permanent magnets [2.51, 221] and magnetic recording devices, e.g. TbFe or FeMn alloys [2.221, 222].

Mechanism (ii), particularly the solution of surface oxygen in the near-surface and bulk region, characterizes the activation of good getter alloys, which are mostly based on Zr [2.46, 47, 198, 224, 225]. The preparation of catalysts from crystalline and amorphous alloys at elevated temperatures in H_2 atmosphere also proceeds along the lines of mechanism (ii) [2.226–229].

Mechanism (iii), finally, applies for the preparation of many specific catalysts of elevated temperature starting with the deposition of a thin film of the catalyst element (Fe, Pd, Rh, Ni, ...) on an oxide support and subsequent annealing in reducing (H_2) and oxidizing (air) atmosphere to obtain clusters of the right size and interface to the oxide support with the appropriate so-called strong metal-support interaction (SMSI) [2.230, 231]. A typical procedure is shown in Fig. 2.17.

Diffusion of surface deposits into the metal or oxide support and their reappearance at the surface due to oxygen adsorption at elevated temperatures has

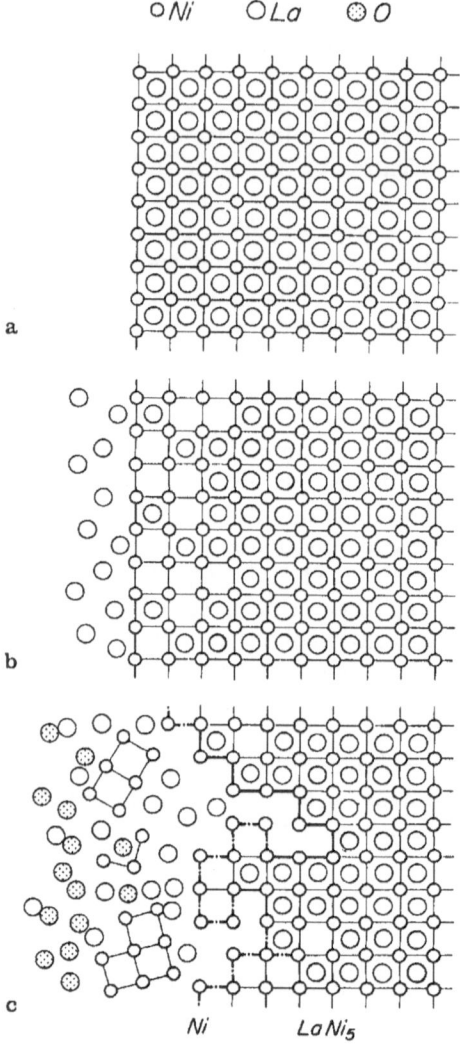

a

b

c

Ni   LaNi$_5$

Fig. 2.16. Surface segregation model and corresponding depth profile (**a**) freshly fractured surface, (**b**) surface enrichment of La to lower the surface energy, (**c**) surface segregation and selective oxidation of La and formation of surface Ni precipitates, (**d**) depth profile according to (**c**)

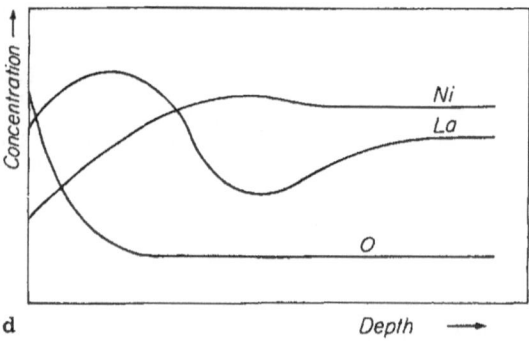

d

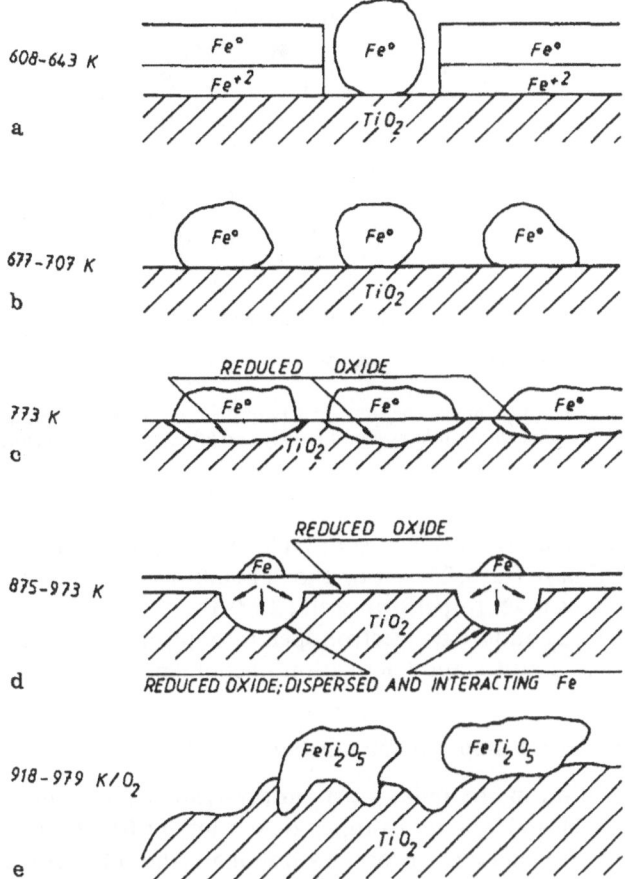

Fig. 2.17 a–e. Schematic diagram of the Fe/TiO_2 system after reduction at progressively higher temperature followed by oxidation [2.230]

also been studied, e.g. for Zr on Pt(100) [2.232] and Fe on MgO [2.233]. This adsorbate-induced restructuring of surfaces was reviewed recently by *Somorjai* and *Van Hove* [2.231].

Further metallic elements are successfully added to interface compounds to enhance [2.234] or diminish [2.222, 235] the rate of oxidation. In TbFe alloys surface oxidation is thought to occur through O diffusion from bulk to surface by way of vacancy sites. Metallic additives such as In were reported to reduce the diffusion by vacancy filling (Fig. 2.18 [2.235]). The influence of the bulk concentration ratio on the oxidation and activation behaviour emerges clearly from studies of Zr-Ni alloys [2.236] and $Zr (Fe_{1-x}V_x)_2$ getter alloys [2.198].

A technically and economically more demanding activation of hydride-forming metals and alloys consists of contacting the surface with a

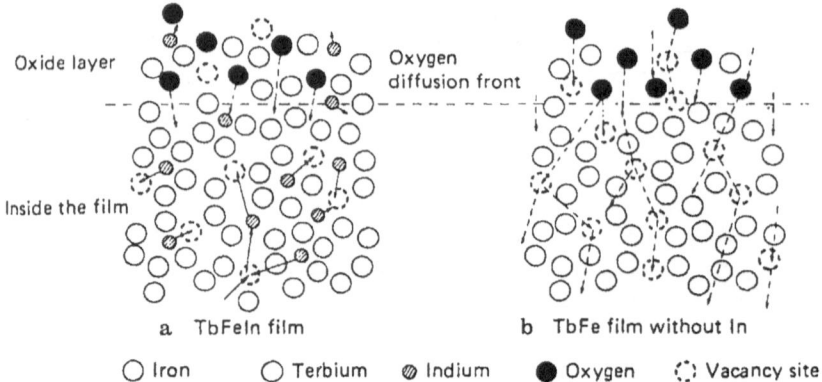

Fig. 2.18 a, b. Oxygen diffusion model involving vacancy-site density in the films. Indium migrates easily due to its small self-diffusion coefficient. It diffuses continuously and fills vacancy sites before oxygen diffusion occurs [2.335]

H_2-dissociating noble-type metal (Pt, Pd; e.g. spot welded foils) and to make use of the atomic H *spilled over* [2.54, 237]. For the "H-spill-over" technique organic complexes were also shown to perform well [2.238].

Photoelectron spectroscopy and Auger electron spectroscopy are the experimental methods most widely used to study activation phenomena.

2.2.8 Deuterium Adsorption and Isotope Effects; Ortho–Para Conversion

In common with H adsorption in the bulk, most surface-related phenomena are studied using H (mass 1) and not its isotopes deuterium D (mass 2) or tritium T (mass 3). Whenever dynamical processes dominate the interaction, isotope effects can occur. They have been reported for the H binding energy, the dynamics of trapping and sticking as well as for surface diffusion [2.21, 239, 240]. None of them point to the extraordinarily short D–D distances that would be required to raise the D–D fusion rate. Isotope effects in bulk diffusion and in the bulk pressure-composition isotherms are well known and are exploited in isotope separation [2.46, 241]; cf. Chap. 5.

In studies of the interaction of Pd(100), Pd(110), and Pd(111) with H and D at low temperature ($\approx 115 \, \mathrm{K}$) an up to 20 times slower rate of filling of α states (bulk H sites) was observed when using D instead of H [2.242, 243]. These results were obtained on the reconstructed surfaces, i.e. after saturation coverage of the chemisorption sites (Fig. 2.19). This isotope effect is normal in the sense that the higher equilibrium pressure of the D_2-Pd bulk isotherms cause a higher driving force for D_2 at equal H_2 and D_2 pressure. The size of the effect, however, is larger than expected and might point to strong quantum phenomena.

Sequential dosing of Pd(111) by D_2 and H_2 and subsequent temperature-programmed desorption recently allowed a distinction to be made between two different pathways for H penetration into the bulk: directly or after equilibrating

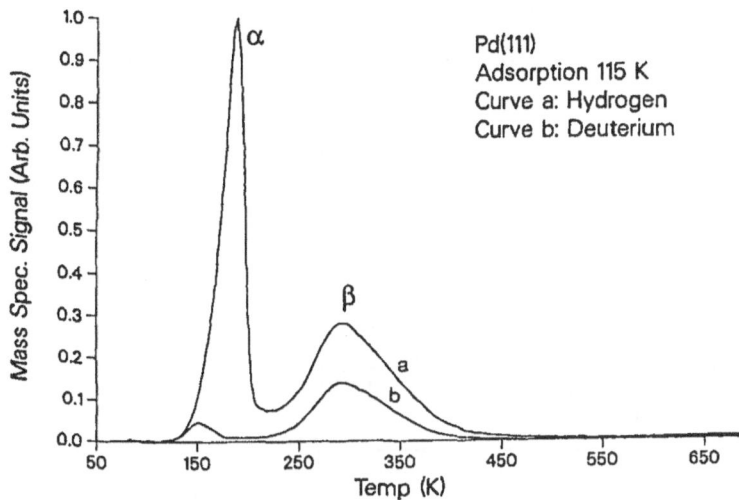

Fig. 2.19. Temperature programmed desorption of H_2 and D_2 from Pd(111) [2.243]. α: desorption from bulk H sites; β: desorption from chemisorption states. After exposing Pd(111) at 115 K to equal doses of H_2 (a) and D_2 (b) a much stronger desorption of α-H than α-D is observed apparently due to faster H than D absorption into α-states

in the chemisorption state [2.243]. In experiment 1 (2) the chemisorption states were saturated by H (D) at 80 K, where negligible bulk penetration occurs. Then 15000 L D_2 (H_2) dosing was added at 115 K. In both experiments the gas which was dosed second appears in the low temperature α-peak, which characterizes bulk states. Only minor HD desorption was observed. Apparently the gas which was dosed second bypassed the chemisorption state.

On the Ni(110) surface, which has a very high initial sticking coefficient for H_2 and D_2 (≈ 0.96), significant differences were found between the elastic scattering of H_2 and D_2 [2.244]: Fig. 2.20 shows the angular distribution for D_2 and H_2 scattering along the (001) direction. The elastically scattered part for D_2 is substantially lower. If energy loss were caused by electron–hole excitations no isotope effect would be expected. In phonon interactions, however, the relative energy transfer to the surface should be about twice as large for D_2 as for H_2.

On the Ni(111) surface an isotope effect was observed to occur at low temperature only [2.245]. Whereas temperature-programmed desorption reveals no difference between H_2 and D_2 at 140 K adsorption temperature, a weaker and kinetically retarded desorption of D_2 from β_1 sites was observed at 87 K adsorption temperature (Fig. 2.21). Apparently the rate of dissociative adsorption at that temperature is already sensitive to zero-point vibrational energy differences between D_2 and H_2.

As with every homonuclear diatomic molecule H_2 and D_2 can exist either in the ortho or the para spin modification with parallel or antiparallel nuclear spins, respectively. At temperatures above ≈ 100 K the ratios are o:p (H_2) = 3

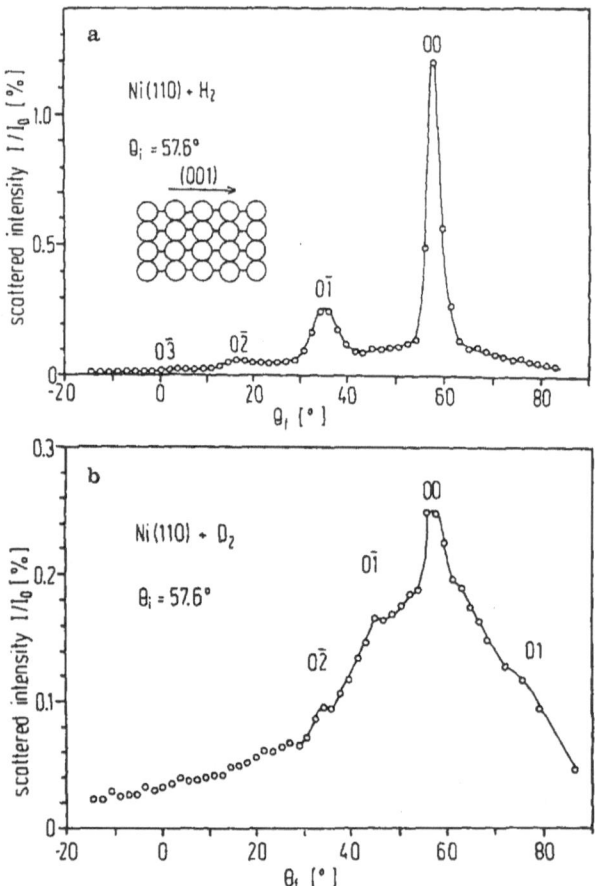

Fig. 2.20 a, b. Top: Angular distribution of H_2 scattered from Ni(110) along the (001) direction, $\theta_i = 57.6°$, $T_S = 400\,K$, $\langle E_\perp \rangle = 43\,meV$ (H_2:D_2 = 1:1 beam), surface periodicity $d = 3.52\,Å$. The large fraction of molecules which underwent dissociative adsorption and subsequent desorption has been separated by using a modulated beam and lock-in techniques [2.244]. Bottom: Angular distribution of D_2 scattered from Ni(110)

and o:p $(D_2) = 2$. At low temperatures para-H_2 and ortho-D_2 are the more stable [2.21]. It was shown that the ortho–para conversion rate is two orders of magnitude faster on a Ag(111) than on a Cu(100) surface. *Ilisca* [2.73] recently suggested that the difference does not characterize Ag and Cu, but is inherently related to the surface orientations.

Low temperature tunneling diffusion of H and D on the W(110) surface shows an isotope effect that is several orders of magnitude smaller than expected, while the higher temperature activated diffusion displays an inverse isotope effect larger than expected. Both results were explained consistently [2.166] by the fact that there is a large time scale difference between the H or D and W motions. Large phonon overlap factors reduce the tunneling. An adsorbate-

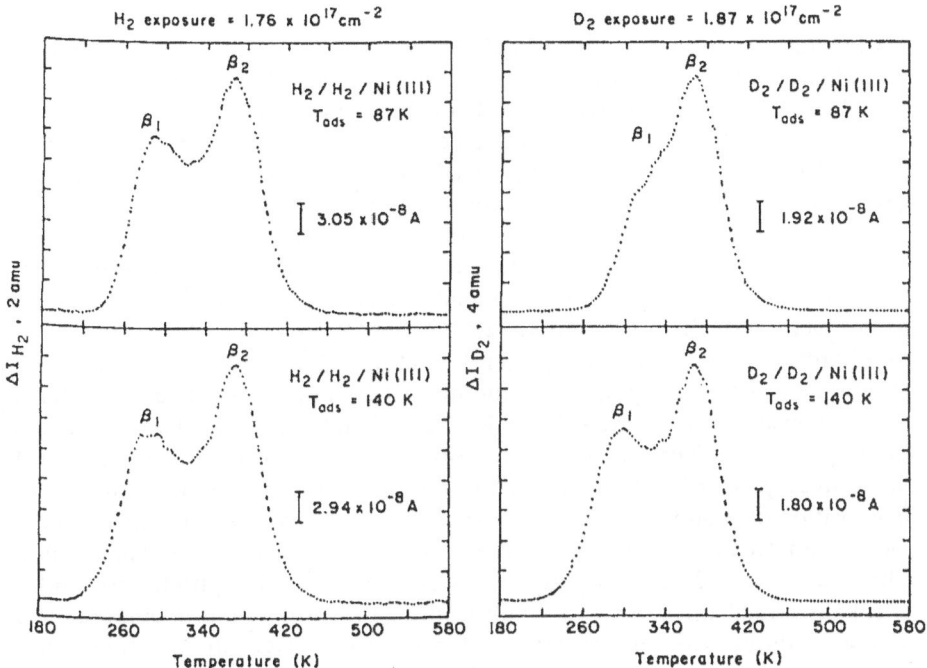

Fig. 2.21. Effect of surface temperature on the dissociative adsorption kinetics of H_2 and D_2 on Ni(111). The H_2 and D_2 saturation coverage spectra for H_2 or D_2 adsorbed at 87 and 140 K are shown in the left and right panels, respectively [2.245]

mass-dependent phonon–vibron coupling parameter describes activated diffusion (a many-phonon process) and tunneling.

Activated D_2 dissociative chemisorption with D-induced surface reconstruction has been found on the Ag(110) surface [2.246].

The influence of a high electric field on the photon stimulated desorption of hydrogen and deuterium from Rh and Ni surfaces is to yield H^+, D^+, H_2^+ and D_2^+ as desorbing species with apparently no isotope exchange [2.247].

For the detection of surface deuterium, direct recoil spectroscopy using 600 eV Li^+ ions [2.248], low energy He^+ scattering in the forward direction [2.249], and laser Raman spectroscopy [2.250] are promising techniques. The H_2–D_2 exchange reaction, i.e. the formation of HD molecules, is an effective and simple probe of the dissociative properties of a surface (Sect. 2.3).

2.3 Experimental Methods

The range of experimental methods for studying surfaces, adsorption, adsorption-induced phenomena and activation is very large. Most of the methods are used for surface studies in general, a few of them, however, are specific to H at metallic surfaces. Due to its low atomic number, the direct detection of H by some methods is difficult or impossible (e.g. in Auger electron spectroscopy); accordingly H-induced variations of the substrate properties are often analyzed instead.

Typical surface-sensitive methods probe one or a few atomic monolayers. As gas kinetics tells us that a monolayer is adsorbed in 1 s at a partial pressure of 10^{-6} Torr on a reactive surface (sticking probability 1), ultrahigh vacuum (10^{-10} mbar) is needed to study reactions on atomically clean surfaces. The preparation of surfaces of the appropriate cleanliness and structure is extremely important and often time consuming. To prepare atomically clean and structurally perfect surfaces of elemental metals repeated outgassing, sputtering and subsequent annealing is required [2.32, 251]. Alternatively, clean surfaces can be obtained by evaporation or sputter deposition of fresh clean films. The cleaning of surfaces of alloys is more demanding. All etching methods (e.g. sputtering, electropolishing) are more or less selective and etch one alloy element preferentially [2.41, 252]. The best results are obtained by cleavage or fracture—if the mechanical properties of the alloys allow it—after outgassing at high temperature. Deposition of fresh clean films by co-evaporation or co-sputtering of all components is another alternative. The surface sensitivity of the methods, which renders them so powerful for studying the early stages of adsorption, turns into a weakness when one wishes to analyze the activation of multicomponent air-exposed alloys with surface layers of some nanometers in thickness. Depth profiling by repeatedly sputtering off the top layers is feasible, although sputtering is selective and may reduce the oxidation state. Methods which probe thicker layers can be applied at higher gas pressure, but the information they provide is less detailed and often suffers from resolution and intensity problems.

About 80% of the surface analysis related to H sorption is performed using LEED (low energy electron diffraction; structure), PES (photoelectron spectroscopy; electronic structure, chemical composition) and EELS (electron energy loss spectroscopy; surface dynamics). Given below is a short description of the frequently used methods and of the H-specific methods. For other methods and for more detailed descriptions the reader is referred to the review papers cited below and to the general references on surface science [2.1–18], catalysis [2.253], and experimental techniques [2.254].

Photoemission [2.6, 15, 255]: The sample is irradiated by monoenergetic photons causing electrons from the valence band and from core levels to be photoemitted. Ultraviolet lamps (mostly He I and He II at 21.2 and 40.8 eV), characteristic X-rays (mostly MgK_α and AlK_α of 1254 and 1486 eV) and synchrotron radiation with a tunable monochromator are used as photon sources. The photoelectrons are detected as a function of their kinetic energy E_K from which their binding energy E_B is determined from

$$E_B = \hbar\omega - E_K - \Phi \tag{2.4}$$

where $\hbar\omega$ and Φ are the photon energy and sample work function, respectively. According to the energy range of the photons, one distinguishes ultraviolet and

X-ray photoelectron spectroscopy (UPS, XPS). The emission from the valence bands measures the occupied region of the density of states and gives, for example, information on the bonding of adsorbates and adsorbate-induced states. The core level emission intensities and core level binding energies serve to probe surface concentrations and the chemical environment of the atom, respectively. By means of core level spectroscopy one can study surface core level shifts and surface structure [2.117, 256, 257]. The surface sensitivity is determined by the mean free path of the photoelectrons which varies between 0.5 and 3 nm for kinetic energies between 50 eV and 1 keV. From angle-resolved photoemission spectroscopy (ARPES, ARUPS) [2.258 260] of the valence bands, energy dispersion curves $E(\vec{k})$ can be evaluated and compared with results from band structure calculations. Angle-resolved photoelectron spectroscopy at a fixed energy of a core level yields structural information through the diffraction of the electrons, e.g. from adsorbates, and is called X-ray photoelectron diffraction (SPD, ARXPD) [2.117, 261].

Measurements of the *work function* and its changes during adsorption provide information about charge transfer and bonding. Analysis of the cut-off energy of the valence band as measured by UPS or capacity measurements between sample and a special probe are currently used techniques [2.14, 254].

Auger electron spectroscopy (AES) [2.6, 15, 262]: Core hole excitations are created either by 3–10 keV incident electrons or by characteristic X-rays (MgK_α, AlK_α) and Auger electrons are emitted at energies characteristic for the element by the two-electron Auger process and analysed as a function of their kinetic energy. The integrated intensity of a peak or the peak-to-peak height in the first derivative of the intensity (background eliminated) give information on the surface concentration. The surface sensitivity is determined by the mean free path of the Auger electrons.

Spin polarized electron emission: The detection of the spin of photo-, Auger- or secondary electrons without or with energy preselection probes the surface magnetism by measuring the polarization of the emitted electrons [2.110–113, 263].

Inverse photoemission or bremsstrahlung isochromat spectroscopy (IPES or BIS) [2.6, 264–266] probes otherwise inaccessible empty states by the time-reversed photoemission process. A breamsstrahlung continuum is produced by electrons impinging onto the surface.

The complementary nature of the various electron-spectroscopic methods for studying adsorbates is clearly seen in the reviews by *Plummer* et al. [2.267] and by *Umbach* [2.268]. The use of electrons as a probe for surface H, including tritium imaging, was reviewed by *Malinowski* [2.269].

Low-energy electron diffraction (LEED), [2.3, 6, 7, 270, 271]): A beam of monoenergetic electrons with energies from 20 to 500 eV is directed at a single crystal surface. The elastically backscattered electrons, which are concentrated along specific directions because of interference, appear as spots on a screen (other detection methods exist, but did not find wider use) and map out the two dimensional reciprocal lattice of the surface. The spots are recorded photographically or by use of a video camera which permits to follow structural transitions simultaneously e.g. with dosing.

Thermal energy *helium atom diffraction* has been very successfully used for structural studies of submonolayers and monolayers of adsorbed H and the related reconstructions [2.272]. The corrugation function which is extracted from the diffracted beam intensities maps an electron charge density contour at low density and thus is very surface sensitive.

Besides LEED and He diffraction, X-ray scattering at grazing incidence [2.273] and scattering of thermal He atoms from ordered and disordered surfaces [2.275] have been used for studying the structure of adsorbed H_2 and H. An illustrative example are the phase transitions of H on W(100) studied with synchrotron radiation at grazing incidence [2.274].

The Fourier transform of the diffraction pattern gives the unit cell dimensions. However, as in bulk diffraction problems, the detailed atomic structure is deduced by trial-and-error iteration to get the minimum in the R-factor. Direct methods in surface crystallography were discussed recently by *Pendry* et al. [2.276]. The bulk structure lying beneath the surface is used as a reference structure.

The diffraction of low energy H_2 beams has been used to study the H_2–Cu surface potential for various surface orientations [2.71].

Time of flight analysis of direct recoils (TOF-DR; Chap. 7): Pulsed rare gas ion beams at grazing incidence are used to bombard the surface. Directly ejected surface particles including light adsorbates like H can be analyzed by time-of-flight measurements. The surface composition and to some degree the structure are probed.

Surface extended X-ray adsorption fine-structure (SEXAFS) [2.256, 277] is based on measurements of the X-ray absorption coefficient of an atom as a function of the energy of incident photons. Core electrons are emitted above threshold. If the absorbing atom is bonded to neighbours, the photoelectron wavefunction goes through a series of maximum and minimum amplitudes which contains information on the distance between absorbing atom and its neighbours. In the energy range 50–500 eV only single scattering of the Photoelectron wavefunctionn is important so that the analysis, although not direct, is at least straightforward.

Scanning tunneling microscopy (STM) [2.278–280]: The electron density of a solid does not drop to zero at the surface, but decays approximately exponentially outside the surface within a short decay length which is related to the electronic or chemical properties of the surface. The tunnel current between the substrate and a counterelectrode sharpened to a pointed tip depends strongly on the decay length and on the distance between tip and substrate. Scanning of the tip over the substrate traces an almost true image of the surface topography and also probes the surface chemistry. The STM is nondestructive, it is a structural and chemical methods, applicable to both periodic and nonperiodic surface features and can be operated at atmospheric pressure. The STM yields less accurate data on surface structures than e.g. low-energy electron diffraction. However, since it resolves space and time it allows the observation of the nucleation of a surface reconstruction. The STM has been used for studying the H-induced reconstruction of the Ni(110) surface (Sect. 2.4.1).

Electron energy loss spectroscopy (EELS) and *high resolution electron energy loss spectroscopy* (HREELS) [2.5, 6, 8, 281–283]: The sample is irradiated with a monochromatic beam of electrons and the energy distribution of the inelastically scattered electrons is measured. The scattering process can involve the excitation of core or valence electrons to unoccupied states and thus provide information for chemical analysis. It may also cause the excitation of surface and bulk phonons and of vibrational states of adsorbates. The high surface sensitivity makes EELS a unique and important tool for surface vibrational spectroscopy particularly of adsorbates including H. The incident beam typically has an energy in the range 1–10 eV with a spread of 5–10 meV. Vibrational losses are of the order of 100 meV.

Incoherent inelastic neutron scattering (IINS) [2.284]: Neutron scattering is known to be a very valuable tool for studying the dynamics of bulk hydrogen. High neutron fluxes (10^{14}–10^{15} neutrons/cm^2 s) allow the investigation of the dynamics of adsorbed hydrogen for samples with a reasonably large surface-to-bulk ratio. The sample can be studied in a reactor vessel over a wide range of pressure and temperature, and therefore the technique is thus complementary to methods such as electron energy loss spectroscopy, which measures e.g. the vibrational spectrum of adsorbates on well-characterized surfaces under ultrahigh vacuum. Neutron methods can cover an energy range from 1 μeV to 300 meV and thus enable studies of the diffusion and rotational dynamics of adsorbates on a time scales of 10^{-8}–10^{-13} s. Figure 3.26 illustrates the inelastic neutron scattering spectra for H (diluted with D) on Raney nickel at coverages below saturation.

In *thermal desorption spectroscopy* (TDS) and *temperature-programmed desorption* (TPD) ([2.21, 285] and Chap. 7) an adsorbate-covered surface is heated at a linear or programmed rate and the desorbing species are detected with a mass spectrometer. These methods are currently used for H and D. Information on adsorption energies and adsorption rates at different sites (surface, subsurface and bulk) is obtained. The clean substrate metal often has to be cooled to low temperatures for adsorption.

Secondary ion mass spectroscopy (SIMS) [2.286, 287]: The surfaces is bombarded with rare gas ions such as Ar$^+$ of 1–4 keV energy. Surface atoms including H are ejected as secondary ions (or neutrals) and are detected with a mass spectrometer. The surface composition can thus be evaluated. Depth profiling is possible by means of continuous bombardment. H concentration profiles have been studied using SIMS by Bastaz [2.287].

Nuclear probes [2.288]: making use of nuclear reactions or of nuclear moments allow the sampling of surface layers and interfaces much thicker than the ultra high vacuum electron-spectroscopic methods. The reaction $^1H(^{15}N, \alpha\gamma)$, ^{12}C, for example, has been used for H depth profiling [2.92, 289, 290].

Conversion-electron Mössbauer spectroscopy (CEMS) [2.291]: A nucleus is excited by γ-absorption and can undergo inverse β-decay, creating a core-hole. The core-hole decays by an Auger process. Electrons emitted from atoms located in a layer of thickness equal to the electron mean free path are detected. Thus 50–150 nm thick layers are probed. Like Mössbauer spectroscopy, the technique is limited to few isotopes. Depth profiling is also possible (cf. FeTi in Sect. 2.5.3).

In the H_2–D_2 *exchange reaction* [2.21, 23, 292] the dissociative and associative character of a surface is probed by means of a mass spectrometer. The surface is exposed to a mixture of H_2 and D_2. If dissociation and associative desorption occurs, masses 2, 3 and 4 are observed for H_2, HD, and D_2, respectively. Otherwise mass 3 is absent. This method has been used to evaluate activation models for FeTi [2.41].

The *Pd-gate metal-oxide-semiconductor device* (Pd-MOS) [2.293, 294] is built of a Pd-SiO$_2$-Si structure and is capable of detecting H_2 at pressures from 1×10^{-10} mbar to 1 bar. H_2 molecules in the ambient, and in ultrahigh vacuum adsorb and dissociate on the Pd gate. H atoms diffuse rapidly through the Pd film to the Pd-SiO$_2$ interface where they form a dipole layer and decrease the effective work function of Pd. The work function shift can be measured as a shift of the capacitance versus voltage curve (Fig. 2.22). The device can be used to detect H in metallic overlayers down to submonolayers thickness [2.294] and to measure sticking coefficients.

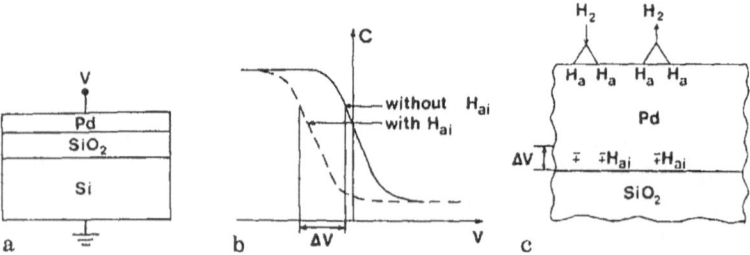

Fig. 2.22. (a) Schematic picture of a hydrogen-sensitive MOS structure. (b) $C(V)$ curve for a structure with and without the presence of hydrogen. (c) dissociation and association take place at the Pd surface. The shift of the $C(V)$ curve in the presence of hydrogen is caused by the hydrogen atoms at the interface, H_{ai}, forming a dipole layer [2.293]

2.4 Hydrogen on Elemental Metals

As mentioned earlier the numerous results on the adsorption of hydrogen on the elemental transition and noble metals, mainly Ni, Pd, Nb, Fe, Rh, Ru, Pt, Ag and W, were reviewed recently by *Christman* [2.21]. That review also contains tables on structure and phases, sticking, heat of adsorption, vibration, work function, and frequency factors. We therefore give only a short summary of these results which have led to the present high degree of understanding of H on clean transition metal surfaces. Far less advanced is our knowledge of hydrogen adsorption by rare earth and simple metals (Sects. 2.4.4–5).

2.4.1 Hydrogen on Nickel

Ni does not form a stable metal hydride. Hydrogen adsorption has been studied mostly on the close-packed Ni(111) surface and on the more open Ni(110) surface of fcc Ni.

On Ni(111) at low temperatures hydrogen forms two adsorption states β_1 and β_2 with thermal desorption peaks at 346 K and 384 K, respectively. β_2-state H atoms occupy threefold coordination sites up to half a monolayer coverage. On smooth surfaces both the β_1 and β_2 states show activated dissociative adsorption with small (≤ 0.1 eV) activation energy and sticking coefficients as low as ≈ 0.05 and 10^{-4} for β_2 and β_1, respectively. The desorption flux is highly peaked towards the surface normal ($\cos^5\Theta$, Fig. 2.23). Defects introduce non-activated adsorption sites which lower the desorption temperature in flash desorption and broaden the desorption flux towards an ordinary cosine distribution [2.244, 2.295]. Photoelectron spectra of one monolayer of H adsorbed on Ni(111) are characterized by a H-induced band split off from the

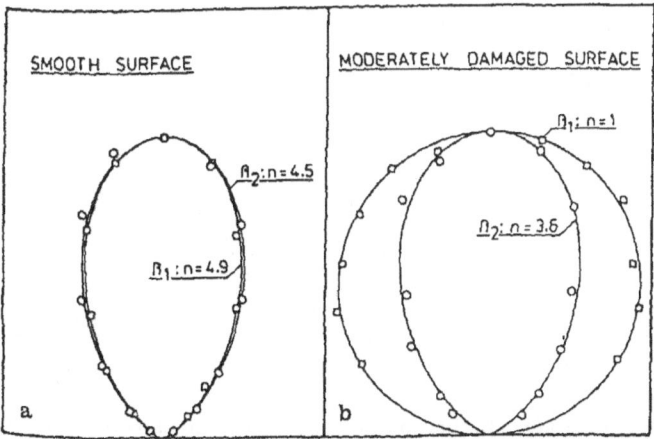

Fig. 2.23 a, b. Polar plot of the normalized desorption flux $D(\theta)/D(0^\circ)$ from the two adsorption states β_1 and β_2 for H_2/Ni(111). Tilt axis for change of angle θ: [0$\bar{1}$1] [2.295]

bulk bands. As the H coverage is decreased, the split-off state moves upwards into the bulk bands and disappears at only about 0.5 saturation coverage. The surface states of the substrate shift continuously with H coverage on Ni(111) [2.296]. The phase diagram for ordered H in the $c(2 \times 2)$-2H structure as evaluated from LEED reflections is shown in Fig. 2.24 [2.21]. Above 270 K no ordered H overlayer is formed. The phase diagram and the change of the heat of H adsorption with coverage can be consistently described by a lattice gas model assuming nearest neighbour exclusion, second- and third-neighbour repulsion and further neighbour interactions [2.297].

On Ni(110) the initial sticking coefficient approaches unity ($S_0 = 0.96$ for a substrate and gas temperature of 300 K). Dissociation occurs without an activation barrier. A simple cosine distribution is observed in the desorption

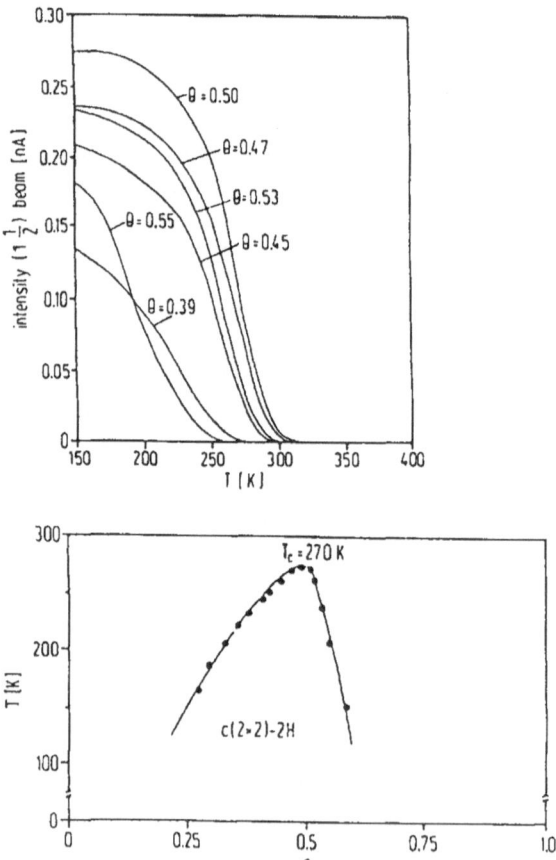

Fig. 2.24. Top: Temperature dependence of the fractional-order LEED beams of the $c(2 \times 2) - 2H$ structure on Ni(111). Parameter of the curves is the coverage θ [2.21]
Bottom: Phase diagram for the $c(2 \times 2) - 2H$ structure on a Ni(111) surface as derived from the inflection points of the curves shown in the top figure. The critical temperature T_c is indicated [2.21]

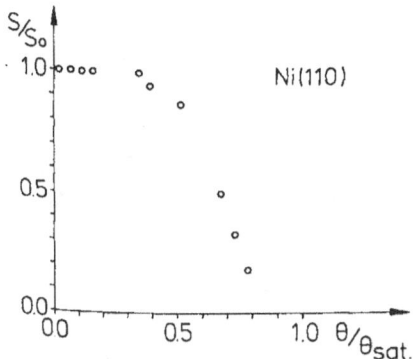

Fig. 2.25. Variation of the relative dissociative sticking coefficient with coverage for D_2 on Ni(110), $T_s = 155$ K [2.244]

of H_2 relative to the surface normal, meaning that the desorbing H_2 are in thermal equilibrium with the surface. The variation of the relative sticking coefficient S/S_0 with coverage is reproduced in Fig. 2.25 [2.244]. There is no detectable variation of the initial sticking coefficient S_0 with temperature in the range 150–800 K.

H on Ni(110) provides a beautiful example of how the interaction of adsorbed H atoms with each other and with the substrate result in a rich variety of ordered phases and phase transitions as a function of coverage including surface reconstruction. He diffraction from H on Ni(110) yields at low temperatures ($T < 130$ K) the following ordered phases [2.298]: $c(2 \times 6)$ for $\theta = 1/3$ ML, $c(2 \times 4)$ for $\theta = 1/2$ ML, $c(2 \times 6)$ for $\theta = 2/3$ and $5/6$ ML, (2×1) for $\theta = 1$ ML and (1×2) for $\theta = 1.5$ ML. Four of these phases are shown in Fig. 2.26. The common structural elements are long zig-zag chains of along the Ni closed-packed rows with H atoms near to threefold sites. Up to one monolayer coverage the substrate Ni atoms do not change their position. Further H, however, induces a substrate reconstruction and at 1.5 ML coverage 1/2 ML H is thought to be adsorbed on the second Ni layer. On the reconstructed surface three distinct desorption states (α, β_1 and β_2) are observed in thermal desorption spectroscopy (Fig. 2.27, [2.299]). The α peak is particularly sharp and can only be observed when the reconstructed 1 $\times$ 2 phase is present. Apparently it corresponds to H desorbing from the second Ni layer. The reconstructed 1 $\times$ 2 phase can be regarded as a surface hydride. The decrease of the diffracted He beam intensity with rising temperature is shown in Fig. 2.28 together with a typical diffraction pattern at low and high temperature.

The question of whether the (1 $\times$ 2) reconstruction at 1.5 ML coverage and low temperature corresponds to a pairing or missing row (Fig. 2.29) was studied in detail. Angle-resolved photoemission studies [2.301] reveal a H-induced state 9.0 eV below E_F, split off from the Ni bulk bands, in addition to the Ni valence-band satellite (Fig. 2.30). The 3d-like bulk states are markedly reduced indicating their participation in the bonding. A theoretical analysis based on experimental photoemission data favours the pairing-row model. The reconstruction is explained by the progressive H-induced lowering of the energy of 3d states [2.300].

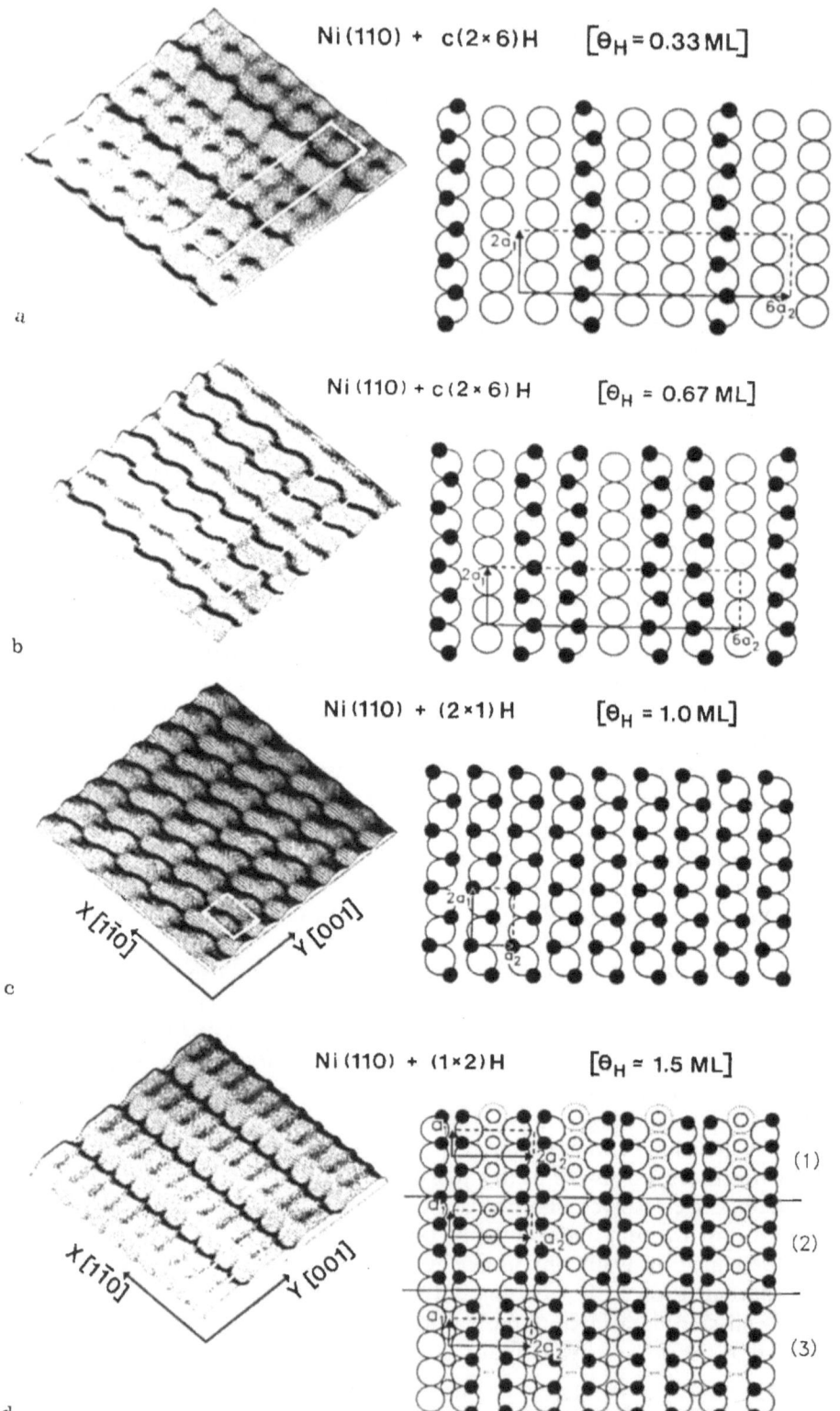

Ni(110) + c(2×6)H [$\theta_H = 0.33$ ML]

Ni(110) + c(2×6)H [$\theta_H = 0.67$ ML]

Ni(110) + (2×1)H [$\theta_H = 1.0$ ML]

Ni(110) + (1×2)H [$\theta_H \approx 1.5$ ML]

X [1 1̄ 0]   Y [001]

a

b

c

d

(1)

(2)

(3)

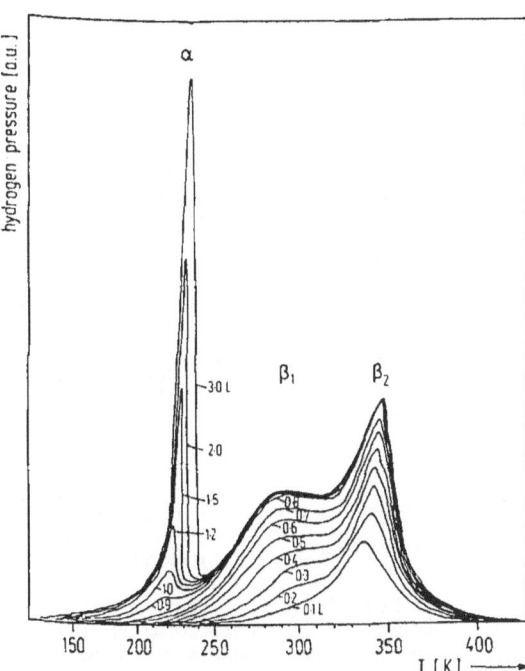

Fig. 2.27. Hydrogen on Ni(110): Series of thermal desorption spectra obtained after adsorption of hydrogen at 120 K. The heating rate was ≈ 9 K/s. The various desorption states are indicated. Note the extreme sharpness and peculiar kinetics of the α-state [2.299]

Further support for the pairing-row model came from the observed phonon softening [2.302] and recently from LEED intensity vs. voltage data which also demonstrated periodic vertical displacements (buckling) of the atoms in the second layer [2.303]. Calculations of the pair interactions showed, in agreement with experiment, that the most stable configuration is the zig-zag chain with H atoms located on either side of the (110) rows [2.304]. A reconstructed streaky (1×2) phase is formed upon H_2 exposure of Ni(110) at room temperature or upon heating of any low temperature .phase above $T = 220$ K [2.21, 305]. The local periodicity was confirmed by the first scanning tunneling microscopy of a H-induced surface reconstruction to be (5×2) [2.306].

Chemisorption of H_2 was observed on stepped Ni(100) surfaces precovered with a dense layer of atomic H. No such state, however, was observable on the flat (100) surface [2.77]. Total energy calculations of H on Ni(100) reveal significantly stronger H binding energy than effective medium theory [2.105]. The most favorable H position is in the centre site, $0.6a$ above the plane of the surface Ni atoms only. The bridge-site minimum lies only 0.1 eV higher, consistent with the high surface mobility of the hydrogen atoms.

Fig. 2.26. He diffraction of H on (Ni(110) [2.298]. Best fit corrugation functions (left) and hard sphere models (right) of (**a**) the first $c(2 \times 6)$H phase with $\theta = \frac{1}{3}$ ML; (**b**) the second $c(2 \times 6)$H phase with $\theta = \frac{2}{3}$ ML; (**c**) the (2×1)H phase with $\theta = 1$ ML; (**d**) the (1×2)H phase with $\theta = 1.5$ ML on the reconstructed substrate. The pronounced hills correspond to H atoms

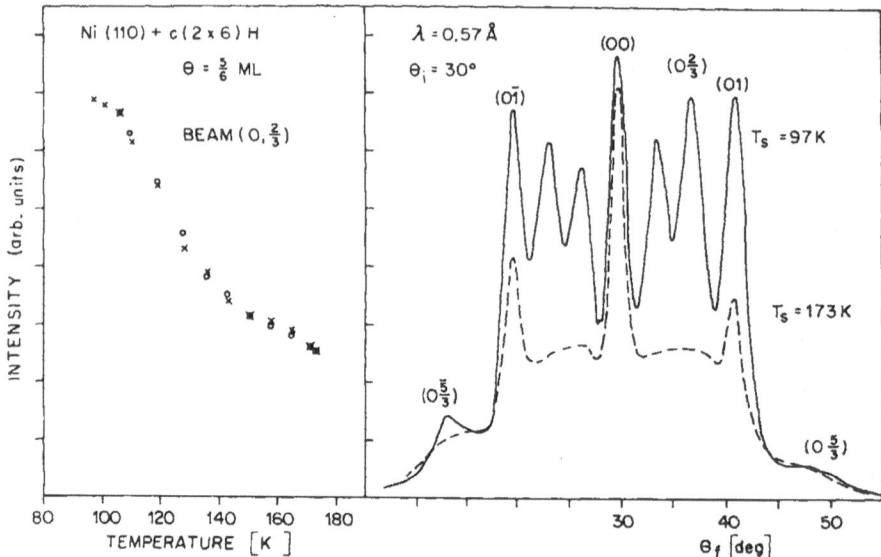

Fig. 2.28. He diffraction from H on Ni(110) [2.298]: Left side: Intensity behaviour of the (0, 2/3) beam of the $c(2 \times 6)$H phase with $\theta = 5/6$ ML as a function of temperature. Crosses refer to heating, and circles to cooling of the sample. The strong intensity change around 140 K is indicative of a reversible order-disorder transition. Right side: Typical diffraction spectra at the lowest (full line) and highest (dashed line) temperature indicated on the left [2.298]

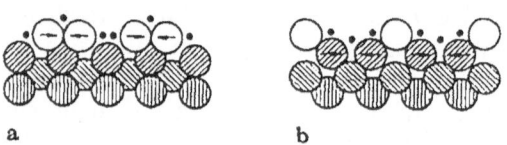

Fig. 2.29. Hard-sphere profiles (view is in the [110] direction) through four layers of (**a**) pairing-row model and (**b**) missing-row model for $(1 \times 2) -$ Ni(110)/H with $\theta_H = 1.5$ ML. Row-pairing is indicated by lateral arrows. Full black circles: H atoms. Note that to improve on the visibility of the H atoms they are shown somewhat displaced from their true position [2.300]

H was calculated to reduce the magnetic moment of Ni(100) surface atoms to $0.2\,\mu_B$ in qualitative agreement with experiment [2.307, 308]. The desorption of H implanted in Ni(111) (> 1 keV) was found to be strongly forward peaked along the surface normal. The effect was tentatively explained by the recombination of an implanted and a chemisorbed H atom [2.309].

2.4.2 Hydrogen on Palladium

H on Pd(110): H_2 adheres to Pd(110) with a sticking coefficient of almost unity and dissociates without activation. He diffraction, LEED and TDS reveal in good agreement with each other the following results: H_2 exposure of Pd(110)

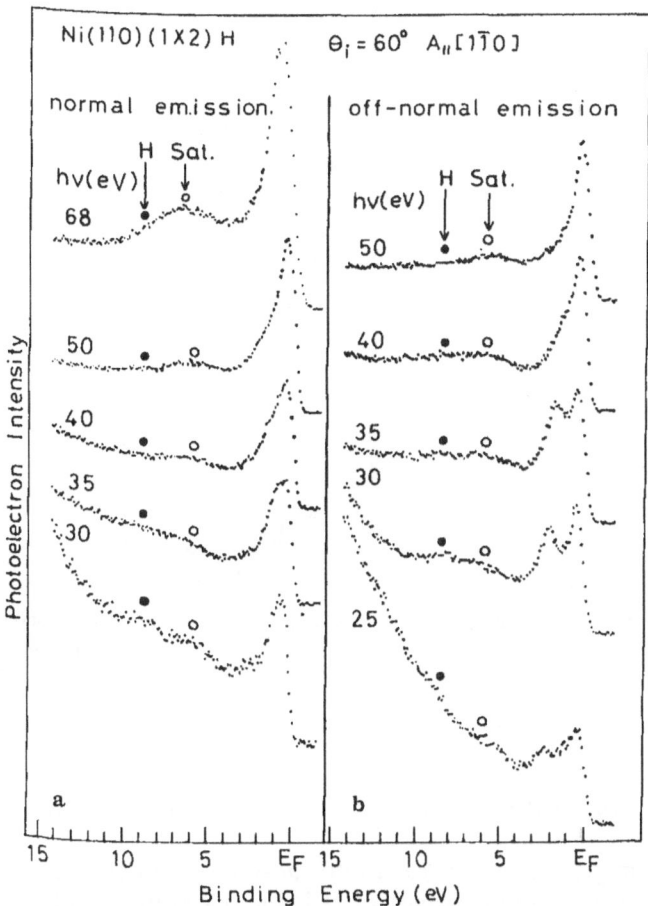

Fig. 2.30. Photon-energy (hv) dependence of angle-resolved photoemission spectra for Ni(110)(1 × 2)-H taken at $\theta_i = 60°$ (surface component of the vector potential A of the light in the [1$\bar{1}$0] azimuth, i.e. $A_{\parallel}$ along [1$\bar{1}$0]: (**a**) normal emission; (**b**) off-normal emission [parallel component of electron momentum [$K_{\perp} = \frac{2}{10}d(\Gamma X)$ at 8.5 eV]. The peaks due to the H-induced bonding state are indicated by solid circles. The open circles indicate the position of the Ni valence-band satellite [2.301]

at 100 K yields two ordered surface phases, a (2 × 1)H phase at monolayer coverage and a (1 × 2)H phase at monolayer coverage and a (1 × 2)H phase with reconstructed Pd substrate at 1.5 ML coverage (Fig. 2.31, inset). Interestingly, the substrate reconstruction enables 0.5 ML of H to enter the near surface region and to populate subsurface sites. TDS spectra (Fig. 2.31) show H desorption from two different subsurface states (α_1, α_2) and from two normal chemisorption states (β_1, β_2). Upon heating the (1 × 2)H phase transforms back into the (2 × 1)H phase and part of the H moves subsurface [2.21, 310, 311]. The H-induced (1 × 2) reconstruction is well described by the pairing-row model with an alternating 0.4 Å lateral displacement of close-packed ⟨110⟩ Pd rows and a

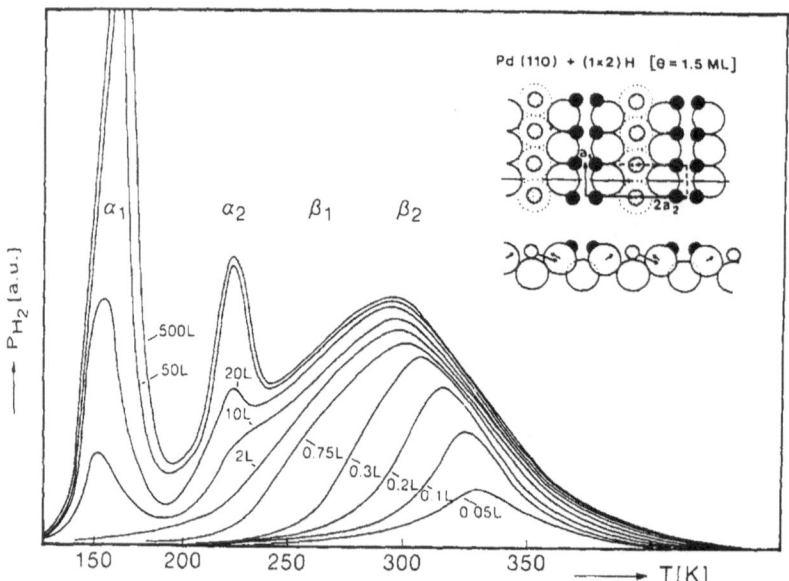

Fig. 2.31. *H* on Pd(110): Series of thermal desorption spectra. Exposures from 0.05 L up to 500 L at 120 K [2.311]. Inset: Top view and side view of a hard-sphere model of the (1 × 2) phase on Pd(111) as deduced from He diffraction. The arrows indicate the motion of the H and Pd atoms upon transformation from (1 × 2)H to (2 × 1)H, where some of the H atoms move into subsurface sites [2.310]

corresponding vertical shift of the same order, as confirmed by ion scattering [2.312].

H on Pd(111): The migration of surface H into subsurface sites was originally suggested to explain the disappearance of H-induced features in photoelectron spectra of H/Pd(111) upon heating to 300 K [2.313]. This suggestion stimulated many further experimental and theoretical studies [2.296, 314–317]: Self-consistent pseudopotential calculations show that H atoms in subsurface tetrahedral sites have electronic properties very similar to those of H in the surface threefold sites; furthermore, the electronic properties of H in subsurface octahedral sites are similar to the clean surface [2.316]. Accordingly, the photoelectron spectra alone do not give clear evidence for subsurface H. *Chubb* and *Davenport* [2.317] calculated the electronic structure of three-layer Pd films with H layers outside and inside and showed that the position of the H induced states depends markedly on concentration. The H-induced bonding of bulk Pd hydride was shown [2.315] to be much weaker and at lower energy than was reported in earlier studies. Indeed, a more detailed analysis by photoelectron spectroscopy as a function of temperature [2.296] revealed that the H-induced band changes position and shape with H concentration. It moves closer to E_F as the H concentration is decreased, merging into the bulk band and finally losing its identity. *Felter* et al. [2.63, 318] present both experimental and theo-

Intensity during adsorption

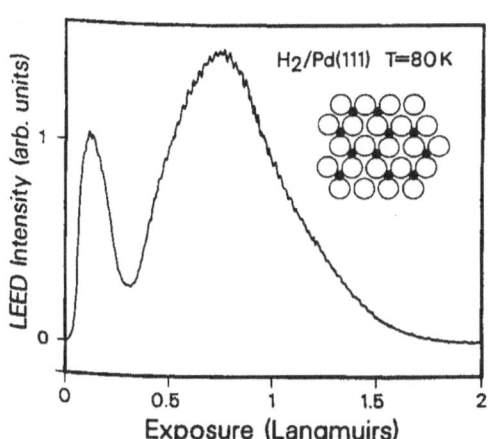

Fig. 2.32. H on Pd(111): Intensity of the $\sqrt{3}$ LEED spot during exposure to H_2 showing transitions from disordered to two ordered (1/3, 2/3) phases. Inset: $\sqrt{3} \times \sqrt{3}$ R30° structure proposed to be responsible for the second peak [2.318]

retical evidence for H bound in the octrahedral site between the first and second planes of Pd(111). They observed order-disorder transformations as a function of temperature and coverage by LEED (Fig. 2.32). Ordered, phases with a $(\sqrt{3} \times \sqrt{3})R$ 30° LEED pattern were observed at coverages of 1/3 and 2/3 of a monolayer at low temperatures. They disorder upon warming to 85 K and 105 K, respectively. Theoretical calculations using the embedded atom method predict the occupation of subsurface sites even at these low temperatures. The H–metal bond in the octahedral subsurface site is comparable to that of the surface threefold site, but H–H repulsion makes the subsurface site more favorable at high coverages. The same method also yields the correct phase transition temperatures. Rapid surface diffusion occurs at temperatures > 55 K. The metal interlayer spacings and H positions above and below the top Pd layer were fitted to reproduce LEED I–V curves at 2/3 coverage [2.319]. One of the two H atoms in the unit mesh is in a three-fold hollow surface site, above third layer atoms (denoted A^+ in Fig. 2.33). The second H resides either in the same type of site or in another three-fold site above second-layer metal atoms B^+, or in subsurface sites between first and second-layer metal atoms A^- or B^-. Good R-factors were achieved for subsurface occupation fractions of up to 60%. Metal interlayer spacings are within the error bars unchanged from those of the clean surface.

Differences in the low temperature desorption of H and D from Pd(111) [2.243] were summarized in Sect. 2.2.8. They could conceivably be applied in an isotope separation process.

H on Pd(100): Band structure and total energy calculations in a tight-binding model for a H impurity near a Pd(100) surface yield bridge geometry as the most stable absorption site, in agreement with experiment [2.320]. Preferential subsurface occupation is favoured by trio H interaction. Pseudopotential local-

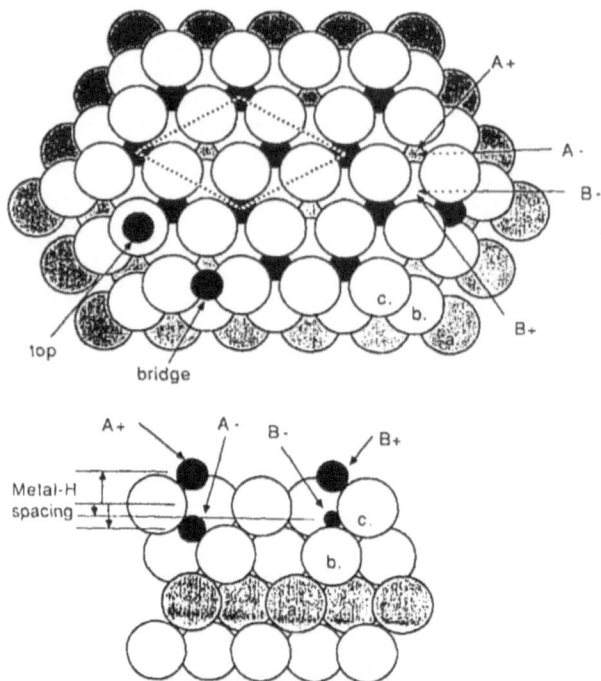

Fig. 2.33. Top: Arrangement of the three outermost layers of Pd(111) in the fcc stacking arrangement, abc. The $(\sqrt{3} \times \sqrt{3})R30^\circ$ unit mesh is outlined. The various sites considered are top, bridge and three-fold. The latter are designated $A+$, $A-$, $B+$, and $B-$, where the $+$ and $-$ symbols designate surface and subsurface (between first and second metal layers) sites and A and B designate above third (*a*) and second (*b*) layer metal atoms respectively. Bottom: As above, except now in side view. The figure is in the (011) plane and contains one of the sides of the $(\sqrt{3} \times \sqrt{3})R30^\circ$ unit mesh

orbit calculations, however [2.321], yield fourfold hollow sites at low coverage $\Theta \leqq 1$ and bridge-bonded H at $\Theta > 1$.

A significant difference between H adsorption on Ni and Pd is due to the fact that Pd dissolves H and forms a bulk hydride at moderate pressure. This allows one to study the transition from surface H via surface hydride to bulk hydride. Indeed *Godowski* et al. [2.322] used TPD to demonstrate the formation of a near surface hydride upon the exposure of Pd(111) to large H doses at temperatures of 90–140 K. Above 140 K the variations in the TPD spectra were characteristics of H solution in the bulk. The H absorption kinetics through clean Pd(110) into the α-phase region were found recently to be rate limited by H bulk diffusion rather than by dissociative H_2 chemisorption. Submonolayer coverages of S, however, cause the kinetics to become surface dominated with dissociative chemisorption as the rate limiting step [2.323]. In analogy, the inhibition of H desorption from Pd hydride by the adsorption of sulphur species [2.207] is probably related to low surface mobility and prevention of recombinative desorption by perturbation of the vibrational properties [3.324].

The poisoning effect of oxygen in the catalytic dissociation of H_2 on Pd has been studied using a Pd MOS structure [3.325].

2.4.3 Hydrogen on Copper

H_2 trapping and physisorption on Cu at low temperatures $\leq 15\,K$ has been studied by a number of authors [2.68, 70, 74, 75, 326]. Trapping and sticking are caused by minute changes of the free H_2 rotational excitations and H–H stretch vibrational excitations. They are hardly detectable by high resolution electron energy loss spectroscopy, but can be studied by molecular beam scattering in resonance modes. Band-like electron states were observed by photoelectron spectroscopy [2.131].

H_2 on Cu is a prototype system for activated adsorption [2.327]. Ways to overcome the activation barrier are either to dissociate H_2 on a hot filament close to the Cu surface [2.35] or the use of H_2 beams of the appropriate energy [2.328]. Appreciable values for the sticking coefficient are obtained only for beam energies above 0.2 eV. The angular variation of the sticking coefficient is very sharply peaked towards the surface normal ($\cos^{10}\Theta$) (Fig. 2.34). The dissociation probability and transition probabilities to excited vibrational states of scattered H_2 or D_2 for various initial energies of the incident H_2 or D_2 can be explained in terms of tunneling through the activation barrier [2.329].

Sticking results for H adsorption on Cu(110) at room temperature are presented in [2.35]. Exposure to 250 L H_2 (dissociated on a W filament) causes a (1 × 2) periodicity in the LEED pattern but does not alter the He and Ne

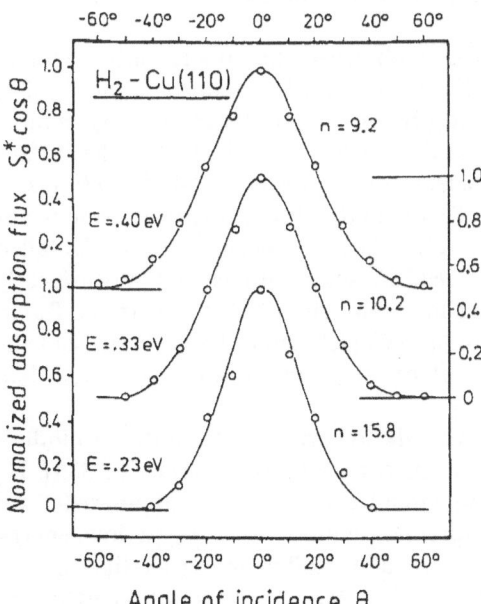

Fig. 2.34. Angular variation of the sticking coefficient plotted as $S_0^* \cos \theta$ for H_2/Cu(110) at three different beam energies. The solid curves are the $\cos^n\theta$ approximations. Tilt axis for the change of angle is the [001] direction

diffraction pattern of the clean surface. As atom diffraction probes the topmost layer only, whereas LEED 'sees' the first few layers, it was concluded that H moves below the surface and induces a reconstruction in deeper layers, in agreement with earlier photoemission results [2.131]. Upon further H exposure the (1×2) periodicity extends to the surface layer. Below 160 K H adsorbs between Cu rows in surface sites [2.330] and the work function is increased. Above 160 K H was found to move to both filled-trigonal surface sites and tetrahedral subsurface sites.

A theoretical analysis of this H-induced surface relaxation in the effective medium theory [2.36] reveals that H induces a local lattice distortion even when chemisorbed. The relaxations are largest for H just under the surface, but are not large enough to make this site more stable than the chemisorption site.

Saturated adsorption of H on Cu(111) below 200 K produces a H bonding band split off from the conduction band 6.2 eV below E_F, and strong charge redistributions within the Cu $3d$-band [2.81]. Subsurface site occupation was also proposed for H on Cu(111) [2.162]. Cluster studies of the chemisorption of H on Cu(100) reveal 51 kcal/mol chemisorption energy, in good agreement with experimental estimates [2.331].

2.4.4 Hydrogen on Rare-Earth Metals

No exhaustive literature exists about H on rare earth metals. Results obtained up to 1985 were suminarized in the reviews of *Netzer* and *Matthew* on "Surfaces of rare-earth metals" [2.32] and of *Netzer* and *Bertel* on "Adsorption and catalysis on rare-earth surfaces" [2.31]. UPS, EELS and work function measurements of the effect on H adsorption on UHV-cleaned surfaces of some rare-earth metals were reported.

Dissociative chemisorption has been observed, but uncertainties remain concerning the sticking probability: Earlier studies by *Atkinson* et al. [2.332] revealed initial sticking coefficients of the order of 0.5–0.7 for H_2 and D_2 on most rare-earth metals, with the exception of those that are divalent at the surface (Sm, Eu, Yb) and have sticking coefficients of the order of 10^{-3}. *Netzer* and coworkers observed in recent UHV EELS and UPS experiments significant H-induced features on Ce(001) at very large exposures only. Detailed angle-resolved UPS studies of H_2 on Ce(001), where surface and subsurface H could be distinguished, revealed rather low reactivity of Ce towards H_2. Rapid diffusion of some H into bulk sites (α-phase CeH_x) cannot be excluded, but the reactivity of clean rare-earth metal surfaces towards H_2 may be lower than generally anticipated [2.333, 334].

The segregation of impurity H from bulk to surface was reported to induce a distorted hexagonal reconstruction of the square Ce(001) surface [2.335]. H_2 adsorption on the Ce(001) surface results in a CeH_x solid solution phase (4.2 eV structure in UPS spectra). At exposures exceeding 500 L H_2 a low energy plasmon feature signals the formation of a (metallic) dihydride-like surface hydride phase. Subsequent annealing to 300–400 °C produces a structure at

3.4 eV which was tentatively ascribed to the population of octahedral sites. Heating above 400 °C leads to a H-induced hexagonal surface reconstruction [2.333].

On UPS spectra of UHV-cleaned Y two features in addition to the expected valence band emission were observed. A comparison of the spectra of Y and of Y exposed to H_2 led *Baptist* et al. [2.336] to suggest that the additional peaks are due to a surface hydride, or subsurface H originating from bulk impurity H, or reaction with the residual gas.

The exposure of polycrystalline La to H_2 at room temperature leads to the formation of a dihydride surface with saturation after 50 and 2000 L H_2 at the top surface and in the outer 20 Å, respectively [2.337]. On evaporated La thin films initial reaction probability $r = 1$ was found for H_2, O_2 and H_2O at room temperature [2.338]. The reactions of H solution and of LaH_2 formation proceeded at $r = 1$ and $r = 0.1$, respectively. Preadsorbed layers of H_2O were found to inhibit the rate of H_2 absorption much more strongly than layers of O_2.

Surface magnetism of evaporated Gd thin films was probed by spin-polarized photoemission [2.144]. Exposure to 0.5 and 1L H_2 at 20 K reduces the polarization from 70% to 45% and 30%, respectively, and lowers the magnetic ordering temperature from about 300 K to 160 K and 120 K, respectively, pointing to a strong interaction of dissociated H with the Gd surface. Upon warming the sample above 200 K the polarization of clean Gd was restored. Apparently, the H was dissolved in the bulk.

Thin Nd overlayers on Cu exhibit very low sticking probability ($\leq 10^{-4}$) for H_2 [2.239]; cf. Sect. 2.5.8. Photoelectron spectroscopic analysis of Yb thin films evaporated onto a $PdH_{0.2}$ substrate reveals trivalent Yb ($3d^9 4f^{13}$ final state) for submonolayer coverage and mixed valent Yb ($4f^{13}/4f^{14}$ final state) above monolayer coverage. The spectra of more than monolayer coverage resemble those of mixed valent $YbH_{2.6}$ pointing to the diffusion of H from the Pd substrate into Yb [2.339].

The growth of Ce submono-layer and monolayer films on SiO_2 and their reaction with H_2 has been studied using the metal-oxide-semiconductor technique. A sticking coefficient of ≤ 0.11 for H_2 on Ce was estimated [2.294].

Using photoelectron spectroscopy a reversible surface semiconductor–metal transition was observed in hydrides of light and heavy rare-earth metals at low temperatures. The reversible diffusion of hydrogen from octahedral-like surface sites into the bulk is thought to push rare-earth $5d$-states from the hydrogen bonding band back to the conduction band, producing a metallic surface dihydride (Fig. 2.35) [2.94].

An interesting surface phase separation phenomena was observed on several rare-earth metals. Upon exposure to H_2O they do not initially form a hydroxide, but a mixture of dihydride and oxide [2.31, 340, 341], which can be considered as a water splitting process [2.51]: Unfortunately, H is very strongly bound in rare-earth hydrides. However, by using a very sensitive Pd–MOS device with monolayers of Ce on top of the Pd we could prove that H diffuses from CeH_x into Pd, whence desorption would be rather easy. The water splitting reaction

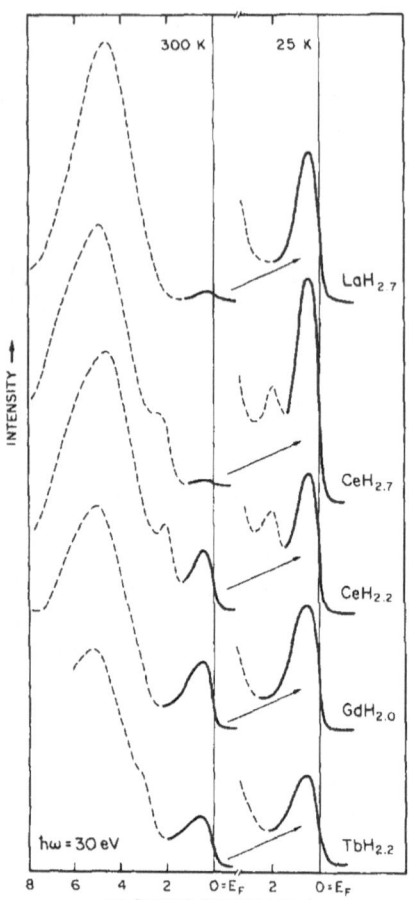

Fig. 2.35. Photoelectron spectra of the valence bands of polycrystalline $LaH_{2.7}$, $CeH_{2.7}$, $CeH_{2.2}$, $GdH_{2.0}$ and $TbH_{2.2}$ taken at a photon energy of 30 eV at 300 and 25 K [2.94]

on the Ce–Pd–MOS stopped when all Ce was oxidized [2.342]. If in a further step photochemical or electrochemical reduction of Ce oxide becomes possible a continuous water splitting process might be feasible. The topochemistry of hydride formation in rare-earth metals, particularly of the effect of surface oxide layers, has been studied by metallographic methods [2.343].

To conclude this section it should be emphasized again that the experimental results on the H_2 interaction with rare-earth metal surfaces at different temperature are rather controversial. A significant effort is needed to understand the fundamental aspects of the transition from adsorbed H_2 and H via near surface H and surface hydrides to bulk hydrides. These transitions are accompanied by valence changes (Yb, Ce, Eu, Sm) of fundamental interest and importance in catalysis.

2.4.5 Hydrogen on Mg, Al, and Ca

Although these light metals are promising hydride formers no consistent picture
of their reaction with H_2 or H can be drawn. Difficulties involved in preparing
clean samples and the low cross section for photoemission are the main reasons.
Furthermore, the interest in chemisorption theory on jellium-like metals only
developed relatively recently. Early results already pointed to an extremely low
sticking probability for H_2 on Mg and Al [2.344, 345] and to the importance
of point or extended defects (paramagnetic centres such as V^-) for the
chemisorption of H_2 and H_2–D_2 exchange reaction on MgO [2.346–350].

Mg, Al and Ca are considered to be among the most free-electron-like
metals and are those best approximated by "jellium", i.e. an electron gas plus
a uniform positive background. Angle-resolved photoemission studies of the
surface electronic structure of Mg(0001) revealed two surface states in band
gaps and peaks due to indirect transitions with high density of occupied states

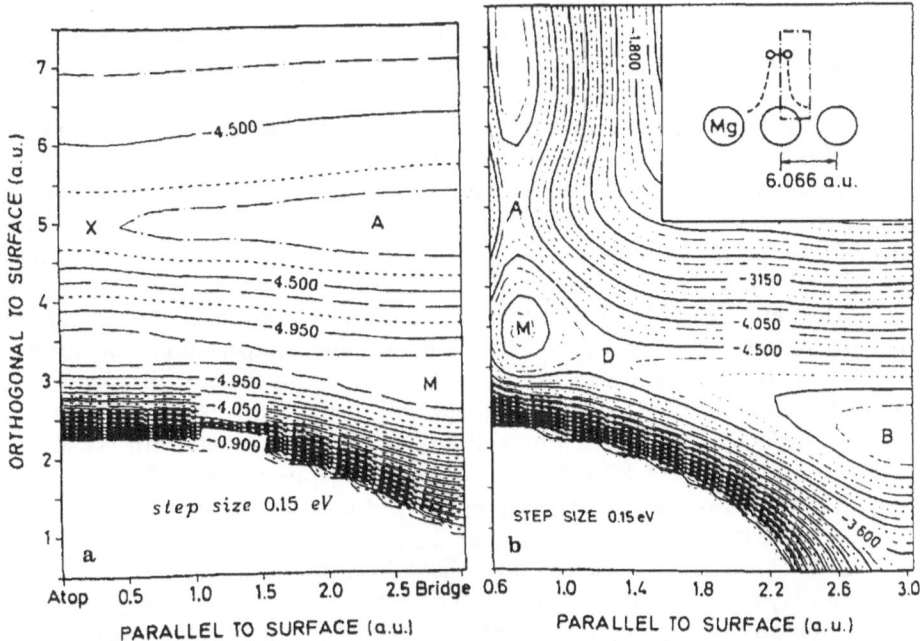

Fig. 2.36. Equipotential-energy curves for H_2 on a Mg(0001) surface, (a) with H_2 parallel to the
surface and a bond length of $R = 1.5$ a.u., and (b) dissociating over the atop site into the bridge
sites (see geometry in inset). The energies in electron volts are relative to those of the free atoms.
The molecular adsorption energy in M is thus only 0.4 eV relative to a free molecule. The distance
from the surface is measured from the first atomic layer, and the distance parallel to the surface,
which in (b) equals one-half the H–H separation R, is measured from the atop site towards the
bridge site. The coordinate of (a) gives the midpoint between the two protons. Note that a molecule
that has surmounted the activation barrier A cannot automatically transfer its kinetic energy
perpendicular to the surface into the H–H vibrations parallel to the surface. Therefore, the molecule
will not dissociate immediately, even though the dissociation barrier D is of the same size as A [2.39]

[2.351]. A self-consistent calculation of the potential energy surface for H_2 on Mg(0001) led to the following picture of the adsorption process (Fig. 2.36) [2.39]: H_2 approaching the surface meets a barrier which is lowest for the atop position. Behind the barrier there is a H_2 chemisorbed state with negligible activation energy for surface migration. A second activation barrier hinders dissociation unless H_2 migrates to a bridge site, where the barrier is small (0.1 eV), and dissociates into adjacent centre sites. The activation energy for dissociation depends strongly on the adsorption site geometry and is too high to allow reasonable dissociation rates. This is in general agreement with newer theoretical considerations [2.27] which reveal a vanishingly small dissociation probability for non-transition metals (cf. Sect. 2.2). Earlier jellium calculations [2.132] had also predicted chemisorbed H_2 as the lowest energy state and H-binding energies below or close to the dissociation energy of H_2 for Al and Na and slightly larger for Mg. Cluster calculations [2.352] show an increased H attraction in $H-Mg_3-M-Mg_3-H$ clusters if the central M atom is Al instead of Mg. The opposite is expected for V and Ti. To what extent sticking and dissociation might be affected by this atom substitution has not yet been studied.

Experimentally it is much simpler to study oxygen adsorption than hydrogen adsorption, particularly for Mg, which reacts 100 times faster with oxygen as compared to Al [2.344, 351].

The exposure of clean Mg and Al to H_2 and H (filament produced) up to doses of 100 L at room temperature did not induce any structure in the photoelectron spectra obtained by Flodström et al. [2.344]. The formation of Al–H species was found by means of HREELS after the interaction of H with the Al(110) surface in a glow discharge [2.353]. No dissociation of H_2 was found on Al [2.354], Al/Na [2.37, 355] or Be [2.356].

Mintz et al. [2.357] studied the interaction of H_2 with clean and oxidized polycrystalline Mg surfaces by TOF–DR at room temperature. The spectra exhibit no surface hydrogen for H_2 exposures up to 2000 L on clean Mg, indicating either that H_2 does not react at all or that adsorbed H diffuses fast into subsurface or bulk sites. Preoxidized Mg, however, reacts with H_2 with a sticking probability of 8.10^{-4} in a one-site adsorption model. Furthermore, these authors also observed the segregation of bulk impurity H in Mg to the surface upon the adsorption of oxygen.

So far no detailed surface studies have been published of the reaction of Mg with H_2 at the elevated temperatures needed for hydride formation, for the simple reason of contamination problems [2.29]. Arguments as to which surface process becomes important thus remain speculative [2.358-361]. Carbonate groups were found on the surface of Mg after H absorption–desorption cycles and subsequent exposure to air [2.362]. Mg readily forms a hydride at the interface with PdH_x by interfacial diffusion of atomic H (Sect. 2.4.8). The effect of organic, metalorganic and metallic surface species on the activation behaviour of Mg is described briefly in Sect. 2.5.5.

2.5 The Hydrogen-Related Surface Properties
of Intermetallic Compounds and Alloys

2.5.1 General Remarks

As already mentioned in Sect. 2.2 very few studies of the interaction of hydrogen (H_2 and H) with clean surfaces of intermetallic compounds and alloys have been made so far. Most surface related work has concerned the activation of previously air-exposed samples for H absorption and gettering. Surface layers on the individual powder grains clearly also affect the heat transfer in metal hydride powder beds and the electric conductivity of metal hydride powder electrodes in electrochemical cells, effects which have never been studied in detail so far.

Thus, the problems which should be looked at in order to characterize the surface properties of intermetallic compounds and alloys concern:

i) *Clean surfaces*: Determination of the composition, crystal structure, electronic structure, vibrational properties of the clean surface, the reaction of that clean surface with hydrogen and the modifications induced by the H adsorption and absorption.

ii) *Air-exposed surfaces* of bulk and powder samples of hydride forming alloys and getter alloys (Sect. 2.2.7): What surface layers are formed (composition, morphology)? Do they passivate the surface for the $H_2 \rightleftharpoons 2H$ reaction so that an *activation* (e.g. at elevated temperature) is needed or is there a self-restoring mechanism which keeps the surface active? Where does the $H_2 \rightleftharpoons 2H$ reaction occur? How serious are the surface reactions for corrosion and cyclic life?

iii) *Previously air-exposed surfaces in a metal hydride powder bed*: What is the composition, thickness and *thermal conductivity* of the surface layers formed on the individual powder grains in the H_2 atmosphere? Could a particular heat treatment lead to surface layers of higher heat conductivity?

iv) Previously air-exposed powder compacted to a metal hydride *electrode* immersed in an electrolyte (e.g. KOH): What surface layers are formed? How good is their electron and *proton conductivity*? How stable are they with reversed polarity (charge/discharge), and do they prevent further corrosion of the underlying alloy, e.g. by oxygen produced at overcharging?

The results obtained over the last decade show that surface segregation induced by selective oxidation is the key mechanism to understanding activation of previously air-exposed samples of those intermetallic compounds which become active around room temperature in a hydrogen atmosphere. For intermetallics needing a high temperature activation such as FeTi the still rather controversial opinions about activation mechanism will be described in Sect. 2.5.3. In the activation of getters at high temperature the key mechanism is the solution of surface and near surface oxygen in the bulk.

Coolings et al. [2.363] noticed as early as 1969, by means of magnetic susceptibility measurements, that the cubic Laves phase compounds $Sc(Ni_{1-x}Co_x)_2$ undergo severe surface damage by crushing at room temperature and heating. The occurrence of metallic Ni and Co was observed and explained by selective oxidation and surface segregation.

2.5.2 LaNi₅ and Related Compounds

Clean LaNi₅ surfaces have the electronic structure of transition metal surfaces. In agreement with the almost temperature independent Stoner-enhanced magnetic susceptibility of bulk samples [2.128], surface analysis by photoelectron spectroscopy [2.364, 365] reveals a Fermi energy close to the top of the Ni 3*d*-derived states. There is no doubt that H_2 adsorbs dissociatively on clean LaNi₅ surfaces.

Air-exposed surfaces of bulk and powder samples of polycrystalline and single crystal LaNi₅ react rapidly with high pressure H_2 (> 20 bar) at room temperature after a short incubation time (minutes). At lower pressure elevated temperature is needed.

Bulk samples fractured in air immediately prior to analysis by photoelectron spectroscopy display a strongly enhanced La surface concentration corresponding to an atomic ratio La:Ni = 1:1. Whereas La is fully oxidized (La(OH)₃) a good proportion of the Ni remains metallic (Fig. 2.37). Powder samples

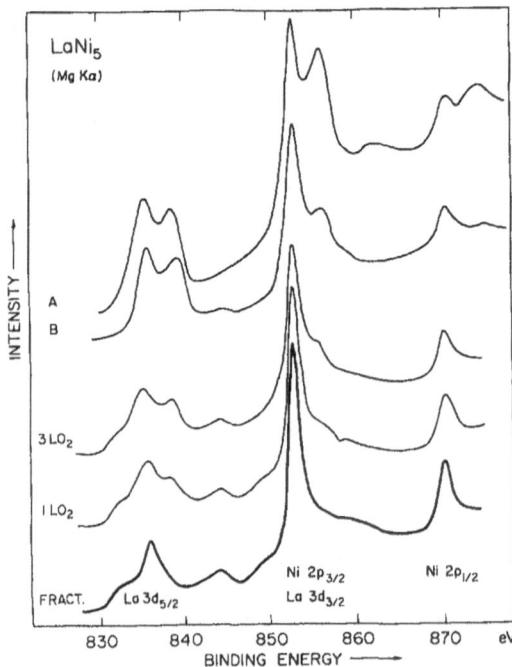

Fig. 2.37. XPS spectra of LaNi₅ (UHV-fractured bulk sample exposed to O_2; A: air-fractured bulk; B: activated powder air exposed) [2.365]

which were also exposed to air just before analysis but had gone through full H absorption-desorption cycles show qualitatively similar behaviour; the amount of metallic Ni is even larger. The lower part of Fig. 2.37 shows La and Ni spectra of UHV-fractured polycrystalline LaNi$_5$, clean and exposed to 1 L and 10 L oxygen. The spectra clearly indicate the selective oxidation of La together with a strong surface enrichment of La. Upon prolonged exposure to air or oxygen more and more surface Ni oxidizes.

The magnetic susceptibility χ_{LaNi_5} of the Pauli paramagnetic LaNi$_5$ was shown [2.128] to increase irreversibly upon grinding, air exposure and H cycling. An increase from $4.6 \cdot 10^{-6}$ emu/g to e.g. $76 \cdot 10^{-6}$ emu/g after 102 cycles was measured. The corresponding magnetization M was interpreted by a super-position of the linearly varying magnetization $\chi_{LaNi_5} \cdot H$ of the bulk in the magnetic field H and to the magnetization M_{sp} of superparamagnetic Ni precipitates each containing about 6000 Ni atoms

$$M = \chi_{LaNi_5} \cdot H + M_{sp} \qquad (2.5)$$

after 102 cycles nearly 1% of the Ni was present in the form of such particles.

With the results of the surface analysis and the magnetic data of LaNi$_5$ we developed the surface segregation model [2.40] to explain the easy activation for H absorption (Fig. 2.16): On a freshly cleaved sample the surface composition is the same as the bulk composition (Fig. 2.16a). Because the surface energy of La is much lower than that of Ni the surface is expected to become La enriched (Fig. 2.16b) in thermodynamic equilibrium. The La surface energy is lowered even further by the chemisorption of oxygen, which enhances the segregation effect. Ni atoms cluster together to form particles containing some thousands of atoms. The segregation continues upon further exposure of the surface to oxygen (air or O_2 impurities in H_2) and prevents the formation of a passivating surface layer. Dissociative H_2 adsorption and associative desorption ($H_2 \rightleftharpoons 2H$) can occur on the metallic Ni particles and on the metallic subsurface of LaNi$_5$. Once the absorption has started, disintegration of bulk samples produces fresh surface and the absorption speeds up. Ni in the outermost layers also oxidizes at high oxygen exposure; it is easily reduced again in hydrogen atmosphere. Ternary and quaternary La–Ni–O–H phases may be formed. Step b), i.e. surface segregation in the absence of oxygen, was recently shown to occur reasonably fast at elevated temperature on a LaNi$_5$(0001) single crystal surface [2.366]. In the schematic Fig. 2.16c some Ni particles are shown epitaxially grown on the LaNi$_5$ substrate. Indeed, the Ni(111) surface fits the LaNi$_5$(0001) surface structure rather well as is illustrated by Fig. 2.38. Photoelectron diffraction of temperature-induced La segregation also points to epitaxial growth of Ni(111) on La(0001) [2.366]. Upon oxygen-induced segregation, however, the hexagonal surface structure is destroyed, in agreement with earlier LEED studies [2.40].

The surface Ni precipitates reveal superparamagnetic behaviour from room temperature down to 77 K; at 4.2 K they are ferromagnetic [2.128]. Their diameter is a factor of two or three smaller than that of the Ni precipitates formed in the bulk upon thermal cycling of LaNi$_5$ [2.367]. An irreversible

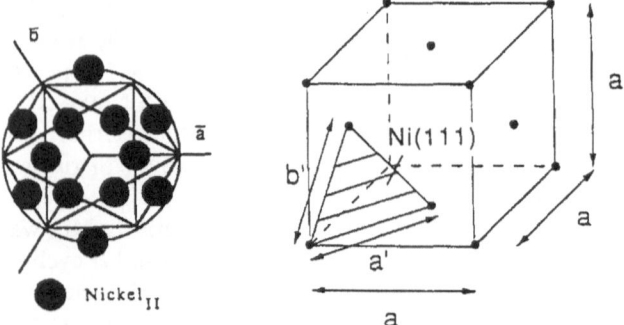

Fig. 2.38. The Ni(111) surface (right side) matches the $LaNi_5(0001)$ surface at $z = \frac{1}{2}$ (left side) rather well [2.366]

increase of the magnetic susceptibility of $LaNi_5$ and $LaNi_{4.9}Fe_{0.1}$ after hydriding was also observed by *Khodosov* and *Linnik* [2.368] and *Oliver* et al. [2.369].

The surface segregation model described above is widely accepted for $LaNi_5$ and many other hydride forming intermetallics (for FeTi see Sect. 2.5.3). Some slightly different models have also been proposed; they are all based on a decomposed surface: *Wallace* et al. [2.217] assume a surface of grains of $La_2O_3/La(OH)_3$ and of Ni. H_2 dissociates on Ni on the outermost surface and H atoms migrate along grain boundaries between Ni and La oxide. The model describes correctly the order of the reaction kinetics. However, it neglects the dynamic aspect of the ongoing segregation (evidenced by capacity losses and increasing magnetization [2.128]). From an analysis of the reaction kinetics *Uchida* and *Ozawa* [2.370] determined the following rate determining steps in the activation process (cf. Chap. 4): (1) diffusion of H in the α-phase for clean surfaces, (2) for air-exposed samples with thin oxide coating dissociation on the "coating" (oxides or hydroxides of La, Ni, or $LaNi_x$), (3) H permeation across coating for thick coating. Dissociation rates were measured to be highest on suboxides of La—although these were many orders of magnitudes lower than on clean Ni or $LaNi_5$. Consequently, activation of heavily oxidized $LaNi_5$ is thought to start with dissociation on top surface suboxides and migration of atomic hydrogen across the oxide coating. According to *Selvam* et al. [2.371] surface Ni oxide is reduced by H_2 in a first step and then acts as a Ni/La_2O_3 supported catalyst for H_2 dissociation. The effect of surface segregation on the loss of storage capacity as a function of the number of absorption cycles and H_2 purity was discussed in [2.40, 128]. The adsorption of H_2O induces a much weaker surface segregation than O_2, and H_2 adsorption has no significant effect on the surface composition at 300 K [2.40, 209]. SO_2 preadsorption blocks further segregation which is apparently related to surface poisoning ([2.372] and Chap. 5).

Ternary phases such as $LaNiO_3$ and La_2NiO_4 were identified after $LaNi_5$ oxidation at 400 °C and above [2.373]. The stability of fine-particle $LaNi_5$ was found to be poor in hardened epoxy resin [2.374].

LaNi$_5$ and LaNi$_5$H$_6$ are relatively resistant to corrosion when used as electrodes in 5–6 M KOH. significant capacity losses, however, occur during cycling. *Boonstra* et al. [2.375] showed that the capacity decrease of the electrode is caused by the oxidation and the formation of La(OH)$_3$ and Ni(OH)$_2$ in the molar ratio 1:5. Hardly any increase in magnetism was observed. XPS analysis of LaNi$_5$ and LaNi$_{4.5}$Al$_{0.5}$ electrodes revealed no severe corrosion in KOH. However, drastic changes of the surface composition after electrolytic charging/discharging as a function of the restpotential (Fig. 2.39 [2.376]). Some surface La is probably dissolved in KOH. Cu encapsulation or special chemical etching were reported to reduce the corrosion [2.377]; see also Sect. 5.3.12.

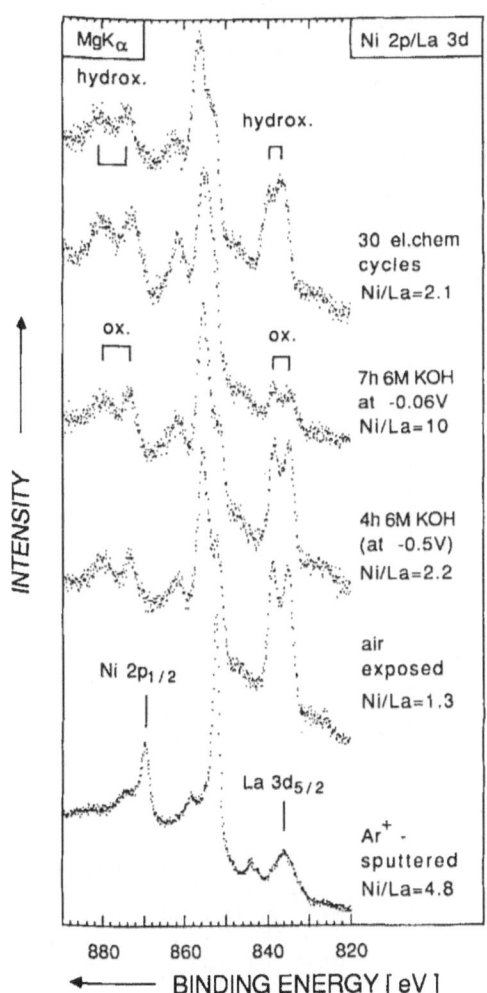

Fig. 2.39. La 3d and Ni 2p XPS core level spectra of LaNi$_5$ powder electrodes (H$_2$ activated) after different pretreatments (Potentials vs. Hg/HgO/6M KOH reference electrode) [2.376]

2.5.3 FeTi and Related Compounds

Angle-resolved photoemission studies [2.378] and band-structure calculations (Chap. 5 of Vol. I) show that the electronic structure of FeTi is comparable to that of other $3d$ transition metals—it resembles that of Cr—which are all known to chemisorb hydrogen dissociatively. The H_2–D_2 exchange reaction occurs fast on FeTi crushed in a H_2/D_2 mixture [2.41]. An ordered surface structure was observed by LEED on a single crystal FeTi surface after repeated heating and cleaning cycles. No superstructure was observed upon H adsorption [2.379]. Preliminary chemisorption studies of H_2 on cleaved, polycrystalline FeTi show that enhanced emission at 5.5 eV below E_F occurs after exposure to as little as 1L H_2 at room temperature [2.378]. UPS spectra of Ar^+-cleaned CoTi [2.380] agree reasonably well with DOS results from band-structure calculations [2.381, 382], although the surface was slightly rich in Co due to preferential sputtering. H_2 exposure at room temperature induces additional emission in the 5–8 eV and 1.5–2 eV regions due to the H 1s–Co $3d$ derived

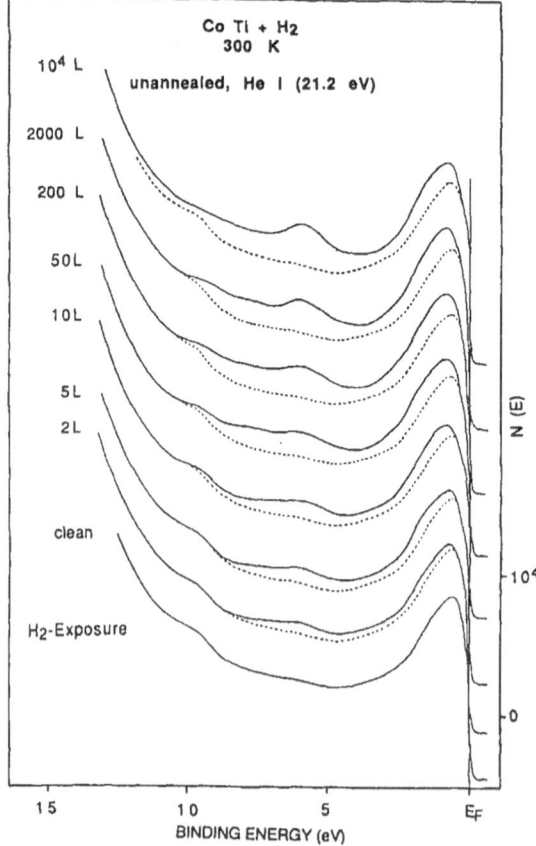

Fig. 2.40. Photoemission spectra of unannealed CoTi exposed to H_2 at 300 K. Exposures are given in Langmuir ($1 L = 10^{-6}$ torr.s). The dashed curves represent the clean surface contribution [2.380]

bonding band and metal p-states lowered by $\approx 3\,\text{eV}$ from $1.5\,\text{eV}$ above E_F, respectively (Fig. 2.40). H_2 dissociates easily, but does not seem to form a surface hydride. H adsorption does not induce a noticeable change in the Co:Ti ratio.

Air-exposed samples of FeTi do not react with H_2 at room temperature. They have to be activated by heating to 350–400 °C in vacuum or H_2; upon cooling to room temperature in H_2 adsorption starts. As the fracture toughness of FeTi is very high compared to other hydride forming intermetallics, disintegration and thus formation of fresh surface is slow. Second phase particles such as $Fe_{3-x}Ti_{3+x}O_y$, Fe_2Ti and α- or β-Ti promote the "bulk"-activation in two ways: First, they lower the fracture toughness of the matrix and facilitate the disintegration. Second, they may provide sites or interfaces for the preferential hydride nucleation and growth [2.41, 383].

What reactions take place on the surface during activation? About a hundred papers and reports have been written on this subject. We gave a critical review of the literature up to 1983 in [2.41]. Some more recent papers are summarized and a somewhat different view of the importance of various steps is given in [2.384]; see also [2.385].

Air-exposed crushed samples are covered by a 10–30 nm thick oxide film which apparently passivates the surface completely [2.41, 214]. XPS analysis shows almost completely oxidized Fe and Ti in a ratio Fe:Ti ≈ 1.0 (Fig. 2.41). During the activation process at temperatures up to 350–400 °C that surface layer has to be either (1) removed, e.g. by dissolution of the surface impurities in the bulk as in getter alloys, or by mechanical separation, or (2) made

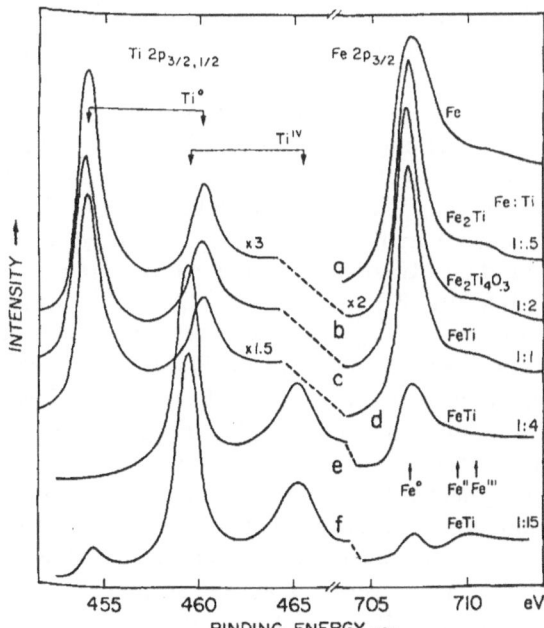

Fig. 2.41 XPS spectra of Fe–Ti compounds (300 K). (a) Fe powder after heating to 650 K, 1 h, 20 bar H_2. (b) Fe_2Ti fractured in UHV. (c) $Fe_2Ti_4O_{0.3}$ fractured in UHV. (d) FeTi fractured in UHV. (e) As (d), but annealed at 650 K, 2 h, 20 bar H_2. (f) Electropolished FeTi as inserted

transparent for H_2 so that H_2 molecules can reach sites for the $H_2 \leftarrow 2H$ reaction, or (3) made reactive for H_2.

By means of photoelectron spectroscopy we have shown that during activation Ti becomes fully oxidized (TiO_2) and enriched at the surface (Fe:Ti ≈ 0.5) whereas the oxidized Fe is reduced to metallic Fe (Fig. 2.41). Results from magnetic susceptibility measurements [2.41, 386] of powder and bulk samples from Pauli paramagnetic FeTi point to the formation of ferromagnetic (superparamagnetic) Fe precipitates at the surface (the bulk magnetism may also rise due to the increased number of Fe atoms on Ti sites, cf. Chaps. 5 and 7 of Vol. I). The selective oxidation and enrichment of Ti together with the formation of metallic Fe shows that a segregation mechanism operates during activation, in full analogy to the one observed in $LaNi_5$ and other compounds, with the difference that high temperature is needed. An analogous decomposition of $Fe_{50}Mn_{50}$ thin films into top surface MnO and underlying α-Fe was observed after annealing at 260 °C [2.223].

As pointed out in previous studies [2.41, 386] these results do not enable one to determine whether H_2 dissociation occurs on surface or near-surface Fe precipitates or on the FeTi subsurface; both are capable causing dissociation. The formation of ferromagnetic surface or near surface Fe precipitates upon heating FeTi to 400 °C, evidenced by magnetic susceptibility measurements, was confirmed by spin-polarized photoemission [2.386] and conversion electron Mössbauer [CEM] spectroscopy [2.387]. Recently *Manchester* et al. [2.384, 388] also found the six-line pattern CEM spectrum typical for α-Fe after heating FeTi to 500 °C, though their earlier measurement [2.389] has given no evidence for surface Fe particles. The Fe particles were also detected by TEM on carefully prepared and not selectively etched samples [2.390, 391]. The six-line CEM pattern of α-Fe disappeared after the removal of a 400 Å surface layer by sputtering [2.392].

The XPS spectra of FeTi activated in a high pressure cell inside the spectrometer at 400 °C in H_2 (Fig. 2.41) and analyzed at room temperature after direct transfer from high pressure H_2 to ultrahigh vacuum gave no evidence for the formation of phases other than TiO_2 and metallic Fe on FeTi, particularly no Fe_2Ti, $Fe_2Ti_4O_x$ nor Ti hydride. A strongly enhanced Ti:Fe ratio in the outer surface layers found by AES depth profiling has been reported in many publications [2.214, 386, 388, 391, 393]. Thus a net result of the surface analysis is the transformation of the passivating Ti–Fe–O layer into a layer of TiO_2 and metallic Fe, which is apparently transparent for H_2. It is not clear from the spectra whether, at the interface between the TiO_2–Fe outermost layer and the FeTi substrate, monolayer amounts of $Fe_2Ti_4O_x$, Fe_2Ti or similar compounds are formed. A sequence beginning with a top layer of TiO_2–Fe thicker than the roughly 3 nm probing depth of XPS and monolayers of $Fe_2Ti_4O_x$ and Fe_2Ti on FeTi would account for a Ti enrichment in the outer surface layers and Fe enrichment in the subsurface layers. AES depth profiles showed such a double layer. The Ti enrichment, however, is always accompanied by an oxygen content much higher than that of Fe_2Ti_4O. Non-destructive depth profiling by XPS with variable X-ray energy ($\approx 1-4\,kV$) could probably settle that point. Fe_2Ti was found at the interface of Fe–Ti thin films after annealing to temperatures above 700 °C [2.394]. After heating FeTi to 800 °C, Fe_2Ti was also found on the surface by means of CEMS [2.388]. In AES analysis, which has a much smaller probing depth than CEMS, Fe vanished completely [2.395]. The Ti-only surface was assumed to form surface Ti hydride, however, experimental evidence is lacking and it is rather unlikely that TiO_2 already formed at temperature below 400 K converts to TiH_x.

The oxygen-stabilized compound $Fe_2Ti_4O_x$ shows surprising activation behaviour [2.396]. Previously air-exposed samples start rapid H absorption and disintegrate into very fine powder at room temperature after a short incubation time. XPS analysis showed interesting differences, even at room temperature in the surface reactivity of UHV-fractured $Fe_2Ti_4O_x$ upon O_2 exposure as compared to FeTi: Whereas almost no selective oxidation and no surface enrichment was found on FeTi, the oxidation of Fe on $Fe_2Ti_4O_x$ surfaces was observed to occur *after* selective oxidation and surface enrichment of Ti. The selective oxidation of Ti seems to be initiated by rapid diffusion of oxygen from the underlying $Fe_2Ti_4O_x$ bulk. Selective interaction of oxygen with Ti atoms leading to the formation of stoichiometric TiO_2 at 300 K was observed in low energy Auger and autoionization emissions on previously sputter-cleaned Fe_2Ti_4O and FeTi [2.397]. $Fe_2Ti_4O_x$ compounds are rather brittle and disintegrate easily, which also facilitates activation. A more detailed analysis of the surface reactions of $Fe_2Ti_4O_x$ might well lead to a deeper understanding of the activation of FeTi.

The effect of contaminating oxide layers on the H uptake kinetics of thin film samples studied by *Fromm* and coworkers [2.194, 199, 200] has already been described in Sect. 2.2.6. The results have been confirmed for thin film FeTi samples [2.398]. Thin suboxide layers formed in the initial stages of oxidation do not prevent the reaction, but do slow it down. Since additional transition metal layers on top of the oxide restore the kinetics one assumes that the oxidation hinders dissociation and not diffusion. The positive effect of particular substitutions, of the presence of second phases and of the non-stoichiometry on the activation observed by many authors [2.399–403] seems to be primarily related to the second stage or bulk activation.

The absence of surface segregation at room temperature explains the high sensitivity of FeTi to poisoning by impurities in H_2 (Chap. 5). Pd coatings protect FeTi from poisoning [2.404].

Ammonia synthesis [2.405], CO hydrogenation [2.406] and decomposition of formic acid [2.399] have been studied on FeTi surfaces. Catalytically active Fe surface particles, delivery of atomic H, and easy dissociation of HCOOH into CO_{ads}, C_{ads} and O_{ads} were found to be the important features. Thin film FeTi has been produced by ion beam mixing of Fe–Ti multilayers [2.408, 409]. Surface studies of Fe-rich single crystal Fe–Ti alloys by the atom–probe method revealed strongly anisotropic Ti segregation upon annealing in the temperature range 550–900 °C [2.410].

2.5.4 AB$_2$ Compounds

Clean surfaces of polycrystalline $ZrMn_2$, $ZrCr_2$ and ZrV_2 [2.360, 411, 412] exhibit a valence band comparable to that of a transition metal and certainly dissociate H$_2$. H adsorption and absorption does not induce any significant surface segregation (cf. Fig. 5.21 in Vol. I).

Many AB$_2$ Laves phase compounds (A = Zr, Ti, V, rare earth; B = Mn, Cr, V, Fe, C, Ni) react readily with H$_2$ even after exposure to air, and form hydrides of convenient stability for H$_2$-storage. Many of these compounds are very brittle and disintegrate easily. All AB$_2$ compounds studied so far show strong selective oxidation and surface segregation. However, in contrast to the AB$_5$ compounds, in which A is always the electropositive component, B or A may be the more electropositive in the AB$_2$ compounds. *Jacob* et al. [2.413] gave a quantitative interpretation of surface segregation of some air-exposed AB$_2$ compounds; it is based on the surface free energy of the constituents.

On $ZrCr_2$ selective oxidation and strong surface enrichment of Zr was found by XPS and scanning AES sputter depth profiling [2.414]. On $ZrMn_2$ and $TiMn_{2-x}$, on the other hand, selective oxidation and surface segregation of Mn was observed upon the exposure of UHV-fractured samples to oxygen [2.415] (Fig. 2.42). Both the lower surface energy of Mn and the larger enthalpy of oxide formation (Mn_3O_4) favour Mn segregation.

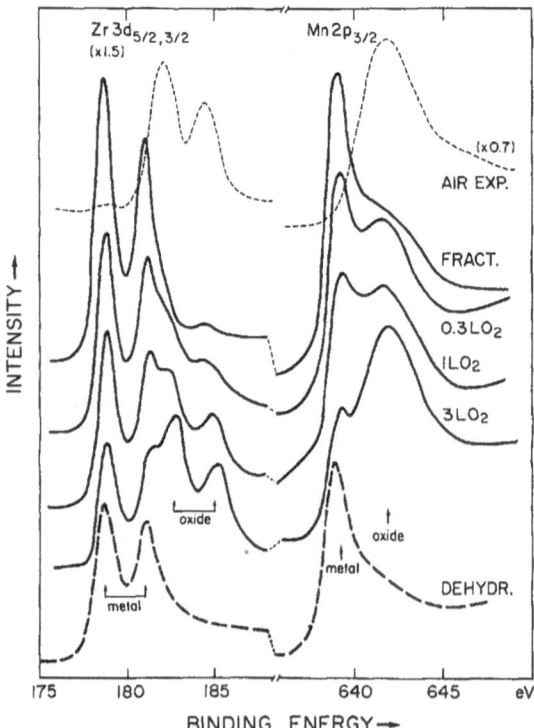

Fig. 2.42. Zr 3d and Mn 2p core level spectra of $ZrMn_2$ (20°C). A bulk sample was fractured in air, then fractured in UHV and finally exposed to increasing doses of oxygen. A further sample was analyzed after disintegration by hydrogen absorption and desorption. Preferential oxidation of manganese and a decrease of the Zr:Mn ratio are observed

Strong Mn enrichment at the surface was also suggested to be responsible for the rapid hydrogenation kinetics of ErB_2, HoB_2, and $ZrMn_2B_{0.8}$ (B = Mn, Fe, Co, Ni) [2.416].

Albers et al. [2.417] recently noticed the presence of the ϕ-phase $TiMn_{1.08}$ in non-stoichiometric $TiMn_{2-x}$ compounds after prolonged annealing under Ar. The ϕ-phase precipitates were found to disappear on H treatment by the formation of α-TiH_n needles observed by TEM. The authors assume that α-TiH_n in overlayers also catalyzes the hydrogen absorption, but give no experimental evidence for this.

2.5.5 Mg-Based Alloys

Detailed studies of the reaction of H_2 with surfaces of Mg alloys and compounds are very scarce, even more scarce than those of elemental Mg.

Part of the reason for the very slow reaction of H_2 with Mg alloy samples is related to the factors mentioned in Sect. 2.4.4 on the surface of Mg: The sticking coefficient and dissociation rate of H_2 on Mg and on possibly formed MgH_2 are extremely low, and the diffusion of H across the MgH_2 skin is very slow.

General reviews of Mg and Mg alloy hydride formation were presented in [2.418, 419]. They contain summaries of the mechanisms thought to be important. Most surface-related mechanisms were deduced indirectly from measurements of the kinetics; some are supported by surface analytical work.

Briefly the following three cases can be distinguished:

i) Some organic compounds formed on the surface of Mg and Mg alloys ("Mg activators", CCl_4-Mg, aromatic molecules bound to surface Mg, tetrahydro-furan-MH [2.238, 418, 420, 421] lead to relatively easy hydride formation due to some kind of H spill-over [2.237]. Details of the mechanism are not known.

ii) Transition metal additives alloyed to Mg are known to form second phase precipitates of intermetallic phases, e.g. Mg_2Ni. These have the electronic structure of a transition metal and probably dissociate H_2. Mg_2Ni and Mg–Mg_2Ni alloys show strong selective oxidation of Mg and surface segregation effects with the formation of near-surface Ni [2.422], comparable, for example, to $LaNi_5$. Thus the transition metal additives favour the dissociation reaction. Furthermore they transform the soft and very ductile Mg into a brittle alloy which readily forms fresh surface by disintegration and provides atomic H by diffusion from the precipitates into Mg at the grain boundaries ([2.423] and Chap. 4).

iii) Alloying with rare-earth or simple metals not only provides less ductile bulk material, which disintegrates easily, but also leads to strong surface segregation phenomena [2.422–424] and herewith sites for H_2 dissociation. These sites have not been identified. Chemisorption energy was mentioned as an important parameter [2.425].

Song et al. [2.426] made a detailed thermodynamic analysis of reaction products on the surface of Mg_2Ni in the presence of oxygen and hydrogen. Chemisorbed oxygen accelerates the surface segregation of Ni and can form water vapour with hydrogen. *Ishido* et al. [2.427] consider H_2O formation as a barrier against hydrogen desorption. Mg–In and Mg–Al alloys also show oxidation-enhanced surface segregation of Mg [2.428].

2.5.6 Other Crystalline Alloys, Getters

Some reactive intermetallic compounds and alloys are widely used as chemical getters (cf. Chap 5) to improve vacuum in sealed tubes and to remove reactive gases such as H_2, O_2, H_2O, N_2 and CO from gas streams and from plasma in fusion technology. Most of the commercially available getter materials are Zr-based alloys, e.g. Zr–V–Fe and Zr–Al. Some of these alloys form stable hydrides and are used to remove or store tritium. After handling in air, the getters have to be activated by heating to high temperature. The getter mechanism and the activation mechanism have been the subject of various surface analytical studies.

As early as in 1978 *Lunin* et al. [2.429] observed strong Ni surface segregation on ZrNi and ZrNi hydride as a function of time and temperature in H_2 and He gas flow. They emphasized that in all cases surface segregation of Ni took place at a temperature 100 °C lower in ZrNi hydride than in ZrNi.

Elemental Zr forms ZrO_2 upon the exposure to O_2 at 310 °C and a Zr suboxide at the metal-oxide interface [2.430]. When heated to 400 °C the ZrO_2 partially decomposes: oxygen ions diffuse through the oxide to the interface to form intermediate suboxides. The suboxide species decompose into Zr and oxygen, the latter diffusing into the bulk Zr. On the surface of Zircalloy 4 (98 wt. % Zr) zrO_2 decomposition and reduction to metallic Zr has been observed at temperatures as low as 200 °C and 300 °C, respectively [2.225].

XPS, AES and SIMS analyses of Zr–Al, Zr–Ni, Zr–Fe and Zr–V–Fe alloy getters show that the getter activation mechanism consists mainly in the diffusion of surface oxygen into the bulk, thereby restoring a reactive metallic surface. A closer look reveals that in a first step the partner element of Zr is reduced to its metallic state and only in a second step at higher temperature Zr also becomes metallic, thus restoring the alloy; surface segregation is of minor importance. The getter process consists of the chemical bonding of reactive gases mostly by selective oxidation of surface components along with surface segregation. The important difference between the various alloys concerns the minimum temperature needed for activation: 800 °C for Zr–Al, 700 °C for Zr–Ni, 400 °C for Zr–V–Fe [2.46] and ≤ 200 °C for $Zr_{91}Fe_9$ [2.431].

In some multiphase Zr–V–Fe alloys surface oxygen diffusion into the bulk was observed to start at 150 °C [2.198]. The economically important question of the lifetime of getters and its limiting factors has not yet been answered. At room temperature getter activity goes to zero after some monolayers coverage. Is partial or full reactivation at elevated temperature possible? There are some

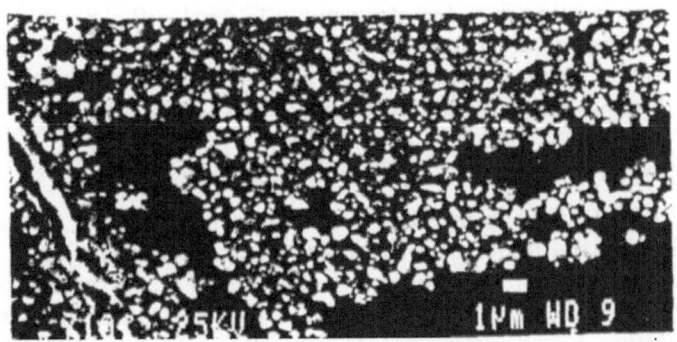

Fig. 2.43. SEM micrograph of a Cu₃Zr alloy (UHV-fractured, 10 h anneal at 200 °C and 1 bar H₂) showing the spherically shaped surface Cu precipitates [2.229]

indications that the surface reactivity goes down as soon as the solubility limit of oxygen in the bulk is reached.

ZrCo behaves differently and compares well with TiFe: samples exposed to air are covered by oxides of both components, easily 100 nm thick. Heating in vacuum reduces Co (or Fe) by the complete oxidation of Zr to ZrO_2 (or TiO_2). At temperatures up to 500 °C Zr (or Ti) remains fully oxidized. The oxygen solubility of ZrCo is known to be very low [2.432]. ZrCo hydride was reported not to be pyrophoric in air.

Crystalline and amorphous Cu–Zr alloys in the concentration range 50–70 at. % Cu slowly decompose at the surface into Zr oxide and Cu and change their colour from silver to reddish. Upon annealing at 150–200 °C in 1 bar hydrogen for a few hours the decomposition is enhanced and macroscopic spherical shaped Cu particles are formed on top of the surface (Fig. 2.43) [2.228, 229]. XPS analysis showed that according to the thermal treatment in hydrogen, the Cu particles can be in electric contact with the metallic alloy substrate or insulated by ZrO_2 [2.229]. Interesting modifications of SMSI can be envisaged.

2.5.7 Amorphous Alloys

Amorphous alloys show a variety of surface segregation effects, which partly account for their potential as catalysts: The as-prepared melt-spun ribbons are covered on both sides by selectively oxidized layers of various thickness, e.g. 10 nm and 60 nm on the shiny and dark side of NiZr, respectively [2.433], with a higher corrosion rate on the dark side. *Spit* et al. [2.434] found a ZrO_2 barrier for H absorption in $Ni_{64}Zr_{36}$ and a Ni-rich layer underneath the surface oxide. Continued cycling of H caused the formation of an additional, internally oxidized layer. After removal by chemical etching of surface ZrO_2 and TiO_2 on amorphous $Ni_{33}Zr_{67}$ and $Ni_{33}Ti_{67}$ alloys, respectively, a rough Ni-rich surface of high electrocatalytic activity was found [2.435]. A 15–35 Å thick layer of ZrO_2 was found by combined AES-XPS analysis on $Ni_{64}Zr_{36}$ ribbons prepared by planar flow casting in a He atmosphere [2.436]. After extensive cleaning by Ar ion bombardment, clean surfaces with compositions close to the nominal bulk composition can be achieved. Analogous surface decomposition effects were observed in Fe–Zr alloys by means of CEMS [2.437, 438].

2.5.8 Metallic Overlayers, Multilayers

As already mentioned in Sect. 2.2.6 metallic overlayers can contribute to the process of H sorption by the underlying substrate in various ways: The H_2 sticking and dissociation rates can be enhanced by preventing the formation of passivating surface layers and by the availability of a surface electronic structure favorable for dissociation. A well-known example is a Pd overlayer [2.91, 92, 170, 190, 191, 194, 199]. Furthermore, as *Pick* et al. pointed out, the barrier for H diffusion from surface into bulk can be reduced [2.190]. Kinetic models allow the determination of the H concentration at the surface, in the surface hydride, and in the bulk.

Butera et al. [2.439] evaporated overlayers of V and Ca onto clean surfaces of the bulk hydride YH_2 and the deuteride $NbD_{0.75}$. Photoelectron spectra showed interfacial diffusion of H from the substrate into the overlayer. Thus the substrate hydride served as a source of atomic hydrogen. In an analogous way, overlayers of Yb hydride were prepared on PdH_x substrate. For low Yb coverage (up to one monolayer) photoelectron spectra indicate trivalent Yb hydride, contrary to the divalent bulk YbH_2 [2.339] (Sect. 2.4.4).

Ce overlayers on a Pd–MOS structure (cf. Sect. 2.3) were exposed to water vapour [2.342]. Some H diffused across the Ce–Pd interface to the gate indicating a water splitting process. Ce overlayers on SiO_2, again a MOS structure, were used to measure H_2 sticking on Ce [2.294] (cf. Sect. 2.4.4).

Thick film Mg samples were covered with a 5–10 nm Pd overlayer as catalyst for H_2 dissociation [2.290, 440]. Unusual Mg hydride formation kinetics were observed probably due to the formation of an interface hydride at the Pd–Mg interface, which was assumed to block hydrogen diffusion.

The preparation of clean surfaces of MgH_2 is a difficult task as the reaction probability of Mg with H_2 is comparable with the oxygen impurity concentration of clean H_2 gas. Thus the impurities may cause effects comparable to those of hydrogen. However, evaporation of monolayers of Mg on polycrystalline PdH_x allowed the observation (XPS, UPS) of MgH_2 formation [2.441]. Above 170 °C H desorbs rapidly and interdiffusion occurs at the Pd–Mg interface.

Hydrogen-induced strain modulation was observed in Nb–Ta multilayers (superlattices) [2.442]. A dramatically increased H solubility as compared to what would be expected for bulk Nb or Ta was reported. Mg–Fe multilayers were studied for their magnetic properties; they certainly show exciting phenomena related to the as yet unexplored H absorption.

Acknowledgement: Careful typing by Marianne Barras and my wife Nina as well as continuous financial support by NEFF (Nationaler Energieforschungs-fonds) is gratefully acknowledged.

References

2.1   A. Zangwill: *Physics at Surfaces* (Cambridge University Press, Cambridge 1988)
2.2   M. Prutton: *Surface Physics* (Oxford University Press, Oxford 1984)
2.3   J.M. MacLaren, J.B. Pendry, P.J. Rous, D.K. Saldin, G.A. Somorjai, M.A. VanHove, D.D. Vvedensky: *Handbook of Surface Structures* (Reidel, Dordrecht 1987)
2.4   L.J. Clarke: Surface Crystallography (Wiley, Chichester 1985)
2.5   M.A. VanHove, S.Y. Tong (eds) *The Structure of Surfaces* I, II Springer Ser. Surf. Sci. Vols. 2, (Springer, Berlin, Heidelberg 1985)
2.6   G. Ertl, J. Küppers: *Low Energy Electrons and Surface Chemistry* (VCH, Weinheim 1985)
2.7   M.A. Van Hove, S.Y. Tong: *Surface Crystallography by LEED* Springer Ser. Chem. Phys. Vol. 2 (Springer, Berlin, Heidelberg 1979)
2.8   H. Ibach, D.L. Mills: *Electron Energy Loss Spectroscopy and Surface Vibrations* (Academic, New York 1982)
2.9   R.F. Willis (ed.): *Vibrational Spectroscopy of Adsorbates*, Springer Ser. Chem. Phys. Vol. 15 (Springer, Berlin, Heidelberg 1980)
2.10 J.T. Yates, T.E. Madey: *Vibrational Spectroscopy of Molecules on Surfaces* (Plenum, New York 1988)
2.11 F. Nizzoli, K.H. Rieder, R.F. Willis (eds.): *Dynamical Phenomena at Surfaces, Interfaces and Superlattices*, Springer Ser. Surf. Sci. Vol. 3 (Springer, Berlin, Heidelberg 1985)
2.12 W. Schommers, P. von Blanckenhagen (eds.): *Structure and Dynamics of Surfaces* I, II, Topics Curr. Phys. **41, 43** (Springer, Berlin, Heidelberg 1986, 1987)
2.13 R.L. Park, M.G. Lagally: *Solid State Physics: Surfaces* (Academic, New York 1985)
2.14 D. Langreth, H. Suhl (eds.): *Many-Body Phenomena at Surfaces* (Academic, New York 1984)
2.15 D.P. Woodruff, T.A. Delchar: *Modern Techniques of Surface Science* (Cambridge University Press, Cambridge 1986)
2.16 D.A. King, D.P. Woodruff (eds.): *The Chemical Physics of Solid Surface and Heterogeneous Catalysis*, Vols. 1–5 (Elsevier, Amsterdam 1981–1987)
2.17 R. Vanselow, R. Howe (eds.): *Chemistry and Physics of Solid Surfaces* IV, Springer Ser. Chem. Phys. Vol. 20, (Springer, Berlin, Heidelberg 1982); R. Vanselow, R. Howe (eds.): *Chemistry and Physics of Solid Surfaces* VI, Springer Ser. Surf. Sci. Vol. 5, (Springer, Berlin, Heidelberg 1985)
2.18 G.A. Somorjai: *Chemistry in Two Dimensions: Surfaces* (Cornell University Press, Ithaca 1981); S.R. Bare, G.A. Somorjai: "Surface Chemistry", in Encyclopedia of Physical Science and Technology 13 (Academic, New York 1987) p. 526
2.19 Z. Paàl, P.G. Menon: *Hydrogen Effects in Catalysis* (M. Dekker, New York 1988)
2.20 K.R. Christmann: Hydrogen Sorption on Pure Metal Surfaces, in Ref. [2.19]
2.21 K.R. Christmann: *Interaction of Hydrogen with Solid Surfaces*, Surf. Science Rep. **9**, 1 (1988) and references therein
2.22 R. Burch: The Adsorption and Absorption of Hydrogen by Metals, in *Chemical Physics of Solids and Their Surfaces* (Chem. Soc., London 1979) Vol. 9, p. 1
2.23 M.P. D'Evelyn, R.J. Madix: Surf. Science Rep. **3**, 413 (1984)
2.24 J.K. Nørskov: Physica B **127** 193 (1984); Phys. Rev. B **26**, 2875 (1982); J.K. Nørskov, F. Besenbacher: J. Less-Common Met. **130**, 475 (1987)
2.25 D.R. Hamann: J. Electron Spectroscop. Relat. Phen. **44**, 1 (1987); P.J. Feibelmann, D.R. Hamann: J. Vac. Sci. Technol. A **5**, 424 (1987)
2.26 M.S. Daw, M.I. Baskes: Phys. Rev. B **29**, 6443 (1984)
2.27 J. Harris: Appl. Phys. A **47**, 63 (1988)
2.28 J.H. Sinfelt: J. Phys. Chem. **90**, 4711 (1986) *Bimetallic Catalysts* (Wiley, New York 1983); Ref. [2.17] p. 19, Ref. [2.14], p. 551
2.29 L. Schlapbach: In *Hydrogen in Disordered and Amorphous Solids*, ed. by G. Bambakidis, R.C. Bowman, NATO ASI B **136**, (Plenum, New York 1986) p. 397
2.30 J.P. Muscat, D.M. Newns: Progr. Surf. Science **9**, 1 (1978)
2.31 F.P. Netzer, E. Bertel: Adsorption and Catalysis on Rare Earth Surfaces in *Handbook on the Physics and Chemistry of Rare Earths*, vol. 5, ed. by K.A. Gschneidner, L. Eyring, (North-Holland, Amsterdam 1982) p. 217
2.32 F.P. Netzer, J.A. Matthew: Rep. Prog. Phys. **49**, 621 (1986)
2.33 J.W. Ward: Properties and Trends in Actinide-H Systems, in *Handbook on the Physics and Chemistry of the Actinides*, ed. by A.J. Freeman, C. Keller, (Elsevier, Amsterdam 1985) Chap. 1

2.34   M.H. Mintz, J. Bloch: Progr. Solid State Chem. **16**, 163 (1985); J. Less-Common Met. **81**, 301 (1981)
2.35   K.H. Rieder, W. Stocker: Phys. Rev. Lett. **57**, 2548 (1986)
2.36   K.W. Jacobsen, J.K. Nørskov: Phys. Rev. Lett. **59**, 2764 (1987); K.W. Jacobsen: Comments Cond. Mat. Phys. **14**, 129 (1988)
2.37   R.B. Murphy, J.M. Mundnar, K.D. Tsuei, E.W. Plummer: Bull. Am. Phys. Soc. **33**, 655 (1988)
2.38   J. Paul: Phys. Rev. B **37**, 6164 (1988)
2.39   J.K. Nørskov, A. Houmøller, P. Johansson, B.I. Lundquist: Phys. Rev. Lett. **46**, 257 (1981)
2.40   L. Schlapbach, A. Seiler, F. Stucki, H.C. Siegmann: J. Less-Common Met. **73**, 145 (1980); H.C. Siegmann, L. Schlapbach, C.R. Brundle: Phys. Rev. Lett. **40**, 972 (1978); Th.v. Waldkirch, P. Zürcher: Appl. Phys. Lett. **33**, 689 (1978)
2.41   L. Schlapbach, T. Riesterer: Appl. Phys. **32** A, 169 (1983) and references therein
2.42   B.J. Berkowitz, J.J. Burton, C.R. Helms, R.S. Polizzotti: Scripta Met. **10**, 871 (1976)
2.43   M.S. Daw, M.I. Baskes: Phys. Rev. Lett. **50**, 1285 (1983); S.M. Foiles, M.S. Daw: J. Vac. Sci. Technol. A **3**, 1565 (1985)
2.44   A. Kimura, H.K. Birnbaum: Scripta Met. **21**, 219 (1987); M.F. Ashby, J.P. Hirth (eds.): *Perspectives in Hydrogen in Metals* (Pergamon, Oxford 1986)
2.45   J. Winter: J. Vac. Sci. Technol. A **5**, 2286 (1987) and references therein; M.A. Pick: J. Less-Common Met. **103**, 5 (1984)
2.46   K. Ichimura, M. Matsuyama, K. Watanabe: J. Vac. Sci. Technol. A **5**, 220 (1987); K. Ichimura, K. Ashida, K. Watanabe: J. Vac. Sci. Technol. A **3**, 346 (1985)
2.47   I. Vedel, L. Schlapbach: *Structure and Reactivity of Surfaces*, ed. by C. Morterra, A. Zecchina, G. Costa (Elsevier, Amsterdam 1989) p. 903
2.48   H. Imamura, W.E. Wallace: J. Phys. Chem. **84**, 3145 (1980)
2.49   R.B. Wright, J.G. Jolley, M.R. Hankins, M.S. Owens, D.L. Cooke: J. Vac. Sci. Technol. A **5**, 586, and 593 (1987)
2.50   V. Paul-Boncour, A. Percheron-Guégan, J.C. Achard, J. Barrault, H. Dexpert, R.C. Karnatak: J. Phys. **47**, C 8-305 (1986) K.S. Sim, L. Hilaire, F. LeNormand, R. Touroude, V. Paul-Boncour, A. Percheron-Guégan: In *Structure and Reactivity of Surfaces* ed. by C. Morterra, A. Zecchina, G. Costa (Elsevier, Amsterdam 1989) p. 863
2.51   L. Schlapbach: J. Less-Common Met. **111**, 291 (1985)
2.52   K. Ledjeff: Int. J. Hydrogen Energy **12**, 361 (1987)
2.53   J.E. Lennard-Jones: Trans. Faraday Soc. **28**, 333 (1932)
2.54   W. Lisowski, E. Nowicka, Z. Wolfram, R. Dus: Appl. Surf. Science **31**, 157 (1988)
2.55   K.D. Rendulic: Appl. Phys. A **47**, 55 (1988); D.W. Goodmann: In Ref. [2.17] p. 169
2.56   E.A. Rohlfing, D.M. Cox, A. Kaldor, K.H. Johnson: J. Chem. Phys. **81**, 3846 (1984); P. Jiang, F. Jona, P.M. Marcus: Phys. Rev. B **36**, 6336 (1987)
2.57   D.M. Cox, D.J. Trevor, R.L. Whetten, A. Kaldor: J. Phys. Chem. **92**, 421 (1988) and references therein
2.58   K. Christmann: Z. Phys. Chem. NF **154**, 145 (1987)
2.59   K. Müller: Ber. Bunsenges. Phys. Chemie **90**, 184 (1986)
2.60   Y. Li, J.L. Erskine, A.C. Diebold: Phys. Rev. B **34**, 5951 (1984)
2.61   J.M. Nicol, J.J. Rush, R.D. Kelley: Phys. Rev. B **36**, 9315 (1987)
2.62   M.Y. Chou, J.R. Chelikowsky: Phys. Rev. Lett. **59**, 1737 (1987)
2.63   T.E. Felter, S.M. Foiles, M.S. Daw, R.H. Stulen: Surf. Science **171**, L379 (1986)
2.64   M.S. Daw, S.M. Foiles: Phys. Rev. Lett. **59**, 2756 (1987); M.S. Daw, S.M. Foiles: Theory of H on Metal Surfaces, in Ref. [2.5] p. 41
2.65   J.P. Muscat: Progr. Surf. Science **18**, 59 (1985); Surf. Science **152/153**, 684 (1985)
2.66   M.L. Cohen: Theory of Surface Reconstruction, in Ref. [2.5] p. 4 and references therein
2.67   M.J. Sparnaay: Surf. Science Rep. **4**, 101 (1984)
2.68   P. Nordlander, C. Holmberg, J. Harris: Surf. Science **152/153**, 702 (1985)
2.69   P.K. Johansson: Surf. Science **104**, 510 (1981)
2.70   S. Andersson, L. Wilzen, M. Persson: Phys. Rev. B **38**, 2967 (1988); S. Andersson, L. Wilzen, M. Persson, J. Harris: Phys. Rev. B **40**, 8146 (1989)
2.71   J. Lapujoulade, J. Perreau: Phys. Scripta T **4**, 138 (1983)
2.72   Ph. Avouris, D. Schmeisser, J.E. Demuth: Phys. Rev. Lett. **48**, 199 (1982)
2.73   E. Ilisca: Phys. Rev. Lett. **66**, 667 (1991)
2.74   U. Harten, J.P. Toennies, C. Wöll: J. Chem. Phys. **85**, 2249 (1986)
2.75   J. Harris, A. Leibsch: Physica Scripta T **4**, 14 (1983)
2.76   P. Nordlander, C. Holmberg, J. Harris: Surf. Science **175**, L753 (1986)

2.77   A.S. Mårtensson, C. Nyberg, S. Andersson: Phys. Rev. Lett. **57**, 2045 (1986)
2.78   Z. Knor: In Catalysis **3**, ed. by J.R. Anderson, M. Boudart (Springer, Berlin, Heidelberg 1982)
2.79   I. Toyoshima, G.A. Somorjai: Catal. Rev.-Sci. Eng. **19**, 105 (1979)
2.80   P. Nordlander, S. Holloway, J.K. Nørskov: Surf. Science **136**, 59 (1984)
2.81   F. Greuter, E.W. Plummer: Solid State Commun. **48**, 37 (1983)
2.82   E. Salomons, R. Griessen, D.G. de Groot, A. Magerl: Europhys. Lett. **5**, 449 (1988)
2.83   A.R. Miedema: Z. Metallkunde **69**, 455 (1978)
2.84   P.A. Dowben, A. Miller (eds.): *Surface Segregation Phenomena* (CRC Press, Boca Raton, Florida 1990)
2.85   T. Sakurai, T. Hashizume, A. Jimbo, A. Sakai, S. Hyodo: Phys. Rev. Lett. **55**, 514 (1985); T. Sakurai, T. Hashizume, A. Kobayashi, A. Sakai, S. Hyodo, Y. Kuk, H.W. Pickering: Phys. Rev. B **34**, 8379 (1986)
2.86   S.M. Foiles: Phys. Rev. B **32**, 7685 (1985)
2.87   P.J. Durham, R.G. Jordan, G.S. Sohal, L.T. Wille: Phys. Rev. Lett. **53**, 2038 (1984)
2.88   U. Vahalia, P. Dowben, A. Miller: J. Vac. Sci. Technol. A **4**, 1675 (1986); H.H. Brongersma, P.A. Ackermans, A.D. van Langenveld: Phys. Rev. B **34**, 5974 (1986)
2.89   K.W. Sulston, S.G. Davison, W.K. Liu: Phys. Rev. B **33**, 2263 (1986); G. Treglia, H. Legrand, F. Ducastelle: Europhys. Lett. **7**, 575 (1988)
2.90   S. Modak, B.C. Khanra: Phys. Rev. B **34**, 5909 (1986)
2.91   M. Strongin, J. Colbert, G.J. Dienes, D.O. Welch: Phys. Rev. B **26**, 2715 (1982)
2.92   M.A. Pick, A. Hanson, K.W. Jones, A.N. Goland: Phys. Rev. B **26**, 2900 (1982)
2.93   E. Wicke: Z. Phys. Chem. NF, **143**, 1 (1985)
2.94   L. Schlapbach, J.P. Burger, P. Thiry, J. Bonnet, Y. Petroff: Phys. Rev. Lett. **57**, 2219 (1986); Surf. Science **189/190**, 747 (1987)
2.95   P. Oelhafen, R. Lapka, U. Gubler, J. Krieg, A. DasGupta, H.J. Güntherodt, T. Mizoguchi, C. Hague, J. Kübler, S.R. Nagel: In *Rapidly Quenched Metals IV*, ed. by T. Masumot, K. Suzuki (Japan Institute of Metals, Sendai 1982) p. 1259
2.96   J.P. Burger, L. Schlapbach, I. Vedel, U. Maier: Z. Phys. Chem. NF, **163**, 569 (1989)
2.97   I. Ishida: J. Phys. Soc. Japan **56**, 1427 (1987)
2.98   D.A. Steigerwald, P. Wynblatt: J. Vac. Sci. Technol. A **5**, 1224 (1987) and references therein
2.99   J.E. Inglesfield: Surface Electronic Structure in *Electronic Properties of Surfaces*, ed. by M. Prutton (A. Hilger, Bristol 1984) p. 1; Rep. Prog. Phys. **45**, 223 (1982)
2.100 D.W. Bullet: Phil. Mag. B **51**, 223 (1985)
2.101 O. Gunnarsson: The Density Functional Theory of Metallic Surfaces in *Electrons in Disordered Metals and at Metallic Surfaces* ed. by P. Phariseau et al. (Plenum, New York 1979) NATO ASI B **42**, 1
2.102 V. Dose: Physica Scripta **36**, 669 (1987)
2.103 F.J. Arlinghaus, J.G. Gay, J.R. Smith: Phys. Rev. B **21**, 2055 (1980)
2.104 J. Ihm: Rep. Prog. Phys. **51**, 105 (1988)
2.105 C. Umrigar, J.W. Wilkins: Phys. Rev. Lett. **54**, 1551 (1985)
2.106 H. Hasegawa: J. Phys. F **17**, 679 (1987); H. Hasegawa, F. Herman: Phys. Rev. B **38**, 4863 (1988)
2.107 A.J. Freeman: in Ref. [2.11] p. 162; J. Magn. Magn. Materials **35**, 31 (1983); A.J. Freeman, C.L. Fu, E. Wimmer: J. Vac. Sci. Technol. A **4**, 1265 (1986)
2.108 L.M. Falicov, R.H. Victora, J. Tersoff: in Ref. [2.5] p. 12
2.109 J. Mathon: Rep. Prog. Phys. **51**, 1 (1988)
2.110 H.C. Siegmann, F. Meier, M. Erbudak, M. Landolt: Adv. Electronics and Electron Physics **62**, 1 (1984); H.C. Siegmann, D. Mauri, D. Scholl, E. Kay: J. de Phys. **12**, C8-9 (1988)
2.111 W. Dürr, M. Taborelli, O. Paul, R. Germar, W. Gudat, D. Pescia, M. Landolt: Phys. Rev. Lett. **62**, 206 (1989)
2.112 D. Weller, S.F. Alvarado, W. Gudat, K. Schröder, M. Campagna: Phys. Rev. Lett. **54**, 1555 (1985)
2.113 R. Allenspach, M. Taborelli, M. Landolt, H.C. Siegmann: Phys. Rev. Lett. **56**, 953 (1986)
2.114 A. Rosengren, B. Johansson: Phys. Rev. B **26**, 3068 (1982)
2.115 A. Fäldt, D.K. Kristensson, H.P. Meyers: Phys. Rev. B **37**, 2682 (1988)
2.116 N. Mårtensson, F.U. Hillebrecht, D.D. Sarma: Surf. Science **152/153**, 733 (1985)
2.117 D. Spanjaard, C. Guillot, M.C. Desjonqueres, G. Treglia, J. Lecante: Surf. Science Rep. **5**, 1 (1985)
2.118 D.E. Eastman, F.J. Himsel, J.F. van der Veen: J. Vac. Sci. Technol. **20**, 609 (1982)
2.119 D. Erbudak, P. Kalt, L. Schlapbach, K. Bennemann: Surf. Science **126**, 101 (1983)
2.120 F. Gerken, A.S. Flodström, J. Barth, L.I. Johansson, C. Kunz: Physica Scripta **32**, 43 (1985)

2.121 B. Johansson, N. Mårtensson: Phys. Rev. B **21**, 4427 (1980); A. Rosengren, B. Johansson: Phys. Rev. B **22**, 3706 (1980)
2.122 A Rosengren, B. Johansson: Phys. Rev. B **23**, 3852 (1981)
2.123 W.F. Egelhoff: Phys. Rev. Lett. **50**, 587 (1983)
2.124 A. Bansil, M. Pessa: Physica Scripta T **4**, 52 (1983)
2.125 L.T. Wille, P.J. Durham: Surf. Science **164**, 19 (1985)
2.126 H.C. Siegmann, P.S. Bagus, E. Kay: Z. Phys. B **69**, 485 (1988)
2.127 G.L. Bona, F. Meier, M. Taborelli, H.C. Siegmann, A. Bell, R. Gambino, E. Kay: J. Magn. Magn. Mat. **54–57**, 1403 (1986)
2.128 L. Schlapbach: J. Phys. F **10**, 2477 (1980)
2.129 J. Harris, S. Andersson: Phys. Rev. Lett. **55**, 1583 (1985)
2.130 J. Harris, S. Andersson, C. Holmberg, P. Nordlander: Physica Scripta T **13**, 155 (1986)
2.131 W. Eberhardt, R. Cantor, F. Greuter, E.W. Plummer: Solid State Commun. **42**, 799 (1982)
2.132 H. Hjelmberg: Surf. Science **81**, 539 (1979)
2.133 J.K. Nørskov, S. Holloway, N.D. Lang: Surf. Science **137**, 65 (1984)
2.134 B.I. Lundqvist: Chemisorption and Reactivity of Metals, in Ref. [2.14] p. 93
2.135 M. Scheffler, A.M. Bradshaw: The Electronic Structure of Adsorbed Layers, in Ref. [2.16] Vol. 2, p. 165
2.136 T.B. Grimley: Theory of Chemisorption, in Ref. [2.16] Vol. 2, p. 333
2.137 Y. Boudeville, J. Rousseau-Violet, F. Cyrot-Lackmann, S.N. Khanna: J. de Phys. **44**, 433 (1983)
2.138 R. Pucci, M. Baldo, G. Giansiracusa, A. Grassi, G. Piccitto: Solid State Commun. **52**, 1025 (1984)
2.139 C.M. Varma, A.J. Wilson: Phys. Rev. B **22**, 3795 (1980)
2.140 R.J. Smith: Phys. Rev. Lett. **45**, 1277 (1980); Phys. Rev. B **21**, 3131 (1980); V. Murgai, S.L. Weng, M. Strongin, M.W. Ruckman: Phys. Rev. B **28**, 6116 (1983)
2.141 K. Dückers, K.C. Prince, H.P. Bonzel, V. Cháb, K. Horn: Phys. Rev. B **36**, 6292 (1987); B. Gumhalter: Progr. Surf. Science **15**, 1 (1984)
2.142 P.W. Selwood: *Chemisorption and Magnetism* (Academic, New York 1975)
2.143 M. Streszewski, C. Jedrzejek: Phys. Rev. B **34**, 3750 (1986)
2.144 A. Cerri, D. Mauri, M. Landolt: Phys. Rev. B **27**, 6526 (1983)
2.145 G. Ertl: Surf. Science **152/153**, 328 (1985)
2.146 P.J. Feibelman, D.R. Haman: Phys. Rev. Lett. **52**, 61 (1984)
2.147 B. Poelsema, L.K. Verheij, G. Comsa: Surf. Science **152/153**, 496 (1985)
2.148 H. Ibach, S. Lehwald, B. Voigtländer: J. Electron Spectrosc. Relat. Phen. **44**, 263 (1987)
2.149 W. Brenig: Physica Scripta **35**, 329 (1987)
2.150 N.V. Richardson, N. Sheppard: in Ref. [2.10] Chap. 1
2.151 D.R. Hamann, P.J. Feibelman: Phys. Rev. B **37**, 3847 (1988) and references therein
2.152 A.G. Eguiluz: Physica Scripta **36**, 651 (1987) and references therein
2.153 J.W. Gadzuk: in Ref. [2.10], Chapt. 2; H. Metiu, J.W. Gadzuk: J. Chem. Phys. **74**, 2641 (1981)
2.154 F. Sols, N. Garcia, F. Flores: Surf. Science **146**, L577, (1984)
2.155 B. Hellsing, M. Persson: Physica Scripta **29**, 360 (1984)
2.156 P. Avouris, B.N. Persson: J. Phys. Chem. **88**, 837 (1984)
2.157 S. Andersson, L. Wilzen, J. Harris: Phys. Rev. Lett. **57**, 1603 (1986)
2.158 N.L. Liu, B.H. Choi, X. Shen: Surf. Science **198**, 79 and 99 (1985)
2.159 M.D. Stiles, J.W. Wilkins: Phys. Rev. **54**, 595 (1985)
2.160 P. Nordlander, J.C. Tully: Surf. Science **211/212**, 207 (1989)
2.161 R.I. Hall, I. Cadez, M. Landau, F. Pichou, C. Schermann: Phys. Rev. Lett. **60**, 337 (1988)
2.162 G. Comsa, R. David: Surf. Science **117**, 77 (1982)
2.163 S. Yucel: Phys. Rev. B **39**, 3104 (1989)
2.164 H. Kaarmann, H. Hoinkes, H. Wilsch: Phys. Rev. B **30**, 424 (1984)
2.165 B.M. Geerken, I.A.M. Corbière, R. Griessen: J. Phys. Chem. Solids **44**, 793 (1983)
2.166 A. Auerbach, K.F. Freed, R. Gomer: J. Chem. Phys. **86**, 2356 (1987)
2.167 C.H. Mak, J.L. Brand, B.G. Koehler, S.M. George: Surf. Science **188**, 312 (1987)
2.168 T.E. Felter, R.H. Stulen, M.L. Koszykowski, G.E. Godowski, B. Garett: J. Vac. Sci. Technol. A **7**, 104 (1989)
2.169 M. Lagos, G. Martinez, I.K. Schuller: Phys. Rev. B **29**, 5979 (1984)
2.170 J.W. Davenport, G.J. Dienes, R.A. Johnson: Phys. Rev. B **25**, 2165 (1982); M.A. Pick: Phys. Rev. B **24**, 4287 (1981)
2.171 P.M. Richards: J. Nucl. Mater. **152**, 246 (1988)

2.172 G.L. Powell, J.R. Kirkpatrick, J.W. Conant: J. Less-Common Met. 172–174, 867 (1991); J.W. Hanneken. D.M. Vaught: J. Less-Common Met. (1991) in press
2.173 V.E. Henrich: Rep. Prog. Phys. 48, 1481 (1985); Prog. Surf. Science 9, 143 (1979); Ref. [2.16] Vol. 3, Chap. 4
2.174 A.B. Kunz: Phil. Mag. B 51, 209 (1985)
2.175 M. Tsukada, N. Shima: Phys. Chem. Minerals 15, 35 (1987)
2.176 J.B. Malherbe, S. Hofmann, J.M. Sanz: Appl. Surf. Science 27, 355 (1986)
2.177 W. Göpel, G. Rocker, R. Feierabend: Phys. Rev. B 28, 3427 (1983)
2.178 W.J. Lo, Y.W. Chung, G.A. Somorjai: Surf. Science 71, 199 (1978)
2.179 J.B. Bates, J.C. Wang, R.A. Perkins: Phys..Rev. B 19, 4130 (1979)
2.180 O.W. Johnson, S.H. Paek, J.W. Deford: J. Appl. Phys. 46, 1026 (1975)
2.181 J.G. Nelson, G.T. Murray: Met. Transactions A 15, 597 (1984)
2.182 E.A. Colburn, W.C. Mackrodt: Surf. Science 117, 571 (1982)
2.183 S.A. Pope, M.F. Guest, I.H. Hillier, E.A. Colburn, W.C. Mackrodt, J. Kendrick: Phys. Rev. B 28, 2191 (1983)
2.184 M. Boudard, A. Delbouille, E.G. Derouane, V. Indovina, A.B. Walters: J. Am. Chem. Soc. 94, 6622 (1972)
2.185 C.L. Fu, A.J. Freeman, T. Oguchi: Phys. Rev. Lett. 54, 2700 (1985); M.B. Brodsky: J. de Phys. 45, C5-349, (1984); J. Magn. Magn. Materials 35, 99 (1983)
2.186 O. Paul, M. Taborelli, M. Landolt: Surf. Science 211/212, 724 (1989)
2.187 A. Amiri-Hezaveh, G. Jennings, D.J. Joyner, R.F. Willis: J. de Phys. 45, C5-371 (1984); T. Shinjo, N. Hosoito, K. Kawaguchi, T. Takada, Y. Endoh: J. de Phys. 45, C5-361 (1984)
2.188 N. Mårtensson, A. Stenborg, O. Björneholm, A. Nilsson, J.N. Andersson: Phys. Rev. Lett. 60, 1731 (1988)
2.189 C. Uher, J.L. Cohn, P.F. Miceli, H. Zabel: Phys. Rev. B 36, 815 (1987)
2.190 M.A. Pick, J.W. Davenport, M. Strongin, G.J. Dienes: Phys. Rev. Lett. 43, 286 (1979); M. El-Batanouny, M. Strongin G.P. Williams, J. Colbert: Phys. Rev. Lett. 46, 269 (1981)
2.191 H.P. Bonzel: Surf. Science Rep. 8, 43 (1987)
2.192 M.P. Kiskinova: Surf. Science Rep. 8, 359 (1988)
2.193 J. Paul, F.M. Hoffmann: Surf. Science 194, 419 (1988)
2.194 H. Uchida: "Einfluss von Kontaminations- und Metalldeckschichten auf die Kinetik der H-Aufnahme von Ti-Filmen" VDI 5/124, Düsseldorf (1987)
2.195 E. Pörschke, D. Shaltiel, K.H. Klatt, H. Wenzl: J. Phys. Chem. Solids 47, 1003 (1986); R. Sherman, H.K. Birnbaum: J. Less Common. Met. 105, 339 (1985)
2.196 M.C. Burell, N.R. Armstrong: Surf. Science 160, 235 (1985)
2.197 C.S. Ko, R.J. Gorte: Surf. Science 161, 597 (1985)
2.198 F. Meli, Z. Sheng, I. Vedel, L. Schlapbach: Vacuum 41, 7 (1990)
2.199 E. Fromm: Z. Phys. Chem. NF 147, 61 (1986) and references therein; H. Uchida, E. Fromm: J. Less-Common Met. 95, 139 and 153 (1983); H.G. Wulz, E. Fromm: J. Less-Common Met. 118, 293 (1986)
2.200 H.G. Wulz, H. Cichy, E. Fromm: J. Less-Common Met. 118, 303 (1986)
2.201 J. Golczewski, T. Schober, N. Iniotakis: Scripta Met. 18, 829 (1984)
2.202 P.M. Richards, S.M. Meyers, W.R. Wampler, D.M. Follstaedt: J. Appl. Phys. 65, 180 (1989)
2.203 L.G. Petersson, H. Dannetun, J. Fogelberg, I. Lundström: Appl. Surf. Science 27, 275 (1986)
2.204 K.D. Rendulic, A. Winkler: Surf. Science 74, 318 (1978)
2.205 P. Marcus: Ann. Chim. Fr. 11, 417 (1986); E. Protopopoff, P. Marcus: Surf. Science 169, L237 (1986); P. Marcus, E. Protopopoff: Surf. Science 161, 533 (1985)
2.206 J.M. Maclaren, J.B. Pendry, R.W. Johner: Surf. Science 178, 856 (1986)
2.207 R.V. Bucur: J. Catalysis 70, 92 (1981)
2.208 D.M. Gualtieri, K.S. Narasimhan, T. Takeshita: J. Appl. Phys. 47, 3432 (1976)
2.209 L. Schlapbach, C.R. Brundle: J. de Phys. 42, 1025 (1980)
2.210 M. Yamawaki, T. Kiyoshi, T. Namba, M. Kanno: Z. Phys. Chem. NF 147, 115 (1986)
2.211 E. Tuscher, P. Weinzierl, O.J. Eder, E. Lanzel: Proceedings 6th World Hydrogen Energy Conference, Vienna 1986 (Pergamon, Oxford 1986) p. 958
2.212 C. Oshima, M. Aono, S. Otani, Y. Ishizawa: Solid State Commun. 48, 911 (1983)
2.213 L. Schlapbach: Bull. Am. Phys. Soc. 35, 1378 (1990)
2.214 G.D. Sandrock, J.J. Reilly, J.R. Johnson: Proc. 11th Intersociety Energy Conversion and Engineering Conf. Stateline Nevada (1976) p. 965; G.D. Sandrock: Report BNL 352410S, Brookhaven Nat. Lab., 1976

2.215 L.C. Beavis, R.S. Blewer, J.W. Guthrie, E.J. Nowak, W.G. Perkins: THEME, Miami 1974, p. S4-38

2.216 W.E. Wallace, R.S. Craig, V.U. Rao: Adv. Chem. Ser. **186**, 207 (1980)

2.217 W.E. Wallace, R.F. Karlicek, H. Imamura: J. Phys. Chem. **83**, 1708 (1979)

2.218 M. Erbudak, F. Stucki: Phys. Rev. B **32**, 2667 (1985); T. Gouder, M. Chtaib, R. Caudano, J.J. Verbist: Surf. Science **162**, 272 (1985); D.D. Sarma, F.U. Hillebrecht, M. Campagna: Surf. Science **162**, 563 (1985)

2.219 H.F. Bittner, C.C. Badcock: J. Electrochem. Soc. **130**, 193C (1983); H.F. Bittner, M.V. Quinzio, C.C. Badcock: *H-Energy Progress V*, ed. by T.N. Veziroglu, J.B. Raylor (Pergamon, New York 1984) p. 1371

2.220 R.L. Chin, A. Elattar, W.E. Wallace, D.M. Hercules: J. Phys. Chem. **84**, 2895 (1980); K. Soga, K. Otsuka, M. Sato, T. Sano, S. Ikeda: J. Less-Common Met. **71**, 259 (1980)

2.221 G. Wiesinger, G. Hilscher, R. Grössinger, H. Kirchmayr: *Hydrogen Energy Progress VI*, ed. by T.N. Veziroglu, N. Getoff, P. Weinzierl (Pergamon, Oxford 1986) Vol. 2, p. 887

2.222 R.B. vanDover, E.M. Gyorgy, R.P. Frankenthal, M. Hong, D.J. Siconolfi: J. App. Phys. **59**, 1291 (1986); D.C. Miller, E.E. Marinero, H. Notarys: Appl. Surf. Science **35**, 153 (1988)

2.223 H. Lefakis, T.C. Huang, P. Alexopoulos: J. Appl. Phys. **64**, 5667 (1988)

2.224 T.A. Giorgi, B. Ferrario, B. Storey: J. Vac. Sci. Technol. A**3**, 417 (1985)

2.225 R. Kaufmann, H. Kleve-Nebenius, H. Moers, G. Pfennig, H. Jenett, H.J. Ache: Surf. Interface Anal. **11**, 502 (1988)

2.226 D.L. Cocke, M.S. Owens, R.B. Wright: Langmuir **4**, 1311 (1988)

2.227 H.C. zur Loye, A.M. Stacy: Langmuir **4**, 1261 (1988)

2.228 F. Vanini, S. Büchler, M. Erbudak, L. Schlapbach, A. Baiker: Surf. Science **189/190**, 1117 (1987)

2.229 X.N. Yu, L. Schlapbach: Z. Phys. Chem. NF **164**, 1171 (1989)

2.230 B.J. Tatarchuk, J.A. Dumesic: J. Catalysis **70**, 308, 323, 335 (1981)

2.231 G.A. Somorjai, M.A. van Hove: Prog. Surf. Science **30**, 201 (1989)

2.232 U. Bardi, P.S. Ross, G.A. Somorjai: J. Vac. Sci. Technol A**2**, 40 (1984)

2.233 R. Hubert, J. Darville, J.M. Gilles: Physica Scripta T**4**, 179 (1983)

2.234 N.A. Braaten, S. Raaen, J.K. Grepstad: Physica Scripta **37**, 778 (1988)

2.235 T. Iijima: J. Appl. Phys. **64**, 5170 (1988)

2.236 M.C. Deibert, R.B. Wright: Appl. Surf. Science **35**, 93 (1988)

2.237 A. Küssner, E. Wicke: Z. Phys. Chem. NF **24**, 152 (1960); P.A. Sermon, G.C. Bond: Catal. Rev. **8**, 211 (1973)

2.238 H. Imamura, T. Nobunaga, S. Tsuchiya: J. Less-Common Met. **106**, 229 (1985)

2.239 R.M. Nix, R.M. Lambert: Surf. Science **220**, L657 (1989)

2.240 K. Christmann, M. Ehsasi: Appl. Phys. A**44**, 87 (1987)

2.241 N. Mitsuishi, S. Fukada, K. Kuroiwa: J. Less-Common Met. **113**, 23 (1985); F.T. Aldridge: J. Less-Common Met. **108**, 131 (1985); F. Doni, C. Boffito, B. Ferrario: J. Vac. Sci. Technol. A**4**, 2447 (1986); K. Nakamura, T. Hoshi: J. Vac. Sci. Technol. A**3**, 34 (1985)

2.242 J.W. He, D.A. Harrington, K. Griffiths, P.R. Norton: Surf. Science **198**, 413 (1988); R.J. Behm, V. Penka, M.G. Cattania, K. Christman, G. Ertl: Chem. Phys. **78**, 7486 (1986)

2.243 G.E. Godowski, R.H. Stulen, T.E. Felter: J. Vac. Sci. Technol. A**5**, 1103 (1987)

2.244 H.J. Robota, W. Vielhaber, M.C. Lin, J. Segner, G. Ertl: Surf. Science **155**, 101 (1985)

2.245 J.N. Russell, S.M. Gates, J.T. Yates: J. Chem. Phys. **85**, 6792 (1986)

2.246 P. Cantini, L. Mattera, M.F. de Kieviet, K. Jalink, C. Tassistro, S. Terreni, U. Linke: Surf. Science **211/212**, 872 (1989)

2.247 S. Jaenicke, J. Dösselmann, A. Ciszewski, W. Drachsel, J.H. Block, D. Menzel: Surf. Science **211/212**, 804 (1989)

2.248 H.K. Schmidt, L.R. Anderson, J.A. Schultz: Appl. Surf. Science **35**, 274 (1988)

2.249 R. Bastasz, T.E. Felter, W.P. Ellis: Phys. Rev. Lett. **63**, 558 (1989)

2.250 D.H. Carstens: Z. Phys. Chem. NF **164**, 1185 (1989)

2.251 R.G. Musket, W. McLean, C.A. Colmenares, D.M. Makowiecki, W.J. Siekhaus: Appl. Surf. Science **10**, 143 (1982)

2.252 P.C. Zalm: Surf. Interface Anal. **11**, 1 (1988); G. Betz: Surf. Science **92**, 283 (1980)

2.253 M.L. Deviney, J.L. Gland (eds): *Catalyst Characterization Science*, Surface and Solid State Chemistry, ACS Symp. Series **288**, ACS Washington 1985

2.254 R.L. Park, M.G. Lagally (eds): *Surfaces*, Methods of Experimental Physics, Vol. **22**, ed. by R. Celotta, J. Levine, (Academic, New York 1985)

2.255 G.K. Wertheim: *X-Ray Photoelectron Spectroscopy*, in *Microscopic Methods in Metals*, ed. by U. Gonser (Springer, Berlin, Heidelberg 1986) p. 193

2.256 R.L. Park: *Core Level Spectroscopies*, in [2.235] p. 187
2.257 W.F. Egelhoff: Surf. Science Rep. **6**, 253 (1987)
2.258 F.J. Himpsel: Adv. Phys. **32**, 1 (1983)
2.259 C.S. Fadley: Prog. Surf. Science **16**, 275 (1984)
2.260 E.W. Plummer, W. Eberhardt: *Angle Resolved Photoemission as a Tool for the Study of Surfaces*, Adv. Chem. Physics Vol. XLIX, I. Prigogine, S.A. Rice, eds. (Wiley, New York 1982) p. 533
2.261 C.S. Fadley: *Recent Developments in Photoelectron Diffraction*, in *Core Level Spectroscopy in Condensed Systems*, ed. by J. Kanamori, A. Kotani (Springer, Berlin, Heidelberg 1988)
2.262 M.P. Seah: *Auger Electron Spectroscopy*, in *Microscopic Methods in Metals*, ed. by U. Gonser (Springer, Berlin, Heidelberg 1986) p. 219
2.263 M. Landolt, R. Allenspach, D. Mauri: J. Appl. Phys. **57**, 3626 (1985)
2.264 T. Fauster, V. Dose: *Inverse Photoemission Spectroscopy*, in [2.17b] p. 483
2.265 N.V. Smith: Rep. Prog. Phys. **51**, 1227 (1988)
2.266 Y. Gao, M. Grioni, B. Smandek, J.H. Weaver, T. Tyrie: J. Phys. E. **21**, 489 (1988); V. Dose: Prog. Surf. Science **13**, 225 (1983)
2.267 E.W. Plummer, C.T. Chen, W.K. Ford, W. Eberhardt, R.P. Messmer, H.-J. Freund: Surf. Science **158**, 58 (1985)
2.268 E. Umbach: Physica **127B**, 240 (1984)
2.269 M.E. Malinowski: J. Less-Common. Met. **89**, 1 (1983)
2.270 L.J. Clarke: *Surface Crystallography*, (Wiley, New York 1985); J.B. Pendry, K. Heinz, W. Oed: Phys. Rev. Lett. **61**, 2953 (1988)
2.271 K. Heinz: Prog. Surf. Science **27**, 239 (1988)
2.272 K.H. Rieder: *Structural Determination of Surfaces and Overlayers with Diffraction Methods*, in [2.11] p. 2; T. Engel, K.H. Rieder: Surf. Science **109**, 140 (1981)
2.273 S. Brennan: Surf. Science **152/153**, 1 (1985); F. Grey, R. Feidenhans: Europhys. News **19**, 94 (1988)
2.274 J.W. Chung, K. Evans-Lutterodt, E.D. Specht, R.J. Birgenau, P.J. Estrup, A.R. Kortan: Phys. Rev. Lett. **59**, 2192 (1987)
2.275 G. Comsa, B. Poelsema: Appl. Phys. **A38**, 153 (1985)
2.276 J.B. Pendry, K. Heinz, W. Oed: Phys. Rev. Lett. **61**, 2953 (1988)
2.277 R.J. Behm, W. Hösler: In Ref. 2.17b, p. 361
2.278 G. Binnig, H. Rohrer: Surf. Science **152/153**, 17 (1985)
2.279 D. Norman: J. Phys. C **19**,.3273 (1986)
2.280 R.J. Celotta: Appl. Surf. Science **31**, 59 (1988)
2.281 H. Ibach: In Ref. 2.11, p. 109; J. Vac. Sci. Techn. **A5**, 419 (1987)
2.282 N.R. Avery: In Ref. 2.10, p. 223
2.283 J.L. Erskine: Crit. Rev. Solid State Mater. Sci. **13**, 311 (1987)
2.284 R.R. Cavanagh, J.J. Rush, R.D. Kelley: In Ref. 2.10, p. 183
2.285 J.T. Yates: In Ref. [2.235], p. 425
2.286 H. Züchner, B. Hüser, Z. Phys. Chem. NF **147**, 35 (1986)
2.287 R. Bastaz: J. Vac. Sci. Technol. **A3**, 1363 (1985)
2.288 D. Fick: Appl. Phys. A **49**, 343 (1989); E. Recknagel, J.C. Soares, eds: *Nuclear Physics Applications on Material Sciences* (Kluwer, Dordrecht 1988); P.K. Khabibullaev, B.G. Skorodumov: *Determination of Hydrogen in Materials*, Springer Tracts in Modern Physics 117, (Springer, Berlin, Heidelberg 1989)
2.289 A. Leiberich, R. Wolfe: Appl. Phys. Lett. **47**, 241 (1985); M. Pilakouta, A. Neskakis, K. Papastaikoudis, A.A. Katsanos, H. Wenzl, K.H. Klatt: J. Less-Common. Met. **130**, 525 (1987)
2.290 J. Rydén: Hydrogen in Thin Metallic Layers Studied by Nuclear Technics, Ph.D. Thesis **263** Univ. of Uppsala, Sweden, 1990; B. Hjövarsson: Metal Hydrogen Interactions Studied by Nuclear Technics, Ph.D. Thesis **270**, Univ. Uppsala, Sweden, 1990
2.291 D. Liljequist, M. Ismail: Phys. Rev. B **31**, 4131, 4137 (1985)
2.292 T. Engel, G. Ertl: "The H_2-D_2 Exchange Reaction", in ref. 2.16, Vol. **4**, Chap. 6, p. 195
2.293 L.-G. Petersson, H.M. Dannetun, I. Lundström: Phys. Rev. B **30**, 3055 (1984); F. Enquist, M. Armgarth, I. Lundström: J. Appl. Phys. **60**, 4297 (1986); R. Jansson, H. Arwin, M. Armgarth, I. Lundström: Appl. Surf. Science **37**, 44 (1989)
2.294 P.A. Gröning, T. Greber, J. Osterwalder, L. Schlapbach: Vacuum **41**, 4 (1990); P.A. Gröning, Diploma Thesis Univ. of Fribourg, unpublished
2.295 K.D. Rendulic, A. Winkler, H.P. Steinrück: Surf. Science **185**, 469 (1987); K.D. Rendulic, A. Winkler, H. Karner: J. Vac. Sci. Technol. **A5**, 488 (1987)

2.296 F. Greuter, I. Strathy, E.W. Plummer, W. Eberhardt: Phys. Rev. B 33, 736 (1986)
2.297 K. Nagai, Y. Ohno, T. Nakamura: Phys. Rev. B 30, 1461 (1984)
2.298 K.H. Rieder, W. Stocker: Surf. Science 164, 55 (1985)
2.299 K. Christmann, F. Chehab, V. Penka, G. Ertl: Surf. Science 152/153, 356 (1985)
2.300 H.J. Brocksch, K.H. Bennemann: Surf. Science 179, L91 (1987)
2.301 T. Komeda, Y. Sakisaka, M. Onchi, H. Kato, S. Masuda, K. Yagi: Phys. Rev. B 36, 922 (1987)
2.302 S. Lehwald, B. Voigtländer, H. Ibach: Phys. Rev. B 36, 2446 (1987)
2.303 G. Kleinle, V. Penka, R.J. Behm. G. Ertl: Phys. Rev. Lett. 58, 148 (1987)
2.304 J.P. Muscat: Phys. Rev. B 34, 8863 (1986)
2.305 M. Jo, M. Onchi, M. Nishijima: Surf. Science 154, 417 (1985); K. Griffiths, P.R. Norton, J.A. Davies, W.N. Unertl. T.E. Jackman: Surf. Science 152/153, 374 (1985)
2.306 Y. Kuk, P.J. Silverman, H.W. Nguyen: Phys. Rev. Lett. 59, 1452 (1987)
2.307 H. Huang, J. Hermanson: Surf. Science 154, 614 (1985); M. Weinert, J.W. Davenport: Phys. Rev. Lett. 54, 1547 (1985)
2.308 M. Landolt, M. Campagna: Phys. Rev. Lett. 30, 424 (1977)
2.309 I. Chorkendorff, J.N. Russell, J.T. Yates: Surf. Science 182, 375 (1987)
2.310 K.H. Rieder, M. Baumberger, W. Stocker: Phys. Rev. Lett. 51, 1799 (1983)
2.311 R.J. Behm, V. Penka, M.G. Cattania, K. Christmann, G. Ertl: J. Chem. Phys. 78, 7486 (1983)
2.312 H. Niehus, Ch. Hiller, G. Comsa: Surf. Science 173, L599 (1986)
2.313 W. Eberhardt, F. Greuter, E.W. Plummer: Phys. Rev. Lett. 46, 1085 (1981)
2.314 M. Lagos: Surf. Science 122, L601 (1982)
2.315 L. Schlapbach, J.P. Burger, J. Phys. Lett. 43, L273 (1982); T. Riesterer, J. Osterwalder, L. Schlapbach: Phys. Rev. B 32, 8405 (1985)
2.316 C.T. Chan, S.G. Louie: Phys. Rev. B 30, 4153 (1984)
2.317 S.R. Chubb, J.W. Davenport: Phys. Rev. B 31, 3278 (1985)
2.318 T.E. Felter, R.H. Stulen: J. Vac. Sci. Technol. A3, 1566 (1985), T.E. Felter, R.H. Stulen, M.L. Koszykowski, G.E. Godowski, B. Garrett: J. Vac. Sci. Technol. A7, 104 (1989)
2.319 T.E. Felter, E.C. Sowa, M.A. VanHove: Phys. Rev. B 40, 891 (1989); M.S. Daw, S.M. Foiles: Phys. Rev. B 35, 2128 (1987)
2.320 L. Stauffer, R. Riedinger, H. Dreyssé: Vacuum 41, 188 (1990); A. Haroun, L. Stauffer, H. Dreyssé, R. Riedinger: Phys. Rev. B 38, 12150 (1988)
2.321 D. Tomanek, S.G. Louie, C.T. Chan: Phys. Rev. Lett. 57, 2594 (1986)
2.322 G.E. Godowski, T.E. Felter, R.H. Stulen: Surf. Science 181, L147 (1987)
2.323 B.D. Kay, C.H. Peden, D.W. Goodman: Phys. Rev. B 34, 817 (1986); Surf. Science 175, 215 (1986)
2.324 L. Schröter, H. Zacharias. R. David: Appl. Phys. A 41, 95 (1986)
2.325 L.G. Petersson, H. Dannetun, J. Fogelberg, I. Lundström: Appl. Surf. Science 27, 275 (1986)
2.326 S. Andersson, J. Harris: Phys. Rev. B 27, 9 (1983); Phys. Rev. Lett. 55, 2591 (1985)
2.327 M. Balooch, M.J. Cardillo, D.R. Miller, R.E. Stickney: Surf. Science 46, 358 (1974)
2.328 G. Anger, A. Winkler, K.D. Rendulic: Surf. Science 220, 1 (1989)
2.329 M. Hand, S. Holloway: Surf. Science 211/212, 940 (1989)
2.330 A.P. Baddorf, E.W. Plummer: Bull. Am. Phys. Soc. 33, 654 (1988); A.P. Baddorf, I.W. Lyo, E.W. Plummer: J. Vac. Sci. Technol. A5, 782 (1987)
2.331 A. Mattson, I. Panas, P. Siegbahn, U. Wahlgren, H. Akeby: Phys. Rev. B 36, 7389 (1987)
2.332 G. Atkinson, S. Coldrick, J.P. Murphy, N. Taylor: J. Less-Common Met. 49, 439 (1976)
2.333 G. Rosina, E. Bertel, F.P. Netzer: Phys. Rev. B 34, 5746 (1986)
2.334 G. Rosina, E. Bertel, F.P. Netzer, J. Redinger: Phys. Rev. B 32, 2364 (1986)
2.335 G. Strasser, G. Rosina, E. Bertel, F.P. Netzer: Surf. Science 152/153, 765 (1985)
2.336 R. Baptist, A. Pellissier, G. Chauvet: Z. Phys. B 73, 107 (1988)
2.337 R. Kumar, M.H. Mintz, J.W. Rabalais: Surf. Science 147, 37 (1984)
2.338 N. Hosoda, H. Uchida, Y. Ohtani, T. Takahashi E. Fromm: Z. Phys. Chem. NF 164, 1129 (1989)
2.339 T. Greber, St. Büchler, L. Schlapbach: Vacuum 41, 556 (1990)
2.340 J. Dexpert-Ghys, C. Lorier, Ch.H. La Blanchetais, P.E. Caro: J. Less-Comm. Met. 41, 105 (1975)
2.341 H.K. Smith, A.G. Moldovan, R.S. Craig, W.E. Wallace, S.G. Sankar: J. Solid State Chem. 32, 239 (1980)
2.342 T. Greber, L. Schlapbach: Z. Phys. Chem. NF 164, 1213 (1989)
2.343 J. Bloch, Z. Hadari, M.H. Mintz: J. Less-Common Met. 102, 311 (1984)

2.344 S.A. Flodström, L.-G. Petersson, S.B. Hagström: J. Vac. Sci. Technol. 13, 280 (1976)
2.345 O. Gunnarsson, J. Hjelmberg, B.I. Lundquist: Phys. Rev. Lett. 37, 292 (1976)
2.346 M. Boudart, A. Delbouille, E.G. Derouane, V. Indovina, A.B. Walters: J. Am. Chem. Soc. 94, 6622 (1987)
2.347 V. Lee, H. Wong: J. Phys. Soc. Japan 45, 1657 (1978)
2.348 V.E. Henrich, G. Dresselhaus, H.J. Zeiger: Phys. Rev. B 22, 4764 (1980)
2.349 E.A. Colburn, W.C. Mackrodt: Surf. Science 117, 571 (1982)
2.350 H. Fujioka, S. Yamabe, Y. Yanagisawa, K. Matsumura, R. Huzimura: Surf. Science 149, L53 (1985)
2.351 U.O Karlsson, G.V. Hansson, P.E. Persson, S.A. Foldström: Phys. Rev. B 26, 1852 (1982)
2.352 A.L. Companion: J. Less-Common Met. 98, 121 (1984)
2.353 P.A. Thiry, J.J. Pireaux, M. Liehr, R. Caudano: J. Vac. Sci. Technol. A3, 1439 (1985)
2.354 J. Paul, F.M. Hoffman: Surf. Science 194, 419 (1988)
2.355 P.T. Sprunger, E.W. Plummer: Bull. Am. Phys. Soc. 35, 386 (1990)
2.356 K.B. Ray, E.W. Plummer: Bull. Am. Phys. Soc. 33, 655 (1988)
2.357 M.H. Mintz, J.A. Schulz, J.W. Rabalais: Surf. Science 146, 457 (1984); Phys. Rev. Lett. 51, 1676 ((1983); M.H. Mintz, J.A. Schulz: J. Less-Common Met. 103, 349 (1985)
2.358 R. Fromageau, J. Hillairet, E. Ligeon, C. Mairz, G. Revel, P. Tzanétakis: J. Appl. Phys. 52, 7191 (1981)
2.359 A.S. Pedersen, J. Kjøller, B. Larsen, B. Vigeholm: Int. J. Hydrogen Energy 10, 851 (1985)
2.360 L. Schlapbach, J. Osterwalder, T. Riesterer: J. Less-Common Met. 103, 295 (1985)
2.361 K.B. Gerasimov, E.L. Goldberg, E.Y. Yvanov: J. Less-Common Met. 131, 99, 143 (1987)
2.362 E. Akiba, S. Ono: J. Less-Common Met. 124, L1 (1986)
2.363 E.W. Collings, R.D. Smith, R.G. Lecander: J. Less-Common Met. 18, 251 (1969)
2.364 J.H. Weaver, A. Franciosi, W.E. Wallace, H.K. Smith: J. Appl. Phys. 51, 5847 (1980)
2.365 L. Schlapbach: Solid State Commun. 38, 117 (1981)
2.366 H. Schafer, J. Osterwalder, L. Schlapbach: to be published
2.367 R.L. Cohen, R.C. Sherwood, K.W. West: Appl. Phys. Lett. 41, 999 (1982); H. Rummel, R.L. Cohen, P. Gütlich, K.W. West: Appl. Phys. Lett. 40, 477 (1982)
2.368 E.F. Khodosov, A.I. Linnik: Sov. Phys. Solid State 22, 1452 (1980)
2.369 F.W. Oliver, T. Kebede, K. Thompson, J. Gilchrist: Solid State Commun. 46, 837 (1983)
2.370 H. Uchida, M. Ozawa: Z. Phys. Chem NF 147, 77 (1986); J. Less-Common Met. 131, 143 (1987); H. Uchida, Y. Ohtani, M. Ozawa, T. Kawahata, T. Suzuki: J. Less-Common Met. (1991) 172–174, 983
2.371 P. Selvam, B. Viswanathan, V. Srinivasan: Int. J. Hydrogen Energy 14, 687 (1989); P. Selvam, B. Viswanathan, C.S. Swamy, V. Srinivasan: Int. J. Hydrogen Energy 16, 23 (1991)
2.372 G.D. Sandrock, P.D. Goodell: J. Less-Common Met. 73, 161 (1980); F.R. Block, H.J. Bas: J. Less-Common Met. 89, 77 (1983); S. Yu, C. Zhang, K. Xie, G. Kang, Z. Lin: Chinese Physics 3, 842 (1983)
2.373 J. Christopher, M.P.S. Kumar, C.S. Swami: J. Mat. Sci 23, 4263 (1988)
2.374 F.T. Parker, H. Oesterreicher: J. Less-Common Met. 79, 297 (1981)
2.375 A.H. Boonstra, G.J.M. Lippitis, T.N.M. Bernards: J. Less-Common Met. 155, 119 (1989) and references therein
2.376 F. Meli, L. Schlapbach: J. Less-Common Met. 172–174, 1252 (1991)
2.377 T. Sakai, A. Takagi, T. Hazama, H. Miyamura, N. Kuriyama, H. Ishikawa C. Iwakura: Proc. 3rd Int. Conf. "Batteries for Utility Energy Storage", Kobe, Japan, 1991
2.378 J.H. Weaver, D.T. Peterson: Phys. Rev. B 22, 3624 (1980)
2.379 T.E. Felter, S.A. Stewart, F.S. Uribe: Surf. Sci. 122, 69 (1982)
2.380 F.P. Netzer, J.C. Rivière, G. Rosina: Phys. Rev. B 38, 7453 (1988)
2.381 G. Schadler, P. Weinberger: J. Phys. F 16, 27 (1986)
2.382 R. Eibler, J. Redinger, A. Neckel: J. Phys. F 17, 1533 (1987)
2.383 T. Hirata: J. Less-Common Met. 113, 189 (1985)
2.384 F.D. Manchester, D. Khatamian: Mechanisms for activation of intermetallic hydrogen absorbers, in Hydrogen Absorbing Compounds, ed. by R.G. Barnes (Trans. Tech., 1989)
2.385 T. Schober: J. Less-Common Met. 89, 63 (1983); T. Schober, C. Dieker: J. Less-Common Met. 104, 191 (1984)
2.386 L. Schlapbach, A. Seiler, F. Stucki: Mat. Res. Bull. 13, 697, 1031 (1978); L. Schlapbach, A. Seiler, F. Stucki, P. Zürcher, P. Fischer, J. Schefer: Z. Phys. Chem. NF 117, 205 (1979)
2.387 A. Bläsius, U. Gonser: Appl. Phys. 22, 331 (1980)
2.388 D. Khatamian, F.D. Manchester: Surf. Science 159, 381 (1985); Surf. Science 186, 309 (1987)

2.389 D. Khatamian, G.C. Weatherly F.D. Manchester: Acta Metall. **31**, 1771 (1983)
2.390 Th.von Waldkirch, R. Wessiken. H.-U Nissen, L. Schlapbach: Proc. 3rd Int. Cong. on Hydrogen and Materials, Paris (1982), C9
2.391 H. Züchner, G. Kirch: J. Less-Common Met. **99**, 143 (1984)
2.392 D. Finkler, K. Blaes, U. Gonser, S.J. Campbell: *Mössbauer Studies of H in TiFe* in *Titanium, Science and Technology*, ed. by G. Lütjering, U. Zwicker, W. Bunk (Deutsche Ges. f. Metallkde, Oberursel, 1985) p. 2579
2.393 M. Polak, Y. Ben-Shoshan: Surf. Science **146**, L601 (1984)
2.394 H. Harada, S. Ishibe, R. Konishi, H. Sasakura: Jap. J. Appl. Phys. **24**, 1141 (1985)
2.395 G. Sicking, B. Jungblut: Surf. Science **127**, 255 (1983)
2.396 L. Schlapbach, T. Riesterer: J. Less-Common Met. **101**, 453 (1984) and ref. therein
2.397 P.H. McBreen, M. Polak: Surf. Science **179**, 483 (1987); Surf. Science **163**, L666 (1985)
2.398 H.G. Wulz, E. Fromm: J. Less-Common Met. **118**, 315 (1986); H. Uchida, E. Fromm: J. Less-Common Met. **131**, 125 (1987)
2.399 H. Nagai, M. Nakatsu, K. Shoji, H. Tamura: J. Less-Common Met. **119**, 131 (1986)
2.400 Y. Shenzhong, Y. Rong, H. Tiesheng, Z. Shilong, C. Bingzhao: Int. J. Hydrogen Energy **13**, 433 (1988)
2.401 S.H. Lim, J.-Y. Le: J. Less-Common Met. **97**, 59 and 65 (1984)
2.402 J. Suzuki, M. Abe, T. Yamaguchi, S. Terezawa: J. Less-Common Met. **131**, 301 (1987)
2.403 Z. Zhou, H. Zhang, D. Tang, D. Zhou, X. Ping, J. Ni: In Hydrogen Energy Progress VI, T.N. Veziroglu, N. Getoff, P. Weinzierl, eds. (Pergamon, Oxford 1986) Vol. II, p. 898
2.404 J.H. Sanders, B.J. Tatarchuk: J. Phys. F **18**, L267 (1988)
2.405 G. Kirch, R. Hempelmann, E. Schwab, H. Züchner: Z, Phys. Chem. NF **126**, 109 (1981)
2.406 T. Hirata: J. Less-Common Met. **136**, 217 (1988)
2.407 Ph. McBreen, S. Serghini-Monim, D. Roy, A. Adnot: Surf. Science **195**, L208 (1988)
2.408 R. Brenier, A. Perez, P. Thevenard, M. Treilleux, T. Capra: Mater. Sci. Eng. **69**, 83 (1985)
2.409 J.P. Hirvonen, M.A. Elve, J.W. Mayer, H.H. Johnson: Mater. Sci. Eng. **90**, 13 (1987)
2.410 A. Jimbo, T. Hashizume, T. Sakurai, K. Al-Saleh, H.W. Pickering: J. de Phys. **12**, C9-417 (1984)
2.411 L. Schlapbach: Phys. Lett. **91**, 303 (1982)
2.412 T. Riesterer: Z. Phys. B **66**, 441 (1987)
2.413 I. Jacob, M. Fischer, Z. Hadari: Solid State Commun. **49**, 1161 (1984)
2.414 I. Jacob, M. Polak: Mater. Res. Bull. **16**, 1311 (1981)
2.415 L. Schlapbach: J. Less-Common Met. **89**, 37 (1983)
2.416 H.K. Smith, W.E. Wallace: J. Less-Common Met. **115**, 97 (1986); H.K. Smith, W.E. Wallace, R.S. Craig: J. Less-Common Met. **94**, 85 (1983)
2.417 P.W. Albers, G.H. Sicking, I.R. Harris: Z. Metallkde. **79**, 24 (1988)
2.418 P. Selvam, B. Viswanathan, C.S. Swamy, V. Srinivasan: Int. J. Hydrogen Energy **11**, 169 (1986)
2.419 P. Selvam, B. Viswanathan, C.S. Swamy, V. Srinivasan: Thermochimica Acta **125**, 1 (1988); Bull. Mater. Sci. **9**, 21 (1987)
2.420 H. Imamura, M. Kawahigashi, S. Tsuchiya: J. Less-Common Met. **95**, 157 (1983)
2.421 B. Bogdanovic, S.C. Huckett, B. Spliethoff, U. Wilczok: Z. Phys. Chem. NF **163**, 337 (1989)
2.422 L. Schlapbach, D. Shaltiel, P. Oelhafen: Mat. Res. Bull. **14**, 1235 (1979); A. Seiler, L. Schlapbach, Th. von Waldkirch, D. Shaltiel, F. Stucki: J. Less-Common Met. **73**, 193 (1980)
2.423 J. Genossar, P.S. Rudman: Z. Phys. Chem. NF **116**, 215 (1979)
2.424 B. Darriet, M. Pezat, A. Hbika, P. Hagenmüller: Int. J. Hydrogen Energy **5**, 173 (1980)
2.425 M. Khrussanova, M. Terzieva, P. Peshev: Int. J. Hydrogen Energy **11**, 331 (1986)
2.426 M.Y. Song, M. Pezat, B. Darriet, P. Hagenmüller: J. Mater. Sci. **20**, 2958 (1985)
2.427 Y. Ishido, S. Ono, E. Akiba: J. Less-Common Met. **120**, 163 (1986)
2.428 N. Shamir, M.H. Mintz, J. Bloch, U. Atzmoni: J. Less-Common Met. **92**, 253 (1983)
2.429 V.V. Lunin, P.A. Chernavskii, B.Y. Rakhaminov, N.G. Chulkov: Inorganic Mat. **14**, 1326 (1978)
2.430 P.E. West, P.M. George: J. Vac. Sci. Technol. **A5**, 1124 (1987)
2.431 B. Walz: Ph.D. Thesis, University of Basel (1989)
2.432 M. Devillers, M. Sirch, S. Bredendiek-Kämper, R.-D. Penzhorn: Chem. Mat. **2**, 255 (1990)
2.433 X.N. Yu, L. Schlapbach: Phys. Rev. B **37**, 6215 (1988)
2.434 F. Spit, K. Block, E. Hendriks, G. Winkels, W. Turkenburg, J.W. Drijver, S. Radelaar: Proc. 4th Int. Conf. on Rapidly Quenched Metals, Sendai Japan 1981, p. 1635
2.435 K. Machida, M. Enyo, I. Toyoshima, K. Kai, K. Suzuki: J. Less-Common Met. **96**, 305 (1984)
2.436 J.C. Bertolini, J. Brissot, T. Le Mogne, H. Montes, Y. Calvayrac, J. Bigot: Appl. Surf. Science **29**, 29 (1987)

2.437 S.M. Fries, H.-G. Wagner, U. Gonser, L. Schlapbach, R. Montiel-Montoya: J. Magn. Magn. Mat. **45**, 331 (1984)

2.438 Z.S. Wronski, X.Z. Zhou, A.H. Morrish, A.M. Stewart: J. Appl. Phys. **57**, 3548 (1985)

2.439 R.A. Butera, J.-H. Weaver, D.J. Peterman, A. Franciosi, D.T. Peterson: J. Chem. Phys. **79**, 2395 (1983); R.A. Butera, E. Franz, J.J. Joyce, J.H. Weaver: Solid State Commun. **55**, 1089 (1985)

2.440 A. Krozer, B. Kasemo: J. Vac. Sci. Technol. **A5**, 1003 (1987); J. Ryden, B. Hjörvarsson, T. Ericsson, E. Karlsson, A. Krozer, B. Kasemo: J. Less-Common Met. **152**, 295 (1989)

2.441 A. Fischer, H. Köstler, L. Schlapbach: J. Less-Common Met. **172–174**, 808 (1991)

2.442 P.E. Miceli, H. Zabel, J.E. Cunningham: Phys. Rev. Lett. **54**, 917 (1985)

3. Dynamics of Hydrogen in Intermetallic Hydrides

Dieter Richter, Rolf Hempelmann, and Robert C. Bowman, Jr.

With 28 Figures and 8 Tables

3.1 Survey

If a hydrogen atom is dissolved in a metallic compound, generally it may perform motional processes on very different time scales. At very short times the H atom vibrates against its metallic neighbours which due to their much heavier masses do not participate in this high frequency vibrations. Depending on H concentration and H–H interaction they can be considered either as local or optical modes. At the time scale of the acoustic vibrations of the host lattice, the H atoms move more or less adiabatically according to the distortion pattern imposed by the host phonons and mirror the host density of states. This type of motion is also called a band mode. At much longer times the hydrogen is able to leave its interstitial site and to perform jumps to other sites. For a complicated host structure or an amorphous metal a broad spectrum of different jump rates is expected leading to complex diffusion mechanisms. The experimental investigation of these motional processes on the one hand bears scientific importance, since it elucidates the basic interactions between proton and host and facilitates an understanding of the elementary transport processes of light interstitials. On the other hand it also is of consequence for application. For instance H diffusion is an important step in the kinetics of hydrogenation and phase separation. While e.g. for the storage compound $Ti_{0.8}Zr_{0.2}CrMnH_x$ diffusion was found not to influence hydrogenation, in activated FeTi it appears to be the rate limiting process. The vibrational properties of H contribute strongly to the thermodynamic potentials of a metal–hydrogen system. Their investigation helps us to understand the macroscopic properties on a fundamental level.

In the second section we give a brief overview of methods which have been used to study H-dynamics and discuss in some detail anelastic relaxation, nuclear magnetic resonance (NMR) and quasielastic neutron scattering (QNS) which in our opinion have the greatest potential for diffusion studies. Perturbed angular correlation measurements and their advantages for studying the hydrogen mobility in dilute systems are reviewed in Chap. 6. With respect to vibrational spectroscopy inelastic neutron scattering is unique. The third section is concerned with diffusion studies. After a brief overview of diffusion results which is mainly presented in the form of tables, we discuss particular NMR and neutron results in an exemplary way. First we present NMR investigations of $TiCr_2H_x$, thereafter we turn to a QNS study of $Ti_{1.2}Mn_{1.8}H_3$ and finally we take a critical look at NMR and neutron data on $LaNi_5H_x$. A short account of diffusion results on H in amorphous metals concludes this section. The fourth section is dedicated to H vibrations. We start with neutron investigations of the H potential. As an example we present recent results on PdH_x and discuss their relation to solubility data. A measurement of H vibrations also reveals the local symmetry of the H sites and thus yields structural information. Applications of this approach are presented for the cases of intermetallics, amorphous metals, metal oxides and H on metal surfaces. The intensity of an H mode is directly related to the occupation number of the corresponding site. A temperature dependent intensity measurement can thus tell us about thermal repopulation among H sites. As an example we discuss a study on H trapping in $Nb_{99}Cr_1H_1$. In the final section we give a critical outlook on future prospects in this field.

3.2 Methods

3.2.1 General Methods for Diffusion

In their reviews of hydrogen diffusion behaviour in metals, *Völkl* and *Alefeld* [3.1, 2] described essentially all of the experimental techniques that have been used to deduce diffusion parameters. Many different physical phenomena (e.g. electrical resistivity, hydrogen absorption/desorption rates, mechanical relaxations, magnetic properties, etc.) can be more or less directly related to hydrogen diffusion coefficients. Since these authors [3.1, 2] have provided good summaries of the various techniques as well as many references to the original literature, only some recent developments in methods that particularly pertain to hydrogen diffusion in the intermetallic hydrides and metallic glasses will be addressed in this subsection.

Both hydrogen permeation and the absorption/desorption kinetics can be related to hydrogen diffusion coefficients when bulk diffusion (and not surface-dependent effects) is the rate-controlling mechanism. In addition to the relatively common volumetric and weight change methods to monitor the time dependence of hydrogen transport, some more novel approaches such as neutron radiography [3.3], ^{15}N nuclear reaction with protons [3.4], and various tritium tracer methods [3.5, 6] have been attempted. The primary difficulty with most of these macroscopic methods is in the identification and subsequent control of surface phenomena on the kinetic rates. While deposition of thin films of hydrogen permeable metal (i.e. usually Pd) on cleaned surfaces of bulk samples may provide an acceptable solution in some cases [3.7], this approach is not generally practical in brittle intermetallics that undergo large volume expansion and microcracking upon hydriding. However, *Arnold* and *Welter* [3.8] have recently evaluated the hydrogen and deuterium diffusion properties in the α-phase (i.e. very low hydrogen contents) of Pd-coated $TiFeH(D)_x$ samples. Nevertheless, the role of the interface between the bulk host metal and the film in the observed rates is often obscure. Effective particle size (i.e. the surface area actually involved for hydrogen exchange between the solid and gaseous phases) as well as irregular particle shapes can also complicate the analysis of kinetic data since rather simple models (i.e. spheres or infinite planes) are often assumed. Consequently, diffusion parameters derived from permeation and kinetic data may not be as reliable as many authors have claimed and a healthy scepticism of detailed interpretations is often advisable.

Selected metal nuclei can also act as local probes of hydrogen diffusion through the Mössbauer effect [3.9–13] or quadrupolar parameters from the metal nuclear magnetic resonance (NMR) [3.14–16] studies. Although the Mössbauer studies are restricted to only a few nuclei (e.g. ^{57}Fe, ^{181}Ta), *Wagner* et al. [3.13] obtained reasonable diffusion activation energies for $VD_{0.78}$ and $NbH_{0.78}$ doped with small quantities of isotopically enriched Fe solutes. Consequently, this approach could be useful for a very large variety of hydrides that could include intermetallic hydrides. Diffusion studies with the metal NMR method are possible (in principle with many different nuclei) although practical difficulties abound. Not only must the proton motion be sufficient to influence the electric quadrupolar field gradients responsible for the lineshapes and relaxation times, but also the detection of the very broad resonance peaks can be difficult [3.14–16], and the material cannot be ferromagnetic or strongly paramagnetic, which excludes many intermetallic hydrides of practical interest. Nevertheless, metal nucleus NMR offers the possibility of direct comparison of hydrogen isotope effects (as does the Mössbauer technique) with a presumably invariant probe atom and local environment.

3.2.2 Anelastic Relaxation

When samples are available in suitable bulk form (i.e. wire, ribbon, foils, rods, discs, etc.), electrical and mechanical relaxation techniques can be used to determine diffusion behaviour. For intermetallic hydrides, which are generally extremely friable and are often only available as powders, these methods have limited applicability. The most significant exceptions are the metallic glasses (i.e. the amorphous transition metal alloys) that can usually retain their structural and mechanical integrity upon hydrogenation, although amorphous hydrides tend to be much less ductile than the initially hydrogen-free glasses. In the following we give a brief account of mechanical relaxation methods [3.17, 18], whereas for electrical (i.e. resistometric or electrochemical) methods we refer to the review by *Boes* and *Züchner* [3.19] and to recent improvements and/or modifications published by *Pine* and *Cotts* [3.20], *Schöneich* and *Züchner* [3.21], and *Kirchheim* et al. [3.22].

The introduction of a hydrogen atom into a metal is accompanied by an expansion of the metal lattice. If the sites are orientationally distinguishable— e.g. like the interstitial sites in bcc metals—then single atoms create orientationally distinguishable defects which can produce anelasticity by a reorientation mechanism (Snoek relaxation [3.23]). This relaxation mechanism is not available to hydrogen atoms at sites with cubic symmetry. However, hydrogen atoms in all metals, regardless of their site symmetry, if exposed to a macroscopic strain gradient, may produce a diffusional relaxation (Gorsky effect [3.24, 25]). A useful macroscopic approach to these phenomena is to start with a sphere of defect-free host material and to consider the introduction of orientationally distinguishable defects with a dilute interstitial concentration c. Provided these hydrogen atoms are placed in sites of identical orientation, the sphere distorts to an ellipsoid with principal strains $\varepsilon_{ii}(i = 1, 2, 3)$.

The derivatives $\lambda_{ii} \equiv \partial \varepsilon_{ii}/\partial c$ form the principal components of a tensor (the λ-tensor), which for uniaxial defects is of the form

$$\underline{\lambda} = \begin{pmatrix} \lambda_1 & 0 & 0 \\ 0 & \lambda_2 & 0 \\ 0 & 0 & \lambda_2 \end{pmatrix}. \tag{3.1}$$

Such defects have an ellipticity or shape factor given by the difference $\delta\lambda \equiv (\lambda_1 - \lambda_2)$ and a (volumetric) size factor given by the trace $\operatorname{tr}\{\lambda\} \equiv \lambda_1 + 2\lambda_2$. They can interact with a shear stress through the shape factor $\delta\lambda$ to produce reorientation relaxation (Snoek effect) and can also interact with a dilatational stress through the size factor $\operatorname{tr}\{\lambda\}$ to produce a diffusional relaxation (Gorsky effect).

a) The Snoek Effect

Prior to the application of the stress the hydrogen atoms are equidistributed among the sites of different orientation, because they represent states of equal energy. Depending on orientation, a stress splits the site energies and triggers relaxation to a new equilibrium configuration exhibiting a small degree of directional order, with more hydrogen in sites of lower energy. Thus the reorientation relaxation involves local hydrogen jumps.

If a constant stress is applied, such a system responds with an immediate initial elastic strain ε_e followed by the anelastic strain ε_a which grows with time to a limiting value $\bar{e}_a$. Upon removal of the stress the relaxation process is reversed, and the immediate recovery of the elastic strain is followed by the elastic after-effect or creep recovery which eventually restores the sample to its original form. Since both the equilibrium anelastic strain $\bar{e}_a$ and the elastic strain e_e are proportional to the applied stress, the ratio $\Delta \equiv \bar{e}_a/\varepsilon_e$ is a convenient measure of the magnitude of the relaxation and is called the relaxation strength. For the Snoek effect it is proportional to the square of the shape factor $\delta\lambda$. Because of its high sensitivity to weak relaxation processes, experimentally the Snoek relaxation is usually determined dynamically, for instance by deforming the sample at a constant frequency and measuring the mechanical damping (internal friction) as a function of the temperature. This procedure enables a thermally activated relaxation process to be detected as an internal friction peak: the maximum damping occurs when the external (circular) excitation frequency ω equals the reorientation jump time τ_R. In the simple case of a reorientation relaxation with only one type of reorientation jumps the peak shape has the

Debye form

$$\delta/\pi = \Delta^R \frac{\omega\tau_R}{1 + \omega^2\tau_R^2},$$ (3.2)

where the internal friction is expressed as the logarithmic decrement δ. The activation energy of this process is obtained by tracing out the internal friction peak at several different excitation frequencies over as wide a range as possible.

b) The Gorsky Effect

In contrast to the reorientation relaxation the Gorsky relaxation involves long-range diffusion. It is commonly measured quasi-statically as an elastic after-effect. The relaxation is initiated by bending a sample to induce a spatial gradient in the chemical potential μ of the hydrogen atoms. This gradient causes an "uphill" diffusion of the hydrogen atoms and the development of a concentration gradient. Relaxation is complete when the concentration gradient equalizes the chemical potential across the thickness of the sample. For a strip of thickness d, the Gorsky relaxation time τ_G is related to the (chemical) hydrogen diffusion coefficient D by

$$\tau_G \approx d^2/\pi^2 D.$$ (3.3)

Therefore the main application of the Gorsky effect is to obtain precise measurements of hydrogen diffusion coefficients [3.2]. But the other observable quantity, the relaxation strength—which is proportional to the square of the size factor, $\mathrm{tr}\{\lambda\}$—bears additionally important information on the thermodynamic factor $\partial\mu/\partial c$: thus it allows one to relate the chemical diffusion coefficient as determined by macroscopic methods to the tracer diffusion coefficient as determined by microscopic methods (see Sect. 3.2.4a).

3.2.3 NMR Studies of Diffusion

Nuclear magnetic resonance (NMR) is a very versatile technique for the evaluation of hydrogen isotope diffusion behavior since all three hydrogen isotopes permit good to excellent detection. In fact, NMR may be the most widely used method to assess diffusion behaviour in the high concentration hydride phases since there are relatively few special requirements on the samples as long as they are not strongly magnetic. Although NMR can be used for low hydrogen concentrations (i.e. α-phase solid solutions), the normally low sensitivity of the NMR signals requires rather tedious signal averaging whenever the hydrogen to metal atom ratio is below 0.10 or so. Two extensive reviews by *Cotts* [3.26, 27] treat thoroughly the application of NMR to metal-hydrogen systems. In addition, several somewhat more specialized reviews [3.28–31] that primarily focus upon the diffusion aspects of NMR in metal hydride research have also been published within the past few years. The reader is referred to [3.26–31] for general background and detailed descriptions of the various NMR techniques. The emphasis of the present discussion will be upon the advantages and limitations of the different NMR techniques whose measured parameters are related to the diffusion properties of the hydrogen isotopes. Furthermore, the behaviour of protons will receive the most attention since relatively few NMR studies have been performed on metal deuterides [3.27–30] or tritides [3.32, 33].

The various experimental NMR parameters that are normally used to evaluate proton diffusion characteristics are summarized in Table 3.1. The conventional methods used to generate these parameters are described in the reviews by *Cotts* [3.26, 27] and will not be repeated here. With the exceptions of the line-narrowing experiments, which are only sensitive to very restricted changes in proton mobilities [3.26, 27], each NMR parameter can monitor at least a four decade change in the diffusion rate. From a combination of relaxation time measurements, the proton diffusion rate for an individual hydride sample can be followed (in principle) over nearly ten orders of magnitude. However, the numerous complications that can arise from sample properties (i.e. limited thermal stabilities, phase transitions, etc.), as well as instrumental and analysis limitations, usually prohibit such extensive studies. In many circumstances it can be very difficult to establish whether a change in the temperature dependence of an NMR parameter or the corresponding mean time derived for jumps between interstitial sites (τ_c) is due to a change in diffusion mechanism, or merely reflects an artifact of the experimental procedures or the models invoked in the data analysis.

Theoretical expressions relate the experimental relaxation times to the power spectra $J^{(P)}(\omega)$ of local magnetic dipole fields that are caused by relative motion of the resonant nuclei with respect to other nuclear dipole moments including members of the resonant spin system. To illustrate this relationship,

Table. 3.1. Summary of experimental NMR parameters that can be related to hydrogen isotope diffusion properties. τ_c is the time constant for atomic jump rates and D is the microscopic diffusion coefficient

Experimental NMR parameter	Symbol	Derived quantity	Approximate D range for each NMR parameter [cm²/s]
Narrowing of continuous wave linewidth or free induction decay signal	ΔH	τ_c	$10^{-11}-10^{-9}$
Spin–lattice relaxation time	T_1	τ_c	$10^{-10}-10^{-6}$
Rotating-frame relaxation time	$T_{1\rho}$	τ_c	$10^{-13}-10^{-7}$
Dipolar relaxation time	T_{1D}	τ_c	$10^{-15}-10^{-11}$
Spin–spin relaxation time	T_2	τ_c	$10^{-11}-10^{-7}$
Field-gradient spin-echo	FGSE	D	$10^{-8}-10^{-4}$

the general expressions [3.28] for the proton–proton spin interactions are

$$(T_1)^{-1} = (3/2)\gamma_I^4 \hbar^2 I(I+1)[J^{(1)}(\omega) + J^{(2)}(2\omega)], \tag{3.4}$$

$$(T_{1\rho})^{-1} = (3/8)\gamma_I^4 \hbar^2 I(I+1)[J^{(0)}(2\omega_1) + 10J^{(1)}(\omega) + J^{(2)}(2\omega)], \tag{3.5}$$

$$(T_2)^{-1} = (3/8)\gamma_I^4 \hbar^2 I(I+1)[J^{(0)}(0) + 10J^{(1)}(\omega) + J^{(2)}(2\omega)], \tag{3.6}$$

where γ_I is the proton gyromagnetic ratio; $I = 1/2$; ω is the Larmor frequency in the applied magnetic field B_0 ($\omega = \gamma_I B_0$), and $\omega_1 = \gamma_I B_1$ in the rotating rf field B_1. The $J^{(P)}(\omega)$ are Fourier transforms of the time-dependent correlation functions $G^{(q)}(t)$:

$$J^{(P)}(\omega) = \int_{-\infty}^{\infty} G^{(q)}(t) \exp(-i\omega t)dt. \tag{3.7}$$

These expressions are completely general, but the challenge is to find realistic, yet tractable, models to relate the microscopic motion of the nuclei to the experimental relaxation times. Equations (3.4–7) can be generalized [3.26] to include heteronuclear dipolar as well as quadrupolar interactions.

While the relaxation time experiments are normally quite straightforward with modern transient NMR spectrometers, the measured relaxation times generally contain several contributions. For example, the proton spin-lattice relaxation time T_1 can be partitioned [3.28] as follows:

$$T_1^{-1} = T_{1d}^{-1} + T_{1e}^{-1} + T_{1p}^{-1}, \tag{3.8}$$

where T_{1d} is a dipolar term that corresponds directly to the diffusion-induced time modulation of the proton dipolar interactions [3.26, 27] [as represented in (3.4)], T_{1e} corresponds to the relaxation processes that arise from hyperfine interactions with the unpaired Fermi level conduction electrons, and T_{1p} is the possible contribution from localized paramagnetic centres [3.34]. While the T_{1e} term can often be estimated from experimental T_1 data obtained in temperature regions where T_{1d} contributions are small, this approach may be ineffective if there are intervening phase transitions or if T_{1e} is strongly temperature dependent [3.35]. Although the paramagnetic contributions to proton relaxation times are often assumed to be negligible, *Phua* et al. [3.34] have clearly demonstrated that even a few ppm of paramagnetic rare earth metal impurities (Gd, Ce, etc.) can profoundly influence the relaxation times over wide temperature ranges in some metal hydrides. These paramagnetic relaxation effects can be sufficiently large to produce anomalous temperature behaviour, which has previously been incorrectly attributed [3.34] to the intrinsic diffusion properties. The results of *Phua* et al. [3.34] indicate that paramagnetic effects tend to be larger for the T_1 relaxation times than for the spin–spin (T_2) or rotating-frame ($T_{1\rho}$) relaxation times. However, it is now quite clear that NMR samples may need to contain extremely small quantities of paramagnetic impurities to avoid extraneous contributions to the relaxation times. It is likely that a substantial number of previous NMR studies of diffusion properties in intermetallic hydrides were actually performed on samples of insufficient purity as regards paramagnetic species. Consequently, the reported diffusion characteristics may be erroneous and new NMR studies on samples of better purity are necessary. On the other hand, hydrides which are not contaminated with the more strongly paramagnetic impurities (the rare earths, Mn, etc.) are probably not seriously affected by these paramagnetic relaxation phenomena and their diffusion parameters should be fairly reliable.

The diffusion coefficient D and the interval between jumps (τ_c) are related by the expression

$$D = f l^2 / (6\tau_c), \tag{3.9}$$

where l is the jump distance and f is the tracer correlation factor, which usually varies between 0.5 and 1.0 and is often assumed to be equal to unity. A unique advantage of the NMR method which is shared only by incoherent quasielastic neutron scattering, is its ability to generate independent experimental values for τ_c from the relaxation times (usually T_1), and for D from the attenuation of spin echo signals in applied magnetic field gradients [3.27, 36, 37]. Consequently, it is possible to experimentally deduce l from NMR measurements alone. Figure 3.1 reproduces some results from such a study for a cubic γ-phase TiH$_{1.55}$ sample by *Bustard* et al. [3.38]. A comparison of the D values measured directly by the field-gradient spin-echo (FGSE) method [3.27] with diffusion coefficients calculated from (3.5) for τ_c values derived from T_1 data, show a clear preference for a jump distance that corresponds to the separation between nearest neighbour proton interstitial sites. *Bustard* et al. [3.38] obtained the same results from a γ-phase TiH$_{1.71}$ sample. In principle, similar comparisons to derive l for the intermetallic hydrides should also be possible. However, there are two serious difficulties. First, many intermetallic hydrides are too unstable (i.e. they lose hydrogen gas which changes the stoichiometry) at the temperatures where proton D values have become large enough to be directly measured by the FGSE methods [3.27]. Second, the "background" field gradients that arise from the powdered hydride samples themselves can severely interfere with the FGSE measurements [3.27, 31]. Although revised methods to minimize the "background" effects have been described [3.36, 37], the required instrumentation is rather complex and not always effective when the gradients are large. *Cotts* [3.31] has provided a recent summary of the metal-hydrogen systems whose proton D values were obtained by the various FGSE methods.

The nuclear relaxation times are indirectly related [3.39] to diffusion properties through the motion-induced fluctuations of dipolar and quadrupolar (i.e. only for nuclei with non-vanishing quadrupole moments such as deuterons) correlation-time functions $G^{(q)}(t)$ as shown in (3.4-7). Although the theoretical foundations of this relationship are general and well established [3.39], various

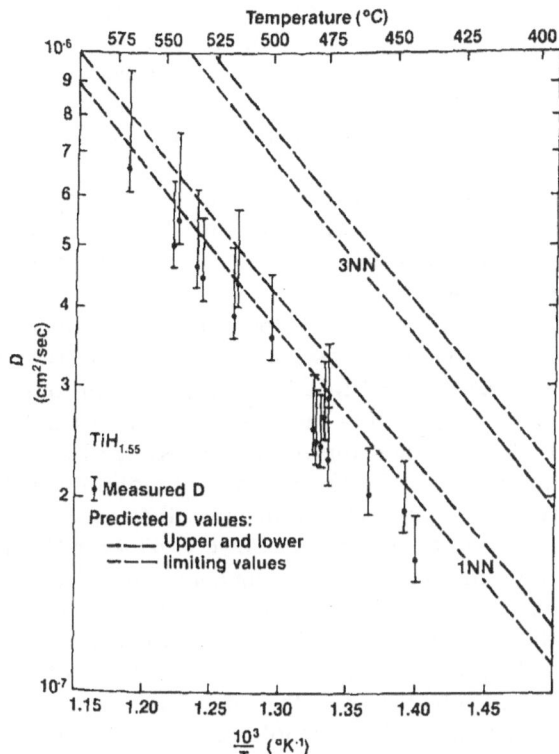

assumptions are always made for the models used to analyse the experimental relaxation-time data even after the (presumed) removal of the non-diffusion contributions. The most common and simplest $G(t)$ expression used to analyse relaxation time data in metal hydrides [3.26–31] is the stochastic exponential correlation-time model for simple hops that was originally proposed by *Bloembergen, Purcell* and *Pound* [3.40] and is widely known as the BPP model. *Cotts* [3.26, 27] provides a very good discussion of the mathematical forms of the BPP relaxation time equations as well as their predicted dependences on τ_c variations. Although the absolute τ_c values derived from BPP model analysis may be in error by as much as a factor of 2, their relative temperature-dependent behaviour is usually far more reliable, giving diffusion activation energies that agree to within about 10% with those obtained by other techniques [3.27]. More accurate $G(t)$ models that attempt to account for orientation-dependent jumps between specific lattice sites in terms of random walk theories are reviewed by *Cotts* [3.31]. Although these improved models can yield more accurate absolute τ_c values, the complexities of the required numerical computations have limited their applications to hydrides with highly symmetric simple cubic crystal structures (e.g., PdH_x, NbH_x, or TiH_x). Furthermore, these lattice-specific models have been found [3.31, 41] to exhibit nearly the same qualitative behaviour as the simpler BPP model. A serious shortcoming of all the lattice-specific models is their inability [3.41] to explain the anomalous temperature and resonance frequency dependences that are experimentally observed in an increasing number of metal-hydrogen systems [3.28, 42–48]. Essentially all of the latter hydrides have either complex crystal structures with several non-equivalent interstitial sites for hydrogen occupancy, or probable diffusion jump paths that include an unstable intermediate site [3.42]. With the exception of some preliminary analyses by *Fedders* [3.49], all the lattice-specific models have only considered the simple atomic hopping processes, which do not provide realistic representations in these situations.

Various approaches have been taken to extract the diffusion parameters when anomalous relaxation-time behaviour is observed. One straightforward method [3.42] is to generalize the BPP model to a linear combination of exponential BPP functions with different activation energies U_i that are weighted by the probability of occupied sites. Another approach is to assume that the diffusion process consists of combinations of different types of motion, which result in an average correlation time $\bar{\tau}_c$ and relaxation times that can reflect different types of motion over a common temperature range [3.43, 46–48]. An example of this process could be localized hopping among a few closely spaced sites plus a much slower long-range motion (perhaps through an intermediate site) to a separate cluster of identical or similar sites. This model can qualitatively account for the non-Arrhenius behaviour, but it is very difficult to obtain quantitative identification of the individual steps in the total diffusion process. A third method [3.44, 50] is to define a phenomenological spectral density function with variable coefficients for τ_c and the frequency dependences of the relaxation times. The experimental data are then fitted assuming an Arrhenius temperature dependence for the diffusion process. Unfortunately, this method provides minimal insight into the actual physical processes if the relaxation-time data show any significant anomalies. Recently, proton relaxation times have been successfully fitted [3.51] with a distribution function of activation energies. The theoretical justification for these methods currently remains incomplete as best. This reflects the great difficulty in developing a "simple, yet accurate" correlation function that can relate multi-step diffusion processes to the nuclear dipolar interactions. Finally, the extraneous contributions of paramagnetic relaxation processes to the observed proton relaxation times are basically still unresolved in most systems.

3.2.4 Quasielastic and Inelastic Neutron Scattering

A simultaneous study of the space and time development of dynamic processes at a microscopic level is possible only by scattering methods. Among these methods neutron scattering plays a particularly important role, since the wavelength of thermal neutrons is comparable to interatomic distances, and at the same time their energy is of the order of typical solid state excitations. Furthermore, the neutron–nucleus interaction is weak. As a consequence neutrons penetrate deeply into matter and are sensitive to bulk properties; multiple scattering processes which, for example, dominate electron scattering because of the strong Coulomb interaction, are only second-order contributions. This facilitates a direct approach to the dynamic response of the system in question. With respect to H in metals, the dynamic range of neutron scattering (10^{-8} eV $\leq \hbar\omega \leq 1$ eV, where $\hbar\omega$ is the energy transfer between neutron and matter) allows an investigation of three different classes of dynamic processes: (i) in the quasielastic regime at low energy transfers, diffusive processes are accessible, (ii) at thermal energies in the regime of the host lattice phonons, the so-called band motions of H occur, where the H atoms move according to the displacements of the host atoms during their vibrations; (iii) at even higher energies the H atoms vibrate against their metal neighbours and we are in the regime of local or optic H vibrations.

With respect to neutron scattering by the hydrogen isotopes, the proton is a special case, since its scattering length depends strongly on the relative orientation of neutron and proton spin. Because at all temperatures of interest the proton spins are oriented randomly, neutrons scattered from different protons have random phase relations and do not interfere: they are scattered incoherently. Thus, neutron scattering by protons in a metal reveals information on the motional behaviour of single protons, while for deuterium and tritium the collective behaviour is observed. Since in the following our main interest is focussed on single particle behaviour, we restrict ourselves to the incoherent scattering by protons. Following the concept of *van Hove* [3.52] the double differential neutron cross-section $\partial^2\sigma/\partial\omega\partial\Omega$ is proportional to the Fourier transform of the proton self-correlation function, $G_s(r, t)$:

$$\frac{\partial^2\sigma}{\partial\omega\partial\Omega} \sim \frac{\sigma_{\text{inc}}^H}{4\pi} \frac{k_f}{k_i} \frac{1}{2\pi} \int\limits_{-\infty}^{\infty} dt \int\limits_{-\infty}^{\infty} d^3r\, G_s(r, t)\, e^{i(qr - \omega t)}$$

$$= \frac{\sigma_{\text{inc}}^H}{4\pi} \frac{k_f}{k_i} S_{\text{inc}}(q, \omega), \tag{3.10}$$

where $\hbar q = \hbar(k_f - k_i)$ is the momentum transfer, which for elastic scattering is related to the scattering angle 2θ by $q = 4\pi/\lambda \sin\theta$ (λ is the neutron wavelength); $S_{\text{inc}}(q, \omega)$ is the incoherent scattering law; the incoherent cross section σ_{inc}^H exceeds the cross-section of typical metals by at least one order of magnitude. Therefore scattering experiments on H in metals are feasible in favourable cases at an H concentration as low as 1 at. ‰ [3.53]. For experimental aspects of neutron scattering we refer the reader to the literature [3.54].

a) Quasielastic Neutron Scattering – H Diffusion

In a quasielastic neutron scattering (QNS) experiment, the neutron spectrum centered around energy transfer zero is analysed as a function of energy transfer $\hbar\omega$ and scattering angle 2θ. Its broadening in energy contains information on the H jump rate and diffusion coefficient. The dependence on the momentum transfer hq and the quasielastic lineshape reveal information on the geometrical details of the diffusive path.

In the following, we treat the self-correlation function for diffusive motion in the classical limit, which is justified as long as phase relations between H wavefunctions at different sites do not exist. Then $G_s(r, t)$ can be interpreted as the conditional probability of finding a proton at time t on a site r if it was at $r = 0$ at time $t = 0$. For the interpretation of QNS results from H in metals, the following further assumptions are commonly made: (i) The diffusion occurs between well-defined sites which are connected by a set of jump vectors S_K. The mean residence time at a certain site is long compared to the jump time, $\tau_i = S/(kT/2m_H)^{1/2} \cong 10^{-13}$ s, between two sites. (ii) Vibrational motion and diffusion are uncorrelated. (iii) The H concentration is small enough that mutual blocking and H–H interaction effects are not important (Chudley–Elliott (CE) model [3.55]). While we maintain the first two conditions throughout this paper, we will briefly discuss the influence of finite H concentration.

Under the assumption of the CE-model, $G_s(r, t)$ can be inferred from a set of rate equations, which for Bravais lattices assume a particularly simple form. In this case, the probability for site occupancy at a site r is given by:

$$\frac{\partial}{\partial t} P(r, t) = \sum_{K=1}^{z} \Gamma^K [P(r + S_K, t) - P(r, t)], \tag{3.11}$$

where z is the number of accessible sites and Γ^K are the corresponding jump rates. The self-correlation function is then $G_s(r, t) = P(r, t)$ with the initial condition $P(r, 0) = \delta(r)$. (Note that we have neglected the vibrational part.) $S(q, \omega)$ is obtained by Fourier transformation and yields a single Lorentzian

$$S(q, \omega) = \frac{1}{\pi} \frac{\Gamma(q)}{\Gamma(q)^2 + \omega^2}, \tag{3.12}$$

with the linewidth

$$\Gamma(q) = \sum_{K=1}^{z} \Gamma^K [1 - \exp(iqS_K)]. \tag{3.13}$$

Thus, a q-dependent measurement of the quasielastic linewidth on single crystal samples provides complete information about the microscopic jump processes including (via the jump vectors) the site occupation. For an example we refer to measurements on α-PdH$_x$ [3.56], where such an experiment revealed nearest neighbour jumps between octahedral sites. At small q, (3.13) can be expanded and we arrive at

$$\Gamma(q) = \sum_{ij} q_i D_{ij} q_j, \tag{3.14}$$

where $D_{ij} = \sum \Gamma^\kappa S_i^\kappa S_j^\kappa$ is the diffusion tensor. Equation (3.14) is also valid for non-Bravais lattices. Thus, at small momentum transfers, QNS yields the macroscopic diffusion coefficient measured over microscopic distances. Since QNS is a true bulk method, effects of inner or outer surfaces do not play any role in such measurements.

Intermetallic compounds do not crystallize in Bravais lattices but, in general, exhibit rather complicated structures with many sites per unit cell. Furthermore, as a consequence of the varying H affinities of the constituent host atoms, chemically different H sites exist which exhibit different binding energies. For example, in $Ti_{1.2}Mn_{1.8}$ tetrahedral sites are surrounded by 4 Mn atoms, by 3 Mn and 1 Ti, or by 2 Mn and 2 Ti, the last combination being the most attractive H site due to the large H affinity of Ti. Thus, compared with the "classical" monatomic metal-hydrogen systems, interstitial diffusion of H in intermetallics is expected to be much more complicated. In the following we briefly outline a procedure leading to the rate equations and finally to the incoherent scattering law for H diffusion in such systems.

As a first step we construct interstitial H sublattices such that for each unit cell there is one H site per sublattice. There may be $v = 1,\ldots,m$ sublattices. Secondly we note all possible jump vectors $S_{v\mu k}$ between sublattices v and μ with a cell distance k. To each jump vector corresponds a jump rate $\Gamma^{v\mu k}$. The condition of detailed balance requires

$$e^{\beta E_v}\Gamma^{v\mu k} = \Gamma^{\mu v - k}e^{\beta E_\mu}, \tag{3.15}$$

where E_v and E_μ are the site energies of the sublattices v and μ, and $\beta = 1/kT$. Defining a probability $P_v(r,t)$ of finding a proton at r and t on sublattice v, the rate Eq. (3.11) becomes

$$\frac{\partial}{\partial t}P_v(r,t) = \sum_{\mu,k}[\Gamma^{\mu v k}P_\mu(r + S_{\mu v k},t) - \Gamma^{v\mu k}P_v(r,t)]. \tag{3.16}$$

Fourier transformation yields

$$\frac{\partial}{\partial t}P_v(q,t) = \sum_\mu P_\mu(q,t)\sum_k \Gamma^{\mu v k}e^{iqS_{\mu v k}} - \sum_{\mu,k}\Gamma^{v\mu k}P_v(q,t), \tag{3.17}$$

or in matrix notation

$$\frac{\partial}{\partial t}P = \underline{\Lambda}P. \tag{3.18}$$

The elements of the jump matrix $\underline{\Lambda}$ are given by

$$\Lambda_{\mu v} = -\delta_{v\mu}\sum_{k,\mu}\Gamma^{v\mu k} + \sum_k \Gamma^{v\mu k}e^{iqS_{v\mu k}}. \tag{3.19}$$

Since $\Gamma^{\mu\nu} \neq \Gamma^{\nu\mu}$ (3.15), $\underline{A}$ is non-hermitian. In order to obtain a hermitian form, we perform a similarity transformation with the thermal occupation numbers [3.57].

$$\underline{\rho} = (\rho_\nu \delta_{\nu\mu}) \quad \text{with} \quad \rho_\nu = \frac{e^{\beta E_\nu}}{\sum\limits_\mu e^{\beta E_\mu}}, \tag{3.20}$$

$$\underline{\tilde{A}} = \underline{\rho}^{-1/2} \underline{A} \underline{\rho}^{-1/2}, \tag{3.21}$$

so that $\underline{\tilde{A}}$ is now hermitian, as can easily be seen from (3.15) and (3.19). The transformation (3.21) does not change the eigenvalues of A. Diagonalization of $\tilde{A}$ leads to the eigenvalues Γ_δ $(q, T, \Gamma^{\nu\mu k})$, and the associated eigenvectors $\tilde{P}^\delta$ $(\delta = 1, \ldots, m)$. We look for solutions G_ν^γ fulfilling the initial conditions

$$G_\nu^\gamma(0) = \rho_\gamma \delta_{\gamma\nu}. \tag{3.22}$$

They are given by

$$G_\nu^\gamma(t) = \sum_\delta \rho_\gamma^{1/2} \tilde{P}_\gamma^{*\delta} \tilde{P}_\nu^\delta \rho_\nu^{1/2} e^{-\Gamma_\delta t}. \tag{3.23}$$

Finally, the self-correlation function is obtained from an average over the initial and a sum over the final states

$$G(q, t) = \sum_{\delta\gamma\nu} \rho_\gamma^{1/2} \tilde{P}_\gamma^{*\delta} \tilde{P}_\nu^\delta \rho_\nu^{1/2} e^{-\Gamma_\delta t}. \tag{3.24}$$

After Fourier transformation with respect to time, the incoherent scattering law becomes

$$S_{\text{inc}}(q, \omega) = \frac{1}{\pi} \sum_{\delta=1}^m w_\delta \frac{\Gamma_\delta}{\Gamma_\delta^2 + \omega^2}. \tag{3.25}$$

Γ_δ are the eigenvalues of the jump matrix, where the weights w_δ are given by

$$w_\delta = \sum_{\gamma\nu}^m \rho_\gamma^{1/2} \tilde{P}_\gamma^{*\delta} \tilde{P}_\nu^\delta \rho_\nu^{1/2} = \left| \sum_\gamma \rho_\gamma^{1/2} \tilde{P}_\gamma^\delta \right|^2. \tag{3.26}$$

Equation (3.25) again contains complete information about all the microscopic details of the proton jump processes in question. However, unlike (3.13), (3.25) is a sum over m Lorentzians exhibiting different widths and weight factors in the spectrum. Only at small q does it reduce to one Lorentzian whose width is solely determined by the diffusion coefficient. At larger q's the interpretation of these spectra requires accurate line-shape analysis which is a difficult task experimentally, and single crystals, which allow an effective use of symmetry relations, are necessary. Unfortunately, in general, intermetallic compounds decompose into powders upon hydrogenation and single crystals are therefore not available. On powder samples a quantitative analysis according to (3.26) is impossible, and simplified semi-phenomenological versions have so far been used in order to interpret scattering data. These will be discussed in Sects. 3.3.3 and 3.3.4. For the dilute phases, which often exhibit the same structure as the concentrated phases, there is some hope for QNS experiments on single crystals. These are expected to establish a detailed microscopic account of proton jump processes in an intermetallic.

So far we have derived the scattering laws for the limit of infinite dilution. For finite but not too high H concentrations, a mean-field approach is often used. It describes the influence of concentration by a blocking factor $(1 - c)$, by which the jump rate and thus the self-diffusion coefficient D_s is reduced compared to the empty lattice: $\Gamma \rightarrow (1 - c)\Gamma$, $D_s \rightarrow (1 - c)D$. In this case the shape of the incoherent spectrum of (3.12) is not changed; $\Gamma(q)$ is merely reduced to $(1 - c)\Gamma(q)$. Even this simple approach gets much more complicated for intermetallics where preferential occupation and blocking of energetically deep sites has to be taken into account. A phenomenological approach to this problem in terms of trapping will be discussed in Sect. 3.3.3.

With increasing concentration, correlation effects between successive jumps gain importance. They are already obvious in the simple blocking picture: viz. if a proton has jumped from a site j to an adjacent site $j + 1$, the probability for a jump back to its starting point is higher than the mean jump probability, since immediately after a jump the original site will certainly be empty. With respect to the self-diffusion coefficient, this backward correlation leads to the correlation factor f, which is well known from self-diffusion in metals. Up to now, microscopic studies of such dynamic correlation effects have been performed mainly theoretically [3.58] by Monte Carlo simulation, and we refer to recent reviews on this subject [3.59, 60]. The first experimental information about this problem was obtained recently [3.61], using the novel muon spin resonance technique. In this, the positive muon is employed as a tracer to monitor dynamic correlations on an atomic level. With respect to quasielastic scattering, correlation effects such as backjumps which are much faster than the average jump process, are expected to lead to broad wings in the QNS spectra.

For a comparison with macroscopic experiments, it is often not the self-diffusion coefficient D_s but the chemical or collective diffusion coefficient, D_c that is of importance. Therefore, we review briefly the relation between these two quantities. For a particle current j, we have

$$j = -cB\frac{\partial \mu}{\partial c}\nabla n = -D_c\nabla n, \tag{3.27}$$

where B is the mobility, μ the chemical potential, and $n(r, t)$ the particle density. For the dilute case $\partial \mu / \partial_c = kT/c$ and $D_c = BkT = D$. If we consider a lattice gas without interaction, the free energy is $F = kTc \ln c + kT(1 - c)\ln(1 - c)$, yielding $\partial\mu/\partial c = kT/c(1 - c)$ and $D_c = D_s/(1 - c)$. Thus for an interaction-free lattice gas, D_c does not depend on concentration. In general the ratio of D_c and D is given by

$$D_c/D = F_{ch}/f,\tag{3.28}$$

where F_{ch} is the normalized chemical factor

$$F_{ch} = \frac{\partial\mu/\partial c}{(\partial\mu/\partial c)_{c=0}}\tag{3.29}$$

and f the correlation factor mentioned above. A recent experiment on TaH_x which compared D_c measured by the Gorsky effect with D observed by QNS demonstrated considerable deviations from the non-interacting model [3.62].

b) Neutron Vibrational Spectroscopy

In addition to their diffusive jumps between different interstitial sites, hydrogen atoms exhibit two other types of motion during their rest time at particular sites, viz. acoustic vibrations (band modes) and optical vibrations (localized modes).

The influence of hydrogen loading on the acoustic host vibrations has been studied widely for binary metal–hydrogen system [3.63, 64]. For these investigations single crystal samples are usually necessary, but these are not available for intermetallic hydrides. This prevents the study of hydrogen band modes in intermetallic hydrides. One exception is the observation of low energy excitations of hydrogen and deuterium in $LaNi_5H(D)_6$ [3.65], which appear to be quasi-optical modes within the acoustic band. Essentially the same features have been observed in NbH [3.66] and in TaH; a discussion of these effects can be found in [3.63] and will not be repeated here.

Instead we pay more attention to the optical or at low H concentration, localized, hydrogen vibrations. In order to keep the representation of the physical phenomena associated with these modes (Sect. 3.4) free from technical details, we now present the scattering function and discuss various approximations and evaluation schemes.

If we disregard hydrogen–hydrogen interactions, which is quite justified for hydrogen in dilute α-phases, and a reasonable approximation for many metal hydrides, then the interstitially dissolved hydrogen atoms can be considered as single, independent, three-dimensional Einstein oscillators, each with three vibrational degrees of freedom. In intermetallic compounds generally, the hydrogen atoms are distributed over several crystallographically nonequivalent sites, and independently contribute to the neutron scattering intensity. This intensity is proportional to the double differential cross-section

$$\frac{\partial^2\sigma}{\partial\omega\partial\Omega} = \frac{\sigma^{tot}}{4\pi}\frac{k_f}{k_i}N_H\sum_{j=1}^{K}z_jf_jS_j(\boldsymbol{q},\omega),\tag{3.30}$$

where σ^{tot} is the total neutron scattering cross-section of hydrogen, k_i and k_f the initial and final neutron wave number, N_H the number of H atoms in the neutron beam, K the number of crystallographically different types of H sites in the unit cell, z_j the number of H sites of type j, and f_j the fractional occupancy of the H sites of type j.

We assume a harmonic form of the hydrogen potential at all sites j and consider first single crystal samples (the polycrystalline average will be performed later on). The scattering function is then given by [3.67]

$$S_j(q,\omega) = \exp[-2W_j(q)]\exp(\hbar\omega\beta/2)\sum_{n,m,l=-\infty}^{\infty} I_n(y_{j1})I_m(y_{j2})I_e(y_{j3})$$

$$\cdot\delta(\hbar\omega - n\hbar\omega_{j1} - m\hbar\omega_{j2} - l\hbar\omega_{j3}), \qquad (3.31)$$

with

$$y_{ji} = \frac{\hbar q_i^2}{2m_H\omega_{ji}}\,\mathrm{csch}(\hbar\omega_{ji}\beta/2).$$

In this equation the first term is the Debye–Waller factor, which will be discussed later, and the second one is the detailed balance term which relates the intensity of energy gain and the energy loss processes. The index i refers to the three normal modes (vibrational directions) of each hydrogen atom.

For the vibrational spectrum we therefore expect three peaks for each type of H sites; depending on the point symmetry of the respective interstitial site, two or even all three peaks can be degenerate. In α-phases the line shape is usually Gaussian with a certain linewidth, which is mainly due to the local distortions around the hydrogen atoms; a Gaussian line shape also appears appropriate for intermetallic hydrides. If, however, the line shape is determined by lifetime effects, it is Lorentzian-like. For an evaluation of the scattering intensities of the different vibrational peaks, mostly simplifications of (3.31) are used. For small arguments y_{ji}, i.e. for $kT \ll \hbar\omega$, the modified Bessel function can be expanded as

$$I_n(y_{ji}) = (1/2y_{ji})^n/n!, \qquad (3.32)$$

which—when we restrict ourselves to fundamental vibrations—results in

$$S_j(q,\omega) = \exp[-2W_j(q)]\sum_{i=1}^{3}\frac{\hbar}{2m_H\omega_{ji}}q_i^2[n_B(\omega) + \tfrac{1}{2}(1 \pm 1)]\delta(\hbar\omega \mp \hbar\omega_{ji}), \qquad (3.33)$$

where the upper (lower) sign refers to neutron energy loss (gain) processes; $n_B(\omega) = 1/[\exp(\hbar\omega\beta) - 1]$ is the Bose occupation number. Note that $n_B(\omega) + 1$ is nothing other than the partition function of a harmonic oscillator, while $n_B(\omega)$ alone represents a reduced partition function (without the occupation probability of the ground state in the sum. It is important to notice that for $k_B T \ll \hbar\omega$, the first factor in the sum of (3.33) is the mean square amplitude of the oscillator (3.35), which is very large for the light hydrogen atom and, together with the total cross-section σ_H^{tot} in (3.30), results in the huge scattering intensity of the localized hydrogen modes. In comparison with this intensity, the scattering from the host lattice is negligible.

In many of the experimental set-ups for localized mode measurements (see [3.54]), a certain energy transfer unequivocally implies a certain q value. This allows a determination of the range of validity of (3.33). For the beryllium filter spectrometer, for instance, this equation holds at 80 K for the whole energy transfer range in question, whereas at 300 K, for example, and at energy transfers below 100 meV, remarkable discrepancies occur. In the latter case one either has to use the modified Bessel functions themselves (3.31), or to include higher order terms in the expansion of (3.32).

The Debye–Waller factor in (3.31) and (3.33) is of course different for each hydrogen site. Because of the comparatively large linewidths of the localized H modes in all intermetallic hydrides investigated so far, possible acoustic side bands are included in the peak area, and therefore the Debye–Waller factor contains only contributions from the optical modes [3.68]. For a single crystal sample we obtain [3.67]

$$2W_j(q) = \langle (qu_j)^2 \rangle = \sum_{i=1}^{3} q_i^2 \langle u_{ji}^2 \rangle, \tag{3.34}$$

where the index i again refers to the normal mode directions and

$$\langle u_{ji}^2 \rangle = \frac{\hbar}{2m_H \omega_{ji}} \coth(\hbar\omega_{ji}\beta/2) \tag{3.35}$$

is the mean square amplitude for the localized vibration in the i direction of a hydrogen atom at site j.

In order to evaluate localized mode measurements of polycrystalline samples, a spatial averaging of the complete scattering function has to be performed, which is difficult to do analytically. Basically two approximations have been suggested in the literature:

i) A separate averaging of the Debye–Waller factor and the phonon form factor in the scattering function [3.69, 70], which transforms (3.33), for example, into:

$$S_j(q, \omega) = \exp[-2W_j(q)] \sum_{i=1}^{3} \frac{\hbar}{6m_H \omega_{ji}} q_i^2 [n_B(\omega) + \tfrac{1}{2}(1 \pm 1)] \delta(\hbar\omega \mp \hbar\omega_{ji}), \tag{3.36}$$

with

$$2W_j(q) = \frac{\hbar^2 q^2}{6m_H} \sum_{i=1}^{3} \frac{\coth(\hbar\omega_{ji}\beta/2)}{\hbar\omega_{ji}}.$$

ii) The use of individual Debye–Waller-factor for each direction [3.71] which leads to

$$S_j(q, \omega) = \sum_{i=1}^{3} \exp[-2W_{ji}(q)] \frac{\hbar}{6m_H\omega_{ji}} q_i^2 [n_B(\omega) + \tfrac{1}{2}(1 \pm 1)]\delta(\hbar\omega \mp \hbar\omega_{ji}).$$

(3.37)

with

$$2W_{ji}(q) = \frac{\hbar^2 q_i^2}{2m_H\hbar\omega_{ji}} \coth(\hbar\omega_{ji}\beta/2).$$

For the Beryllium-filter method (for neutron energy loss) with energy transfers above 100 meV, both the expression in square brackets and the coth term equal unity, and

$$\frac{\hbar^2 q^2}{6m_H\hbar\omega_{ji}} \approx \frac{\hbar^2 k_i^2}{6m_N\hbar\omega_{ji}} = \frac{1}{3}$$

(3.38)

because the proton and neutron mass are equal, and q is mainly determined by k_i. Additionally, one should keep in mind that in this technique the spectra are usually recorded at a constant rate of a monitor counter, which is placed between the monochromator and sample, and which has a counting characteristic proportional to $1/v_N$ and thus to $1/k_i$. Therefore the k_f/k_i correction in (3.30) is automatically done during data-taking. Hence in this approximation the scattering intensity is proportional to the scattering function and is the sum of three peaks of equal intensity.

However, both approximations can only be used if the anisotropies of the hydrogen sites are fairly small. In the case of strongly anisotropic H sites, as e.g. in LaNi$_5$ [3.71], both methods lead to considerable errors in the intensities, as a comparison with a numerically-performed polycrystalline average showed [3.72].

Recently *Tomkinson* et al. [3.73] reported the analytical powder averages for neutron vibrational spectroscopy of anisotropic molecular oscillators. Their results, however, are not valid for general point symmetries and are—compared with a straightforward numerical averaging—of an unwieldy complexity. In addition, a comparison with experimental intensities [3.74] revealed an agreement which was no better than that obtained with the approximations mentioned above.

In summary we conclude that the evaluation of scattering intensities is readily possible if one compares vibrational modes of similar frequencies; if, however, the modes have very different frequencies, then firstly the validity of (3.32) is questionable, and secondly a numerical, spatial averaging procedure has to be inserted into the fitting routine. For the case of strongly anharmonic vibrations where the anharmonicity can no longer be considered simply as a distortion of the harmonic case, the authors know of no appropriate inelastic neutron scattering function. As a result, it has not been possible, to date, to estimate intensity effects due to anharmonicity.

3.3 Experimental Results on Hydrogen Diffusion

3.3.1 General Overview — Long-Range Diffusion

Since the reviews of *Völkl* and *Alefeld* [3.1, 2] were published, studies of the diffusion properties of hydrogen isotopes in metallic systems have remained quite active. Although much theoretical and experimental attention [3.75, 76] has been recently focused upon both the elementary diffusion step for isolated hydrogen isotopes in pure metals and localized motion in the vicinity of various trapping sites such as substitutional or interstitial (e.g. O or N) impurities, hydrogen diffusion properties have also been studied in many ternary or more complex hydrides. These latter efforts have been greatly stimulated by the desire to develop and characterize candidates for various technological applications. Although the diffusion properties of the binary hydride phases are of great fundamental interest, the present discussion will be limited to hydrogen diffusion behaviour in several representative ternary intermetallics and a few of their related alloys. For results on hydrogen diffusion in elemental metals and binary hydrides the reader is referred to [3.63, 75, 76] and Chaps. 1 and 6.

Tables 3.2–7 present summaries of reported hydrogen diffusion parameters for $TiFeH_x$ [3.8, 12, 77–81], Mg_2NiH_x [3.82–90], $LaNi_5H_x$ [3.91–98], $LaNi_{5-x}M_xH_z$ [3.102–105], AB_2H_x hydrides [3.47, 106–110] of the Laves alloys with either the cubic (C15) or hexagonal (C14) structure types, and several A_2BH_x systems [3.44–46, 112–114] with either the cubic E9$_3$ or tetragonal C11b crystal structures. Although the various NMR techniques are by far the most heavily represented in Tables 3.2–7, diffusion parameters obtained with other methods (i.e. QNS, gas absorption/desorption kinetics, and Mössbauer) are also given. A cursory examination of these tables immediately indicates a great diversity and often rather large inconsistencies among the observed diffusion parameters for hydrides with nominally identical compositions. This behaviour contrasts

Table 3.2. Summary of hydrogen diffusion parameters for $TiFeH_x$ and related systems

Sample	Method	Temperature range [K]	U [meV]	D_0 [cm²/s]	Reference
β-TiFeH$_x$	NMR(T_2)	340–420	260(20)	–	[3.77]
β-TiFe$_{0.79}$Mn$_{0.15}$H$_x$	NMR(T_2)	340–410	320(40)	–	[3.77]
γ-TiFe$_{0.79}$Mn$_{0.15}$H$_x$	NMR($T_{1\rho}$)	350–385	800(50)	–	[3.77]
β-TiFeH$_x$	desorption kinetics	273–302	340	1.1×10^{-4}	[3.78]
β-TiFeH$_{1.03}$	QNS	633–759	500(50)	7×10^{-4}	[3.79]
β-TiFeH$_{1.03}$	NMR(T_2)	350–413	330(60)	4.2×10^{-7}	[3.80]
β-TiFeH$_x$	Mössbauer	400–486	540	–	[3.12]
α-TiFeH$_{x(\sim 0)}$	absorption kinetics	380–680	496	10.1×10^{-4}	[3.8]
α-TiFeD$_{x(\sim 0)}$	absorption kinetics	400–670	512	8.2×10^{-4}	[3.8]
TiFeH$_x$	kinetics	293–473	120	–	[3.81]

Table 3.3. Summary of hydrogen diffusion parameters for Mg_2NiH_x

Sample	Method	Temperature range [K]	U [meV]	D_0 [cm^2/s]	Reference
α-MgNiH$_{0.3}$	QNS	423–498	280(40)	$6.7(3) \times 10^{-5}$	[3.82]
α-Mg$_2$NiH$_{0.22}$	NMR($T_{1\rho}$)	300–400	275	4×10^{-7}	[3.83]
α-Mg$_2$NiH$_{0.27}$	NMR(T_2)	290–360	370(30)	–	[3.84]
β-Mg$_2$NiH$_{3.72}$	NMR(ΔH)	210–480	250(50)	–	[3.85]
β-Mg$_2$NiH$_4$	NMR(T_1)	315–465	454(20)	–	[3.84]
β-Mg$_2$NiH$_4$	NMR(T_1)	300–430	350	–	[3.86]
β-Mg$_2$NiH$_{3.85}$	NMR(T_1)	210–460	534(35)	–	[3.87]
β-Mg$_2$NiH$_4$	NMR(T_1)	300–430	480	4.2×10^{-4}	[3.88]
β'-Mg$_2$NiH$_4$	QNS	530–688	158	–	[3.89]
β'-Mg$_2$NiH$_4$	NMR(ΔH)	518–535	860	–	[3.90]

sharply with the usually quite reproducible diffusion parameters obtained from most binary metal–hydrogen systems (providing that the samples are well characterized and experimental artifacts are properly identified and removed) with generally independent methods [3.1, 2, 27].

Numerous factors can potentially cause the often extensive variability in diffusion properties of intermetallic-hydrogen systems. Furthermore, relatively few of these factors are under the direct control of the experimentalist—some may not even be recognized as a possible source of problems. Many of the intermetallics and alloys are difficult to prepare as homogeneous single phases; activation and hydriding reactions may provoke either disproportionations or microscopic phase separations that can be difficult to detect by conventional methods like powder X-ray diffraction. Additional complications arise if there is extensive metal atom disorder or non-stoichiometry thus creating a greater variety of interstitial sites for hydrogen occupancy. Since many ternary hydrides can spontaneously evolve hydrogen, treatments that form more or less hydrogen impermeable surface layers are often used to prepare samples that are deemed suitable for diffusion experiments (for some examples see [3.47, 77, 79, 110]). Nevertheless, temperature-dependent changes in hydrogen content can still occur and may have influenced the apparent diffusivities to give erroneous activation energies in some cases. As mentioned earlier, diffusion parameters that have been obtained with techniques requiring hydrogen transport through particle surfaces (see for example, [3.3–8, 81, 93, 107]) may not correspond to the bulk diffusion mechanisms but instead reflect mainly motion near or through the surface regions. Finally, the generally more complex crystal structures of many intermetallic hydrides (i.e. when several different types of interstitial sites are simultaneously occupied, with substantial differences in occupation factors being rather common) compared with prototype binary hydrides lead to far more complicated diffusion behaviour that can consist of convolutions of several more or less independent processes.

On a microscopic level only FGSE-NMR techniques and QNS at small momentum transfers offer direct access to the H diffusion coefficient. Conven-

Table 3.4. Summary of $LaNi_5H_x$ diffusion parameters

Sample	La purity [%]	Method	Temperature range [K]	U [meV]	D_0 [cm²/s]	$D(300\,K)$ [10^{-8} cm²/s]	Reference
β-LaNi$_{5.3}$H$_6$	99.9	NMR(ΔH)	150–220	220(10)	1.5×10^{-4}	2.1	[3.91]
β-LaNi$_{4.8}$H$_6$	99.9	NMR(T_2, $T_{1\rho}$)	210–300	244(8)	–	–	[3.92]
β-LaNi$_{5.0}$H$_6$	99.9	NMR(T_2, $T_{1\rho}$)	230–320	248(4)	–	1.7	[3.92]
β-LaNi$_{5.2}$H$_6$	99.9	NMR(T_2, $T_{1\rho}$)	220–310	265(18)	–	–	[3.92]
α-LaNi$_5$H$_x$	?	absorption	353–463	410(41)	3×10^{-2}	0.4	[3.93]
β-LaNi$_5$H$_{6.5}$	99.99	NMR(FGSE)	331–375	410(41)	0.14(3)	1.4	[3.43, 94]
β-LaNi$_5$H$_{6.5}$	99.99	NMR(T_1)	250–320	210(21)	–	–	[3.43, 94]
β-LaNi$_5$H$_{6.5}$	99.99	NMR($T_{1\rho}$)	228–350	400	–	–	[3.43, 94]
β-LaNi$_{5.00}$H$_{6.2}$	99.99	NMR(T_2)	230–311	300(10)	1.6×10^{-3}	1.5	[3.95]
β-LaNi$_{5.03}$H$_{4.95}$	99.5	NMR(T_2)	215–301	170(20)	5.0×10^{-6}	0.7	[3.95]
β-LaNi$_5$H$_{6.6}$	?	NMR(T_1)	230–280	230(20)	8×10^{-4}	11	[3.96]
β-LaNi$_5$H$_{6.1}$	99.9	NMR(ΔH)	?	200(30)	–	–	[3.97]
β-LaNi$_5$H$_6$	99.9	QNS	280–400	275(15)	$2.1(7) \times 10^{-4}$	5.0	[3.98]
β-LaNi$_5$H$_6$	99.9	QNS	248–318	242	8.8×10^{-5}	0.8	[3.99]
β-LaNi$_5$H$_6$	?	QNS		–	–	120	[3.100]
β-LaNi$_5$H$_6$	?	QNS	–	110	$0.6 \cdot 10^{-3}$	85	[3.101]

Table 3.5. Effects of alloy substitution on diffusion behaviour in $LaNi_{5-x}M_xH_z$ systems

Sample	Method	U [meV]	Reference
$LaNi_{5.00}H_{6.2}$	$NMR(T_2)$	300(10)	[3.95]
$LaNi_{4.6}Al_{0.4}H_{5.4}$	$NMR(T_2)$	300(10)	[3.95]
$LaNi_{4.5}Al_{0.5}H_{5.24}$	$NMR(T_2)$	280(20)	[3.95]
$LaNi_{4.3}Al_{0.7}H_{4.64}$	$NMR(T_2)$	340(20)	[3.95]
$LaNi_{4.0}Al_{1.0}H_{4.33}$	$NMR(T_2)$	420(10)	[3.95]
$LaNi_{3.8}Al_{1.2}H_{4.04}$	$NMR(T_2)$	450(10)	[3.102]
$LaNi_{3.5}Al_{1.5}H_{3.77}$	$NMR(T_2)$	510	[3.102]
$LaNi_4CuH_{4.7}$	$NMR(T_2)$	220	[3.103]
$LaNi_3Cu_2H_{4.2}$	$NMR(T_2)$	240	[3.103]
$LaNi_2CuH_{3.25}$	$NMR(T_2)$	342	[3.103]
$LaNiCu_4H_3$	$NMR(T_2)$	381	[3.103]
$LaCu_5H_{1.5}$	$NMR(T_2)$	436	[3.103]
$LaNi_{5.0}H_{6.0}$	$NMR(T_{1\rho})$	370(10)	[3.104]
$LaNi_{4.0}Al_{1.0}H_{4.3}$	$NMR(T_{1\rho})$	440(10)	[3.104]
$LaNi_{4.0}B_{1.0}H_{1.4}$	$NMR(T_{1\rho})$	400(10)	[3.104]
$LaNi_{4.5}Al_{0.5}H_{4.8}$	QNS	170(5)	[3.105]

Table 3.6. Diffusion parameters for hydrides of Laves alloys

Sample	Structure type	Method	Temperature range [K]	U [meV]	References
$ZrMn_2H_{2.0}$	C14	$NMR(T_2)$	160–400	185(17)	[3.106]
$ZrMn_2H_{2.45}$	C14	$NMR(T_2)$	160–400	150(17)	[3.106]
$ZrMn_2H_{3.0}$	C14	$NMR(T_2)$	160–400	150(9)	[3.106]
$ZrMn_2H_{3.2}$	C14	$NMR(T_2)$	160–400	165(13)	[3.106]
$ZrMn_2H_{3.4}$	C14	$NMR(T_2)$	160–400	160(20)	[3.106]
$HfV_2H_{2.1}$	C15	$NMR(T_2)$	160–400	190(10)	[3.106]
$TiMn_{1.25}H_x$	C14	absorption kinetics	628–831	442	[3.107]
$TiMn_{1.56}H_x$	C14	absorption kinetics	334–746	188	[3.107]
$TiCr_{1.85}H_{0.8}$	C15	$NMR(\Delta H)$	–	35	[3.108]
$TiCr_{1.85}H_{0.4}$	C15	$NMR(\Delta H)$	–	50	[3.108]
$TiCr_{1.8}H_{0.55}$	C15	$NMR(T_{1\rho})$	190–280	190(20)	[3.47]
$TiCr_{1.8}H_{2.58}$	C15	$NMR(T_{1\rho})$	180–280	250(20)	[3.47]
$TiCr_{1.9}H_{0.63}$	C14	$NMR(T_{1\rho})$	180–290	218(10)	[3.47]
$TiCr_{1.9}H_{2.85}$	C14	$NMR(T_{1\rho})$	200–270	400(20)	[3.47]
$Ti_{0.8}Zr_{0.2}CrMnH_3$	C14	QNS	230–360	220(20)	[3.109]
$Ti_{1.2}Mn_{1.8}H_{2.9}$	C14	QNS	257–376	225	[3.110]

tional NMR relaxation techniques, Mössbauer effect and QNS at large momentum transfers measure individual jump rates which, in the case of multistep mechanisms, do not allow one to draw conclusions on the diffusion coefficient. Given these difficulties, it is not surprising that even on the basis of methods sensitive only to single jumps, contradictory results for diffusion properties have been reported (Table 3.4). In the case of $LaNi_5H_6$ the different relaxation times

Table 3.7. Summary of hydrogen diffusion parameters for A_2BH_x

Sample	Structure type	Method	Temperature range [K]	U [meV]	Reference
$Ti_2NiH_{2.09}$	$E9^3$	QNS	420–523	345(45)	[3.112]
$Ti_2NiH_{3.1}$	$E9^3$	Permeation	–	380	
$Ti_2FeO_{0.23}H_{2.84}$	$E9^3$	NMR(ΔH)	200–290	185	[3.113]
$Ti_2FeO_{0.3}H_{2.44}$	$E9^3$	NMR(ΔH)	200–290	161	[3.113]
$Ti_2FeO_{0.37}H_{1.85}$	$E9^3$	NMR(ΔH)	200–290	122	[3.113]
$Ti_2FeO_{0.45}H_{1.44}$	$E9^3$	NMR(ΔH)	200–290	112	[3.113]
$Ti_2FeO_{0.5}H_{0.82}$	$E9^3$	NMR(ΔH)	200–290	87	[3.113]
$Ti_{64}Co_{32}H_{79.4}$	$E9^3$	NMR(ΔH)	210–290	225	[3.114]
$Ti_{64}Co_{32}H_{89.2}$	$E9^3$	NMR(ΔH)	210–290	225	[3.114]
$Ti_{64}Co_{28}Fe_4H_{91.1}$	$E9^3$	NMR(ΔH)	210–290	225	[3.114]
$Ti_{64}Co_{24}Fe_8H_{87.2}$	$E9^3$	NMR(ΔH)	210–290	281	[3.114]
$Ti_{64}Co_{20}Fe_{12}H_{95.7}$	$E9^3$	NMR(ΔH)	210–290	447	[3.114]
$Ti_{64}Co_{18}Fe_{14}H_{87.4}$	$E9^3$	NMR(ΔH)	210–290	359	[3.114]
$Ti_{64}Co_{16}Fe_{16}H_{93.4}$	$E9^3$	NMR(ΔH)	210–290	363	[3.114]
$Hf_2RhH_{2.2}$	$E9^3$	NMR(T_1, $T_{1\rho}$, T_2)	–	288(3)	[3.44]
$Hf_2CoH_{3.8}$	$E9^3$	NMR(T_1, $T_{1\rho}$, T_2)	–	302(5)	[3.44]
$Ti_2CuH_{1.9}$	$C11b$	NMR($T_{1\rho}$)	280–520	400(30)	[3.45]
$Zr_2PdH_{1.94}$	$C11b$	NMR($T_{1\rho}$)	260–540	370	[3.46]

Table 3.8. Comparison of activation energies U(meV) obtained from different proton NMR parameters

Sample	T_1 (Below minimum)	T_2	$T_{1\rho}$ (Above minimum)	$T_{1\rho}$ (Below minimum)	Reference
$LaNi_5H_{6.2}$	220(20)	300(10)	370(10)	160(10)	[3.102]
$LaNi_{4.5}Al_{0.5}H_{5.32}$	150(30)	300(10)	360(10)	170(10)	[3.102]
$LaNi_{4.0}Al_{1.0}H_{4.33}$	160(20)	420(10)	440(10)	130(10)	[3.102]
$LaNi_5H_7$	217(7)	–	370(10)	210(20)	[3.43]
$LaNi_5H_{6.0}$	240(10)	–	370(10)	160(20)	[3.104]
$LaNi_4BH_{1.4}$	270(10)	–	400(10)	250(20)	[3.104]
$TiCr_{1.8}H_{0.55}$	130(10)	200(10)	190(20)	30(10)	[3.44]
$TiCr_{1.8}H_{2.58}$	135(10)	260(10)	250(20)	100(10)	[3.44]
$TiCr_{1.9}H_{0.63}$	88(10)	205(10)	218(10)	–	[3.44]
$TiCr_{1.9}H_{2.85}$	280(10)	400(20)	400(20)	180(20)	[3.44]

have yielded different activation energies even over a common temperature range.

As an example of a QNS experiment which was designed to reveal the macroscopic diffusion coefficient, we present some results for the hydrogen storage compound $Ti_{0.8}Zr_{0.2}CrMnH_x$ [3.109]. Figure 3.2 displays quasielastic spectra obtained with the high resolution back-scattering spectrometer in Jülich. Figure 3.3 shows the q dependence of the corresponding quasielastic widths.

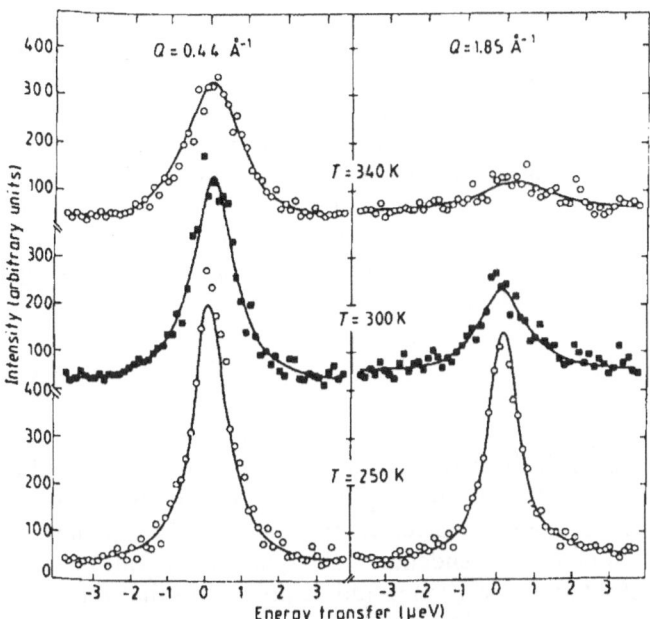

Fig. 3.2. Typical QNS spectra of $Ti_{0.8}Zr_{0.2}CrMnH_3$: the curves represent fits of convoluted Lorentzians to the experimental intensities [3.109]

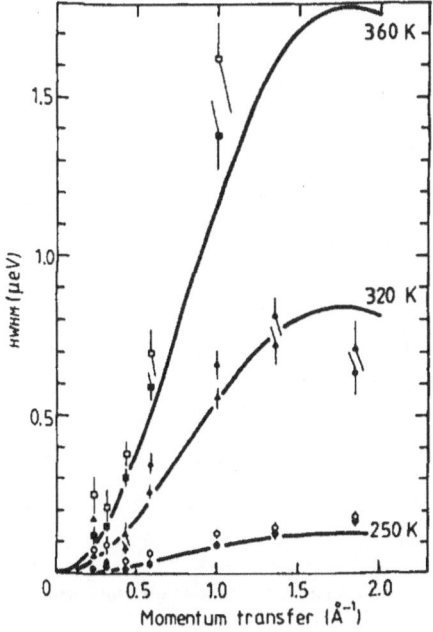

Fig. 3.3. q-dependence of QNS linewidths obtained from $Ti_{0.8}Zr_{0.2}CrMnH_3$. Open symbols before and full symbols after multiple-scattering correction [3.109]

We emphasize that a proper evaluation of such spectra requires multiple scattering corrections which, at small q's in particular, are an important source of systematic errors. Most of the doubly-scattered neutrons, which make the largest contribution to multiple scattering, have been scattered twice at large angles where the scattering function yields large linewidths. Thus they tend to cause a marked false broadening in the small q range. To correct for this effect, knowledge of the scattering function at large q is necessary—often an extremely difficult task. A pragmatic procedure is suggested in [3.109] and [3.98]. In order to correct the linewidth at small q, the spectra at larger q are described by a single Lorentzian with a q-dependent width according to

$$\Gamma(q) = \frac{6\hbar D}{l^2}\left(1 - \frac{\sin(ql)}{ql}\right). \qquad (3.39)$$

This corresponds to a liquid-like isotropic arrangement of jump vectors which is considered to be a reasonable first-order approximation for correction purposes. As shown by Fig. 3.3, the thus corrected linewidths follow a q^2 law. We note that the observation of a q^2 dependence of the quasielastic width at small q is an important test of internal consistency and demonstrates that the experiment actually observes long-range diffusion and is *not* affected by local jumps, or even more importantly by trapping processes [3.115].

Figure 3.4 presents recent QNS results on the H diffusion coefficient in $LaNi_5H_6$ [3.98] and emphasizes another aspect of multiple scattering contri-

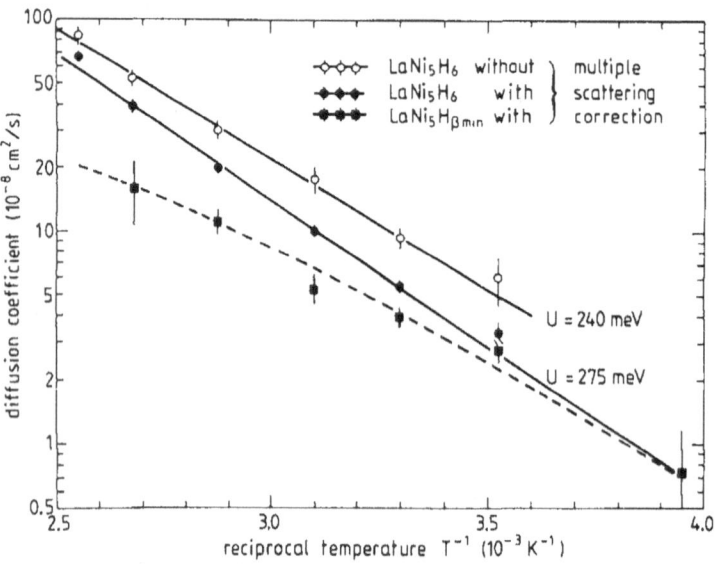

Fig. 3.4. Hydrogen diffusion coefficient in $LaNi_5H_6$ before and after multiple-scattering correction. The results for the β_{min} sample are also included (see text) [3.98]

butions: the more the multiply-scattered neutrons contribute within an energy range defined by the resolution of the spectrometer, the greater their influence on the apparent width. This effect is largest at low temperatures. Therefore, without corrections, the observed activation energy is smaller than the real one. For $LaNi_5H_6$ with a neutron transmission of 80%, which is typical for such experiments, this correction changes the activation energy by 15%. Figure 3.4 also presents diffusion data taken at a lower H concentration, and the most striking result is the concentration dependence of D: the diffusion coefficient decreases with decreasing c. Qualitatively such behaviour can be explained by the coexistence of at least two energetically non equivalent sites in the lattice. The energetically deeper ones saturate first, and the remaining protons are then able to diffuse faster because the trapping probability for the deeper sites is reduced. This effect compensates for the blocking factor $(1 - c)$, which tends to reduce D as c increases (see also Sect. 3.3.2). Table 3.4 lists data on the H diffusion coefficient in $LaNi_5H_6$ from various sources.

It is now quite clear that *Fischer* et al. [3.100] and *Noréus* et al. [3.101], who measured at large momentum transfers where single-jump rates determine the scattering law, have seen a spatially restricted motion of H in $LaNi_5H_6$ and misinterpreted it as hydrogen diffusion; this explains their results.

The diffusion constants of hydrogen can be related to the reaction rate constant $K_r = 1/\tau_r$ for absorption or desorption in a storage powder. For grains of size R we obtain approximately

$$K_r \approx \pi^2 D/R^2. \tag{3.40}$$

For $Ti_{0.8}Zr_{0.2}CrMnH_3$, with $D = 6 \cdot 10^{-8}\,cm^2/sec$ from quasielastic neutron scattering and $R = 0.5\,\mu m$, this yields $K_r = 400\,s^{-1}$, if the absorption and desorption are entirely diffusion-controlled. Actually, the measured reaction rate is only $0.04\,s^{-1}$ [3.116]. Consequently, the reaction is retarded by surface-penetration effects, and its rate is not determined by the relatively fast bulk diffusion. The opposite case occurs for $TiFeH_x$ where the self-diffusion is very slow. Here the reaction seems to be diffusion-determined [3.79].

Finally we point out that diffusion during absorption or desorption is influenced by the driving force related to the hydrogen density gradient, see (3.27). Therefore the chemical diffusion constant $D_c(c)$ comes into play instead of D_s. Rigorously, the problem is non-linear and very complex. The above treatment is only a rough approximation.

3.3.2 Diffusion in $TiCr_yH_x$

As are implied by the generally low activation energies given in Table 3.4, relatively fast hydrogen mobilities have been found for several AB_2H_x systems with the Laves crystal structures of either the C14 or C15 type [3.47, 106–110]. The most detailed NMR studies of Laves-type hydrides have been performed on $TiCr_{1.8}H_x$ (C15) and $TiCr_{1.9}H_x$ (C14). Figure 3.5 illustrates the temperature dependence of

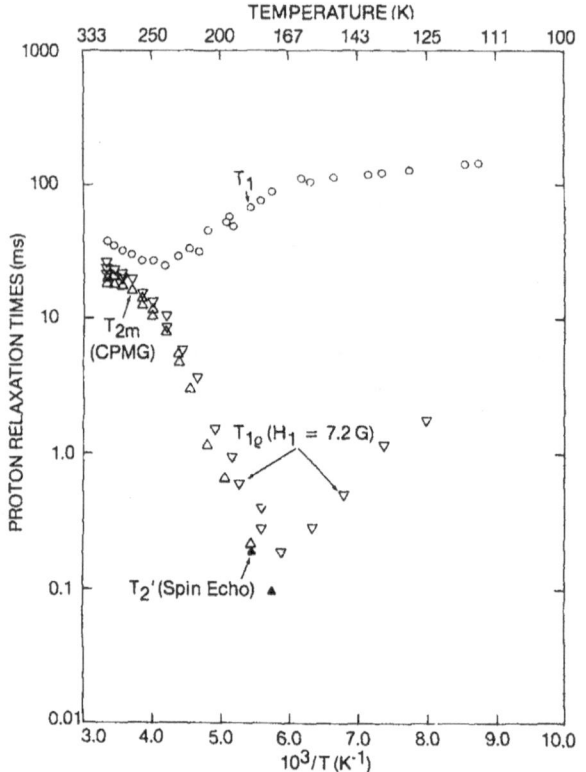

Fig. 3.5. Proton relaxation times T_1, $T_{1\rho}$, T_2' (spin-echo sequence), and T_{2M} (Carr–Purcell–Meiboom–Gill pulse sequence) for α'-phase TiCr$_{1.8}$H$_{2.58}$ with the cubic C15 Laves structure [3.47]

the proton T_1, T_2, and $T_{1\rho}$ relaxation times as they were obtained from the α'-phase sample, TiCr$_{1.8}$H$_{2.58}$. BPP-type analysis (Bloembergen, Purcell, Pound) of the proton relaxation time data has indicated rather complex behaviour for all the TiCr$_y$H$_x$ samples studied. Table 3.8 presents the various activation energies derived from the different relaxation times. *Bowman* et al. [3.47] have postulated that hydrogen motion in AB_2H_x consists of the superposition of two processes: (1) localized motion around the hexagonal rings formed by the preferred A_2B_2 interstitial sites, and (2) hydrogen jumps from one hexagon to another. Host crystal structure is expected to have an important role in the diffusion processes for the C15 and C14 Laves phases. Although all of the hexagons are composed of equivalent A_2B_2 sites in the C15 structure, two distinct types of hexagons occur in the C14 structure. Hence, proton jumps between the nonequivalent A_2B_2 sites for the C14 phases may have higher barriers than jumps between the identical A_2B_2 sites in the C15 hydrides. The U values in Table 3.8 are in qualitative agreement with this prediction. However, the possibilities of hydrogen-ordering effects and trapping at defect Ti$_3$Cr sites in the nonstoichiometric TiCr$_y$H$_x$ samples complicate the situation. More complex descriptions of the microscopic diffusion mechanisms in the C14 and C15 TiCr$_y$H$_x$ phases require detailed information on the temperature dependence of the interstitial site occupancies, and the specific effects of the Ti–Cr substitutional disorder in these nonstoichiometric alloys.

A relatively unusual technique, muon spin rotation, has been applied to study the cubic Laves phase hydride ZrV_2H_x using the positive muon as a radio-active tracer [3.111]. ZrV_2 is an intermetallic compound with energetically different interstitial sites which, upon hydrogenation, are successively filled with hydrogen (thermodynamic distribution). When a positive muon is added as a microscopic probe, it is thermalized at random and thus at low temperatures statistically distributed over the available (unblocked) interstices. By varying the preloading the accessible hydrogen sites were scanned and it was found that at high hydrogen content three kinds of interstices, Zr_2V_2 tetrahedra, ZrV_3 tetrahedra and V_4 tetrahedra, are occupied, in contradiction to published diffraction data. This surprising result was confirmed by detailed neutron vibrational spectroscopy.

3.3.3 QNS Studies of $Ti_{1.2}Mn_{1.8}H_3$

According to (3.13) and (3.26), a q-dependent QNS experiment provides complete information on microscopic H-jump processes in intermetallic hydrides. However, such experiments are an ambitious task and have not yet been undertaken, mainly due to the lack of single crystals. Instead, two different semi-phenomenological approaches have been reported which shall now be discussed. *Hempelmann* et al. [3.110] modelled the main features of the motional mechanisms in such complex structures by adopting a trapping model which was generalized in order to include spatially restricted fast motional processes. Alternatively, *Lartigue* et al. [3.105] started from structural information and interpreted the observed composite spectra as originating from H atoms moving across different groups of nonequivalent interstitial sites. This latter approach will be discussed in Sect. 3.3.4.

Hempelmann et al. based their model on their investigation of $Ti_{1.2}Mn_{1.8}H_3$, where spectra composed of several components exhibiting different q and temperature dependence were observed. This is demonstrated in Fig. 3.6, which displays a lineshape analysis of a spectrum at $T = 374\,K$ and $q = 1.95\,Å^{-1}$. It is evident that the spectrum cannot be described by one Lorentzian but is a super-position of at least two quasielastic lines. Figure 3.7 presents the q dependence of the linewidhs for the two components as observed at 355 K with the back-scattering spectrometer. Also included is the T dependence of the linewidths which is presented in form of an Arrhenius plot.

As $q \rightarrow 0$, only the linewidth of the narrow component tends to zero, whereas at large q the broad component has a pronounced q dependence. The Arrhenius representation of the linewidths reveals a considerably higher activation energy for the narrow component. Figure 3.8 displays the q dependence of the intensities of both components relative to the total scattering intensity at 101 K. Starting from a weight close to 1 at $q = 0$ the weight of the narrow line continuously

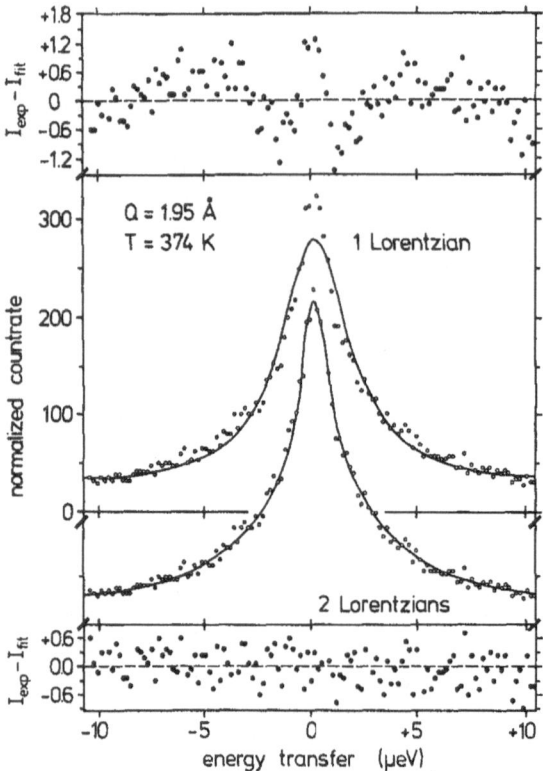

Fig. 3.6. Lineshape analysis of a QNS spectrum obtained from $Ti_{1.2}Mn_{1.8}H_3$ at 374 K. The above spectrum was fitted with 1 Lorentzian. The resulting distribution of errors is shown on top of the diagram. Systematic errors are evident. The lower curve presents a fit with two Lorentzians. The corresponding distribution of errors is displayed below and shows only statistical errors [3.110]

decreases with increasing q. The relative intensity of the broader component on the other hand starts from a weight of zero at $q = 0$, passes through a maximum, and decreases again towards higher q. At large q the total intensity of both components is significantly lower than expected from the reference measurements at low temperature.

Both the q dependence of the widths and the weights of the spectral components resemble closely what was found for H diffusion in the presence of trapping impurities by *Richter* and *Springer* in NbN_xH_y [3.115]. These observations can be interpreted successfully in terms of a two-state model which considers the proton either in a trapped state or in a state of free diffusion. Changes from the diffusive to the trapped state occur with a trapping rate τ_1^{-1}, while escape from a trap is described by an escape rate τ_0^{-1}. Multiple state changes are taken into account at the level of an average T-matrix approximation [3.57]. The scattering law of the two-state model exhibits the following features: At small q's where the scattering experiment averages over large volumes in space $\sim (2\pi/q)^3$, the quasielastic spectrum comprises a single Lorentzian, whose width is determined by the effective diffusion coefficient in the medium, including trapping and detrapping. At larger q's the spatial resolution of the scattering experiment increases and escape processes from

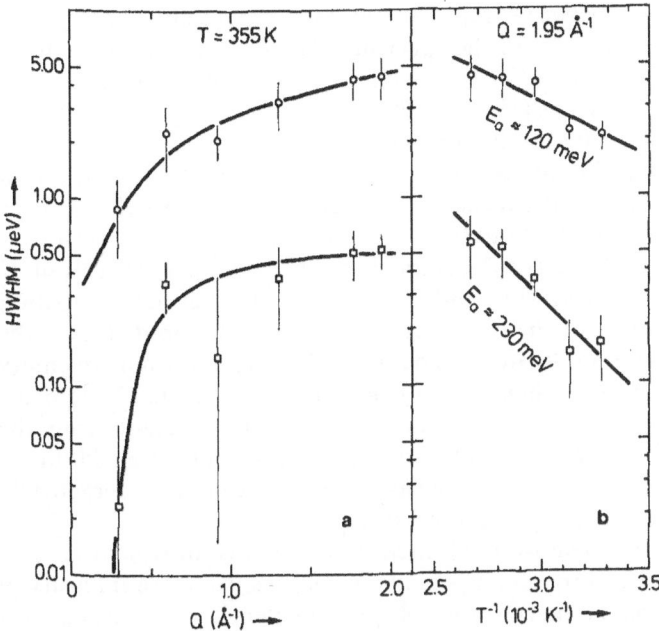

Fig. 3.7 a, b. q and T dependence of the QNS linewidths obtained from a fit with two Lorentzians to the QNS spectra from $Ti_{1.2}Mn_{1.8}H_3$ [3.110]

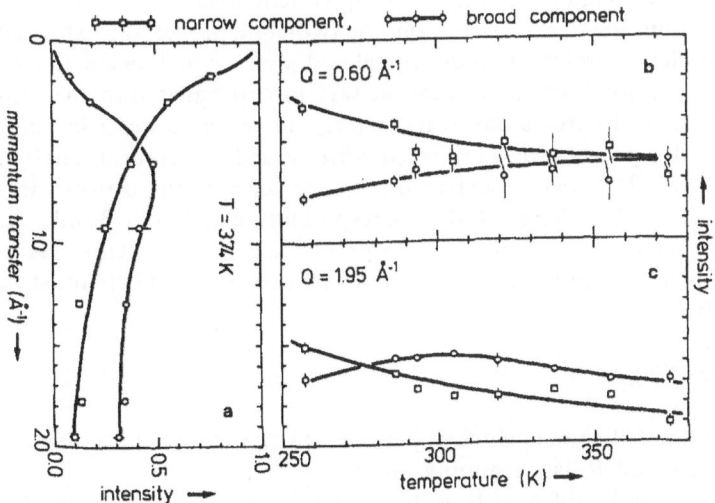

Fig. 3.8 a–c. q and T dependence of the intensities of the two Lorentzians of Fig. 3.7 relative to the total scattering intensity at 101 K [3.110]

traps as well as diffusion between traps are observed separately. We expect a narrow and a broad component in the spectrum. The narrow one stems from escape processes from the trap ($\Gamma \sim 1/\tau_0$) while the broad one originates from diffusive jumps in between traps ($\Gamma \sim 1/\tau$). The activation energy of $1/\tau_0$ should be greater than that of $1/\tau$ by the binding energy at the trap.

As discussed above, in intermetallic compounds a whole spectrum of energetically different H sites is expected. In order to account for this feature, we divide the available sites into "trap sites", comprising the energetically lowest interstitial position, and "free sites" comprising the remainder. Dissolved H preferentially occupies the trap sites and according to the thermal occupation probability saturates most of them. The remaining H is distributed over the free energetically less favourable sites and occasionally gets trapped in an empty trap site. Thus, in spite of the high density of traps, for a single H atom the diffusional process with respect to trapping is not very different from the situation in Nb with dilute N impurities, since the empty traps are still dilute. In order to distinguish that from impurity traps, we call the energetically favourable sites in intermetallics structural traps.

According to Fig. 3.8, a substantial intensity fraction is missing at large q, indicating the existence of broad components in the spectrum, the intensity of which lies mostly outside the energy range of spectrometer ($E = \pm 10\,\mu eV$). Such a third component in the spectrum with a substantial weight only at large q must be related to a rapid motion confirmed to small regions in space. Two kinds of such a local motion are imaginable: (i) correlated jumps as a consequence of the high hydrogen concentration, and (ii) structural effects like local hopping within clusters of energetically equivalent sites.

In order to account for this local motion, the two-state model was extended to a three-state model. In order to model correlated jumps which occur during passages of "free diffusion", the local motion was incorporated into the self-correlation function of the free state. Alternatively, in order to describe local hopping at a trap, this process was connected with the self-correlation function of the trapped state. The spatial extension of the local jump process was considered as a dumbbell of length l; the corresponding rate is $1/\tau_l$. While the significance of parameters $1/\tau_0$ and $1/\tau$ remains unchanged, some care has to be taken as regards the trapping rate $1/\tau_1$, in which saturation effects must be included explicitly:

$$\tau_1^{-1} = 4\pi R_t D c_t^*(T), \tag{3.41}$$

where $D = d^2/6\tau$ is the diffusion coefficient between traps, c_t^* is the concentration of unoccupied traps, and R_t is a trapping radius.

The value of c_t^* can be inferred from thermodynamic arguments using the chemical potentials of free and trapped hydrogen atoms respectively, which in thermal equilibrium must be equal

$$kTln\left(\frac{z_f/Z_f}{1-z_f/Z_f}\right) = -E_b + kTln\left(\frac{z_t/Z_t}{1-z_t/Z_t}\right), \tag{3.42}$$

(E_b is the binding energy at the structural trap, z_f and z_t are the number of protons in the "free" and "trapped" states respectively, and Z_f and Z_t are the numbers of these sites available). After some algebra the concentration of empty traps is obtained as

$$
c_t^* = \frac{1}{2}\left(c_t + c_H + \frac{\exp(-E_b/kT)}{1-\exp(-E_b/kT)} \right)
$$
$$
+\frac{1}{2}\left[\left(c_H + c_t + \frac{\exp(-E_b/kT)}{1-\exp(-E_b/kT)} \right)^2 - \frac{4c_Hc_t}{1-\exp(1-E_b/kT)} \right]^{1/2}, \quad (3.43)
$$

where c_t is the total trap concentration and c_H the hydrogen concentration. For $1/\tau_1 \gg 1/\tau$ the incoherent scattering law can be written as

$$
S_{inc}(q,\omega) = R_1 F \frac{\Gamma_1/\pi}{\Gamma_1^2 + \omega^2} + R_1(1-F)\frac{(2/\tau_1)/\pi}{(2/\tau_1)^2 + \omega^2}
$$
$$
+ (1-R_1)\frac{\Gamma_2}{\Gamma_2^2 + \omega^2}, \quad (3.44)
$$

with

$$
\Gamma_{1/2} = 1/\tau_0 + 1/\tau_1 + 1/\tau f(q) \pm \left[(1/\tau_0 + 1/\tau_1 + 1/\tau f(q))^2 - 4\frac{f(q)}{\tau_0\tau} \right]^{1/2},
$$
$$
(3.45)
$$

$$
R_1 = \frac{1}{2} + \frac{1}{2}\frac{f(q)\dfrac{\tau_1-\tau_0}{\tau_1+\tau_0} - \dfrac{1}{\tau_0} - \dfrac{1}{\tau_1}}{\left[(1/\tau_0 + 1/\tau_1 + f(q))^2 - 4\dfrac{f(q)}{\tau_0\tau} \right]^{1/2}}, \quad (3.46)
$$

$$
f(q) = \frac{1}{\tau}\left(1 - \frac{\sin(qd)}{qd} \right),
$$

and

$$
F = \frac{1}{2}\left(1 + \frac{\sin(ql)}{ql} \right).
$$

where F is the incoherent structure factor of the local motion. It can be understood as the Fourier transform of the asymptotic particle distribution due to the local jumps for infinite times [3.75].

Since the time scales of local hopping and translational diffusion differ by orders of magnitude, hybridization effects between these two different diffusive modes were neglected in the three-state model. Furthermore, it was assumed

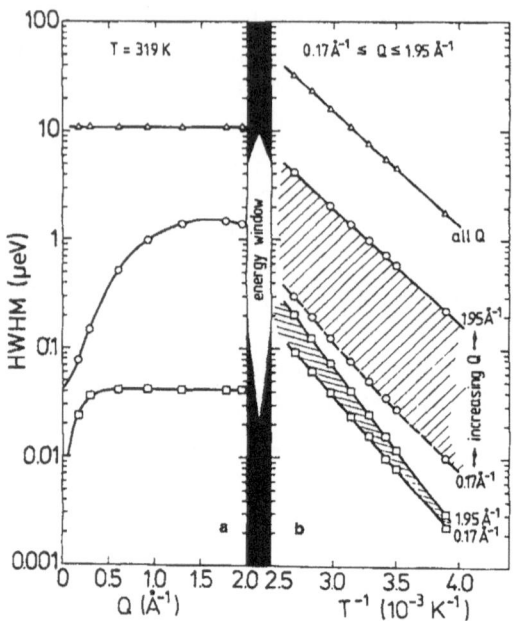

Fig. 3.9 a, b. q and T dependence of the widths of the three lines in the QNS spectra of $Ti_{1.2}Mn_{1.8}H_3$. These widths have been calculated with the parameters resulting from a global fit with the three state model to 48 experimental spectra, including multiple-scattering corrections. The points indicate where the spectra were recorded [3.110]

that the fast motional process occurs during passages of diffusion and not in the trapped state. This view is supported qualitatively by the intensity increase of the broad line with rising temperature (cf. Fig. 3.8). If local motion occurs dominantly during trapping, its spectral fraction should decrease as much as the occupation of traps decreases with increasing temperature. Quantitatively, the fit achieves a significantly better agreement relating local hopping to the free state. The q and T dependence of the resulting linewidth is shown in Fig. 3.9. The activation energies for τ^{-1}, τ_l^{-1} and τ_0^{-1} are 210 meV, 209 meV, and 300 meV, respectively, yielding a binding energy at the traps of 90 meV. While the trap concentration c_t is evaluated at $c_t = 0.24 \pm 0.06$, the concentration of free traps c_t varies between 5% and 9% depending on temperature, which is consistent with the application of the trapping model. The relatively large amount of trapping sites (25%) clearly shows that they do not originate from impurities or surface segregations. An experiment on stoichiometric $ZrMn_2H_3$ revealed analogously composed QNS spectra and thus the structural traps do not appear to be a peculiarity due to the excess Ti atoms. Finally, the jump length l of the local jump process was determined as $l = 1.37$ Å. This length corresponds well to the distance between adjacent sites [3.117] in the hexagonal Laves phase structure. From the similarity of activation energies, the coincidence of local jump length and nearest neighbour distance, and the association of local jumps and diffusional state, it was suggested that the rapid local motion may be connected with correlation due to concentration effects.

3.3.4 LaNi$_5$H$_6$ – NMR and Neutron Scattering

The intermetallic hydride whose diffusion properties have been most extensively measured is β-phase LaNi$_5$H$_{\sim 6}$ which has been examined using essentially all of the NMR techniques [3.43, 91, 94–97, 102, 104] and has been the subject of several QNS studies [3.98–101, 105, 118]. When the reported diffusion coefficients and activation energies for β-LaNi$_5$H$_{\sim 6}$ listed in Table 3.4 are compared, disparities are quite apparent. Nevertheless, a nominal 300 K diffusion coefficient of about 10^{-8} cm^2/s has been independently estimated from several of these studies.

In addition to various sample dependent factors like differences in stoichiometries and possible impurity phases (e.g. precipitated Ni, etc.), difficulties with the experimental techniques themselves have made large contributions to the discrepancies in Table 3.4. The anomalous paramagnetic relaxation effects [3.36, 95] mentioned in Sect. 3.2.3 may be responsible for the small U values typically seen in the less pure LaNi$_5$H$_x$ samples [3.91, 92, 95–97]. *Richter* et al. [3.98, 75] have discussed various instrumental artifacts and data analysis difficulties as well as the data correction procedures necessary to properly determine the proton diffusion characteristics of β-LaNi$_5$H$_x$ with the QNS method. Unfortunately, the extent of these experimental difficulties had not been fully appreciated in most of the previous work and have led to some very questionable interpretations of the observations. However, the combination of recent QNS and NMR results is finally leading to a better understanding (albeit, still rather qualitative) of the complex nature of hydrogen motion in β-LaNi$_5$H$_6$.

Before we discuss the NMR and QNS data in more detail, we first examine the structure of LaNi$_5$H$_6$ which has been a matter of controversy for several years. *A.L. Bowman* [3.119] and later on other groups [3.120], suggested a structure with 2 different H sites in the space group P31 m. The first structure proposal by *Percheron-Guégan* et al. [3.121] contained a distribution of the hydrogen atoms between 5 types of interstitial sites in the unit cell, thereby retaining the space group P6/mmm of hydrogen-free LaNi$_5$. The inherent difficulties in a neutron diffraction study of LaNi$_5$-deuteride are (i) the dominant Ni scattering which obscures the D positions, and (ii) the pronounced anisotropic line broadening. Both difficulties have recently been overcome by *Thompson* et al. [3.122], who used a La^{60}Ni$_5$D$_7$ sample with isotope-pure ^{60}Ni (D is then the dominant scatterer!) and who modelled the anisotropic line broadening in a modified Rietveld refinement procedure. The line broadening is due to strain, particle size and possibly stacking faults (microtwinning); hence the line shape is non-Gaussian and was described by pseudo-Voigt functions. The resulting structure is shown in Fig. 3.10. The main features are: i) three types of D sites (La$_2$Ni$_4$ octahedra, La$_2$Ni$_2$ tetrahedra, and Ni$_4$ tetrahedra), ii) displacement along the c-axis of the intermediate Ni layer, and iii) long range correlations in the D occupancies resulting in a doubling of the unit cell. The observation that the intensity of the (023) superlattice Bragg peak does not change upon substitution of the natural Ni by isotope pure ^{60}Ni shows clearly that only the

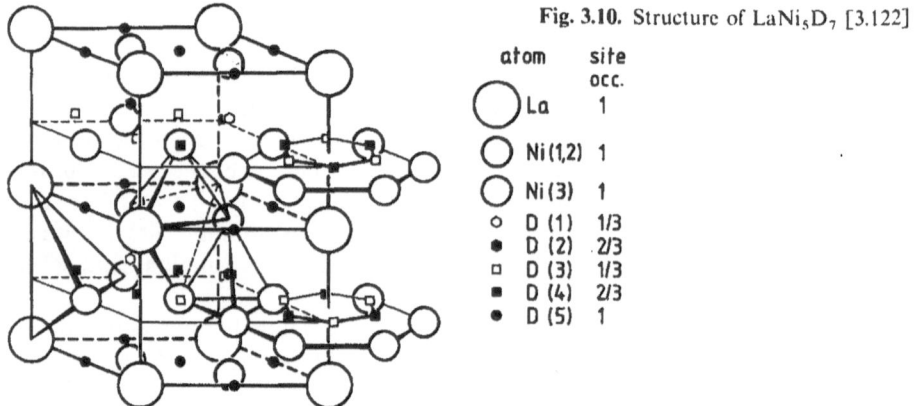

Fig. 3.10. Structure of $LaNi_5D_7$ [3.122]

atom	site occ.
La	1
Ni (1,2)	1
Ni (3)	1
D (1)	1/3
D (2)	2/3
D (3)	1/3
D (4)	2/3
D (5)	1

D atoms are responsible for the superlattice. Further work on the $LaNi_5H_6$ structure by the French group [3.123, 124] finally led to a convergence of the different structural proposals. In their third structural determination [3.124] they obtained the same D-sites as Thompson et al. and thus agreement between the two groups was achieved.

For the identification of the diffusion process such an agreement is not yet at hand. However, from considerations of the more recent NMR [3.43, 94, 102, 104] and QNS [3.98, 105, 118, 120] results important characteristics of the diffusion process can be extracted, though some major features remain contradictory. In order to avoid (or at least to minimize) the paramagnetic relaxation effects [3.36], only NMR data obtained on the highest purity samples should be used to extract the diffusion parameters. The proton relaxation times T_1, $T_{1\rho}$, and T_2 for three high purity (i.e. the starting La metal contained 4 ppm Fe and less than 10 ppm total of all other rare earth metals) β-$LaNi_{5-y}Al_yH_x$ samples [3.102] are shown in Fig. 3.11, while the U values obtained from these parameters are summarized in Table 3.8. A unique activation energy clearly cannot represent the hydrogen diffusion properties of these samples. Similar conclusions were reached from other NMR studies [3.43, 104] of high-purity β-$LaNi_5H_x$ samples. From these results it was suggested [3.43, 102, 104] that the NMR relaxation times reflect at least two types of motions: (1) a short range (i.e. localized) process with an activation energy of about 0.2 eV, and (2) the long range, bulk hydrogen transport with a nominal 0.4 eV activation energy. Although this mechanism is qualitatively consistent with the observed relaxation time behaviour, and can rationalize the effects of Al and B substitution for Ni atoms and the proton diffusion parameters [3.95, 102, 104], the lack of a detailed theoretical model that can directly relate the experimental relaxation time data to the necessarily complex multi-step jump processes have prevented any quantitative assessments.

Compared to NMR quasielastic neutron scattering is in principle in a more favourable position to reveal details of the jump processes, since it leads to a

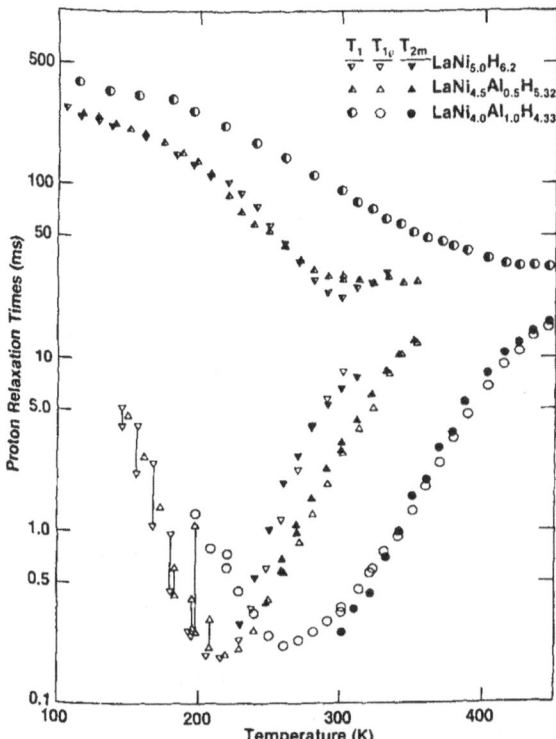

Fig. 3.11. Proton relaxation times T_1, $T_{1\rho}$, and T_2 for high-purity β-LaNi$_{5-y}$Al$_y$H$_x$ [3.102]. The vertical lines connect limiting $T_{1\rho}$ values in regions of non-exponential recovery

simultaneous time and space resolution of the H motion in question. Furthermore, in all circumstances the protons in the sample all contribute to the quasielastic line, and magnetic impurities and small segregated magnetic Ni clusters do not interfere with the scattering result.

Achard et al. performed both high resolution back-scattering and lower resolution time-of-flight experiments on LaNi$_5$H$_{5.8}$ and LaNi$_{4.5}$Al$_{0.5}$H$_{4.8}$ [3.105, 118]. Figure 3.12 presents the lower resolution data obtained from the time-of-flight (TOF) measurements of LaNi$_{4.5}$Al$_{0.5}$H$_{4.8}$. As already reported for the case of LaNi$_5$H$_6$ [3.118], below an intense elastic peak broad quasielastic lines are observed, the intensities of which increase slowly with increasing q. Such a q dependence of the inelastic intensity (see also (3.46) and the explanation thereafter) is indicative of a spatially restricted fast protonic motion. Furthermore, again probably only a fraction of the protons participate in this fast motion. Such short range jump processes appear to be a common feature of many intermetallic hydrides. As explained above, they were observed in the hexagonal Laves phase Ti$_{1.2}$Mn$_{1.8}$H$_3$, and were reported recently also for Mg$_2$NiH$_4$ [3.89]. Typical high resolution spectra obtained from both samples are shown in Fig. 3.13. They show a small quasielastic broadening of the central peak, which appeared to be elastic with the coarser resolution of the TOF spectrometer. The small widths indicate jump processes two orders magnitude

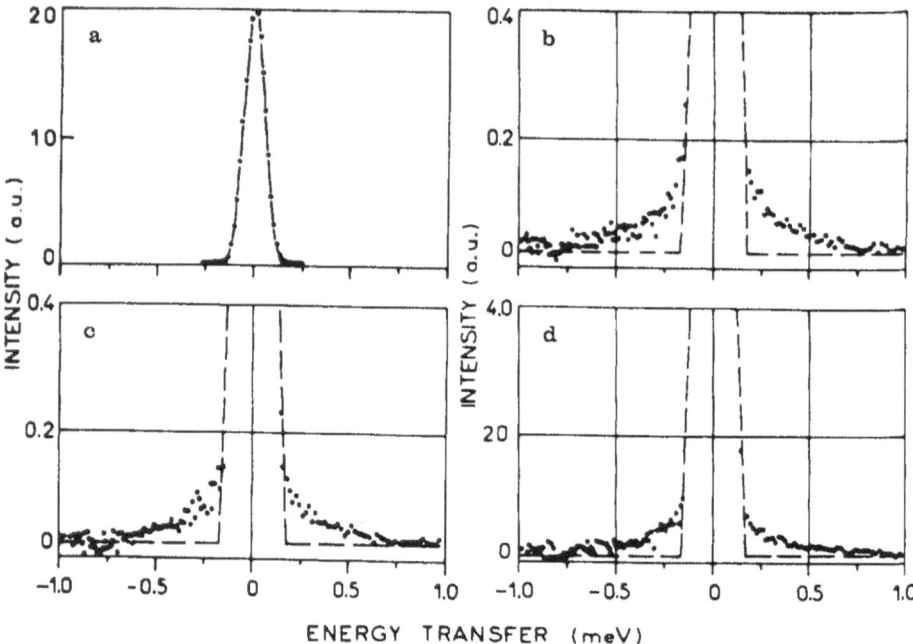

Fig. 3.12. Quasielastic neutron spectra from H in (**a**) and (**b**) $LaNi_5H_{5.8}$, (**c**) $LaNi_4MnH_{5.7}$ and (**d**) $LaNi_4AlH_{4.3}$ ($T = 300$ K, $q = 1.9$ Å^{-1}). In each case the experimental points are compared with the vanadium elastic peak (---) (intensity magnifications: (**b**), (**c**), (**d**) × 50) [3.118]

slower. According to the analysis given by the French group, the spectra can be decomposed into an elastic and a quasielastic component, the spectral weights of which are about 1/2 and do not depend on q. The linewidth of the quasielastic line increases strongly with q and temperature. The q-independent elastic part in the high resolution spectra was taken as evidence that a fraction of the protons are moving too slowly to be detected by neutron scattering and thus appear to be at rest. This interpretation, however, contradicts recent high resolution QNS experiments by *Richter* et al. [3.98], which covered the temperature range 280–400 K.

At small momentum transfers a detailed lineshape analysis shows that only one Lorentzian line appears in the spectrum (Fig. 3.14). As can be seen from the distribution of errors, the experimental lineshape is in perfect agreement with the single Lorentzian line. The existence of only one line or one relaxation process at small q demonstrates that QNS observes true long range diffusion participated in by *all* protons in the sample. The result of an immobile H fraction in $LaNi_5H_x$ at room temperature is therefore most likely related to an insufficient energy resolution in this experiment.

In order to arrive at a more quantitative picture of the fast H motion in $LaNi_{5-x}Al_xH_6$, *Lartigue* et al. [3.105] considered different local jump processes across ring configurations in the $LaNi_5H_6$ structure. In doing so, they used as

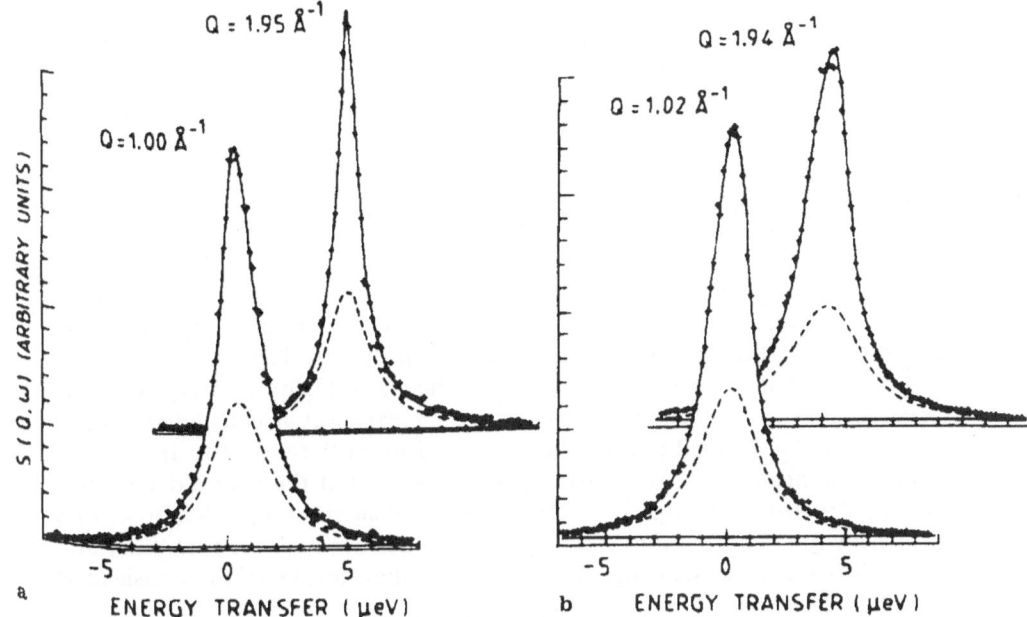

Fig. 3.13. Decomposition of the high resolution spectra from (**a**) $LaNi_5H_{5.8}$ at 295 K and (**b**) $LaNi_{4.5}Al_{0.5}H_{4.8}$ at 353 K into a purely elastic line, a broadened component (---) and a flat background [3.105]

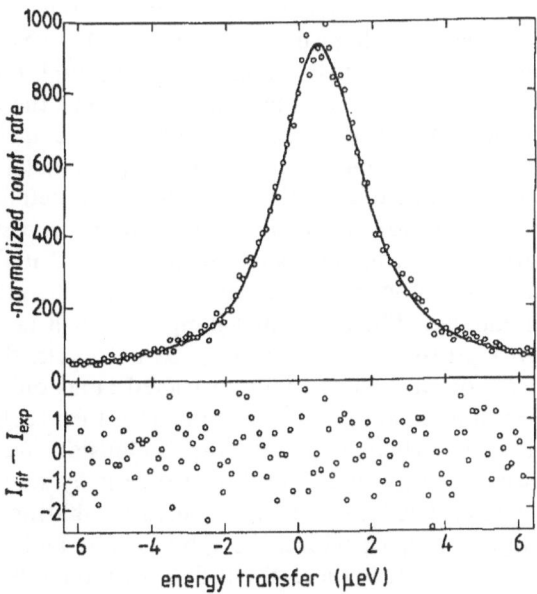

Fig. 3.14. Lineshape analysis of a QNS spectrum from $LaNi_5H_6$ at 373 K and $q = 0.3\,\text{Å}^{-1}$. The data were fitted with one Lorentzian. Unlike the case of Fig. 3.6, no systematic deviations are visible in the error distribution which is shown below the spectrum [3.124]

a basis the structural determination of [3.121], which subsequently turned out to be incorrect. Nevertheless, since their approach is of particular interest, we briefly outline the concept.

For a localized jump mechanism the incoherent scattering function has the general form

$$S_L(q, \omega) = \frac{N_R}{N} \left[B_0(q)\delta(\omega) + \sum_j B_j(q) \frac{1}{\pi} \frac{\tau_j}{1 + (\omega\tau_j)^2} \right] \tag{3.47}$$

and can be calculated from rate equations in a similar way to that shown in Sect. 3.2.4. Here, $B_0(q)$ is the elastic incoherent structure factor (EISF) of the motion and τ_j^{-1} are the eigenvalues of the jump matrix. The spectral weights $B_j(q)$ are combinations of eigenvectors as shown in (3.26), and N_R/N is the fraction of protons participating in the process. *Lartigue* et al. calculated $S_L(q, \omega)$ for several possible local jumping schemes and fitted the model to the experimental data, keeping the respective H fractions fixed at the values determined from the structural investigations. Due to an insufficient q-range, this procedure relied heavily on occupation numbers taken from crystallographic data, and the result is not a consequence of the q and ω dependence of the quasielastic spectra. Considering short jump distances, a determination of the local jump process from dynamical data alone would require an extension of the covered q range by at least a factor of 2. For small q, all EISF decay with $1 - \alpha(ql)^2$ where l is the jump distance, and discriminating between different models is very difficult with no unique solution. It is therefore not surprising that they were able to fit the quasielastic spectra with a model based on the wrong structure. However, independent of the missing geometrical information, the activation energy of 170 meV obtained for the fast process should be reliable. Its value is also in good agreement with NMR results derived from the NMR parameters ΔH, T_1 (below minimum) and $T_{1\rho}$ (below minimum). Although the hydrogen atoms are very mobile within small areas, their motion between these areas occurs on a much longer time scale. Therefore, the large "elastic" contributions are observed within the resolution window of the TOF spectrometer.

Summarizing, in spite of the large efforts to elucidate the microscopic details of H diffusion in β-LaNi$_5$H$_x$, we still do not understand the diffusion process at the atomic scale. Both the neutron scattering results of *Lartigue* et al. [3.105] and *Richter* et al. [3.98] show multicomponent spectra in which the different lines exhibit different activation energies. The associated jump rates will also occur in one or other T_1, T_2 or $T_{1\rho}$ experiments. Neutron scattering clearly shows that the fast jumping modes occur over spatially restricted areas only. Furthermore, from the QNS spectra at small q it is clear that on a time scale of 10^{-8}–10^{-9} s essentially *all* protons participate in long-range diffusion over distances of 20–30 Å. The activation energy of 275 meV is also seen by some T_2 and T_1 experiments [3.95, 102, 104]. The large activation energy of 400 meV found by FGSE methods has no counterpart within the QNS results and at present we do not understand how to reconcile these NMR data with the QNS results. In order to proceed further, not only are more detailed QNS and NMR data on very well characterized samples required, but more rigorous theoretical models (particularly for the NMR relaxation times) must also be developed to sort out the relative contributions of the individual jump processes.

Hydrogen diffusion in $LaCu_5H_{3.4}$ was investigated using pulsed NMR techniques to measure spin–lattice T_1 and rotating frame $T_{1\rho}$ proton relaxation times. Evidence for at least two uncoupled modes of diffusive motion was found. They are substantially less rapid than previously reported for β-LaNi$_5$ hydride [3.125].

3.3.5 Amorphous Metals

A number of papers have appeared on the hydrogen dynamics and diffusion in amorphous alloys and metallic glasses. As summarized by Bowman [3.126], most of this work has involved the metal–metal glasses (e.g., Ti–Cu, Ti–Ni, Zr–Ni, Zr–Pd, etc.), but the amorphous metal–metalloid Pd–Si system has also received much attention [3.127–130].

Computer modelling of amorphous structures reveals that a variety of interstitial sites (octahedral, tetrahedral etc.) exist [3.131], where each type of site has edge lengths fluctuating around an average value. Dissolving hydrogen in these interstices gives rise to a distribution of potential energies for the hydrogen atoms. A continuous Gaussian distribution

$$n(G)\,dG = \frac{1}{\sigma\sqrt{\pi}}\exp\left[-\left(\frac{G-G^0}{\sigma}\right)^2\right]dG \qquad (3.48)$$

is in agreement with electrochemical electromotoric force (e.m.f.) measurements [3.132] and theoretical considerations [3.133], where $n(G)$ is the concentration of interstices of free energy G within the interval dG, and σ and G^0 are the width and mean energy respectively of the energy distribution. Since each site energy is associated with a limited number of sites, Fermi–Dirac statistics are used to obtain the occupancy of the different sites. This relates the hydrogen concentration c to the chemical potential μ of the hydrogen:

$$c = \int_{-\infty}^{\infty} \frac{n(G)\,dG}{1 + \exp\left(\dfrac{G-\mu}{RT}\right)}. \qquad (3.49)$$

Note that for an ideal crystal with $n(G) = \delta(G-G^0)$ this equation reduces to

$$\mu = G^0 + RT\ln\frac{c}{1-c}. \qquad (3.50)$$

Choosing the mean free enthalpy of solution G^0 as a reference level, Kirchheim's theory yields for the H concentration and temperature dependence of the chemical diffusion coefficient

$$D = \frac{D^0 \sigma \sqrt{\pi}}{RT} \exp{(\text{erf}^{-1}|2c - 1|)^2} \cdot \exp\left(\frac{\sigma \, \text{erf}^{-1}|2c - 1|}{RT}\right), \tag{3.51}$$

where D^0, the diffusivity of hydrogen in a hypothetical solid which only has sites of energy G^0, is an adjustable parameter, whereas σ is derived from e.m.f. measurements. Figure 3.15 compares electrochemically measured diffusivities with calculated ones for hydrogen in amorphous $Pd_{77.5}Cu_6Si_{16.5}$. Evidently, (3.51) is able to describe the dependence of the H diffusion coefficient on the H concentration over a range of three orders of magnitude. The second exponential term in (3.51) brings about a decrease of the apparent activation energy with increasing H concentration, which is also observed experimentally. Obviously the macroscopic diffusion of H in amorphous metals can satisfactorily be described under the assumption of a Gaussian distribution of site energies. A proper description of H diffusion in the amorphous metal–metalloid Pd–Si system turned out to be more demanding [3.128, 130]: the H diffusion mechanism in $Pd_{85}Si_{15}H_{7.5}$ was shown by means of quasielastic neutron scattering to be characterized by two well-separated regimes of jump rates, in clear contradiction to existing ideas on diffusion in amorphous materials, in disagreement particularly with the anticipated continuous distribution of jump rates. A semiquantitative interpretation of the data was possible in terms of a two-state diffusion-trapping model. No evidence for a fractal dimension of the network of diffusive paths was found.

Amorphous metals seem to be the first host systems in which a reorientation relaxation associated with hydrogen diffusion (the Snoek effect) has been observed

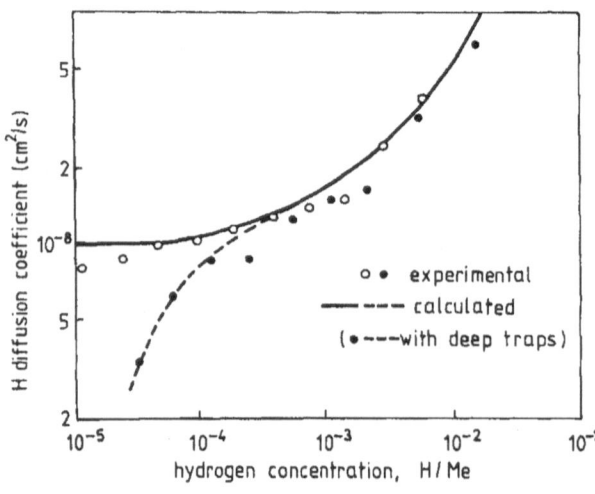

Fig. 3.15. Diffusivity of hydrogen vs concentration for an amorphous alloy of $Pd_{77.5}Cu_6Si_{16.5}$ at 295 K [3.126]

[3.134, 135]. Figure 3.16 shows the internal friction peak exhibited by glassy $Pd_{0.8}Si_{0.2}$ after room temperature equilibration against various pressures of hydrogen gas. The coupling between peak height and peak temperature indicates concentration-dependent relaxation kinetics, with shorter relaxation times τ_R for higher hydrogen contents. For a given H content, from the mean $1/T$ shift between the peaks, the temperature dependence of $\bar{\tau}_R$ is represented by

$$\bar{\tau}_R = \bar{\tau}_{OR} \exp(\bar{E}_R/kT), \qquad (3.52)$$

with values of $1.4 \cdot 10^{-14}$ s for the prefactor $\bar{\tau}_{OR}$, and 320 meV for the mean activation energy of $Pd_{82}Si_{18}H_{1.8}$. For lower H contents higher activation energies were found, for instance 450 meV in the case of $Pd_{82}Si_{18}H_{0.3}$. The widths of the internal friction peaks in Fig. 3.16 are considerably larger than expected for a single relaxation process; this indicates a distribution of relaxation times and thus a distribution of site energies, qualitatively in accordance with Kirchheim's model.

Measurements of the Gorsky effect have also been performed on the same systems [3.136]. Figure 3.17 shows the temperature dependence of the resulting diffusion coefficient. Again a strong concentration dependence of the activation energies for H diffusion (E_D) is observed. In Fig. 3.18 we compare the characteristics of the Gorsky relaxation in Pd–Si alloys with those of the Snoek relaxation. The upper part of this plot compares the values of $\bar{E}_R$ and E_D, the lower part the corresponding prefactors $\bar{\tau}_R^{-1}$ and D_0 (the latter values have been shifted relative to each other to bring the data points onto a common curve). It is immediately evident that both relaxations exhibit a similar concentration

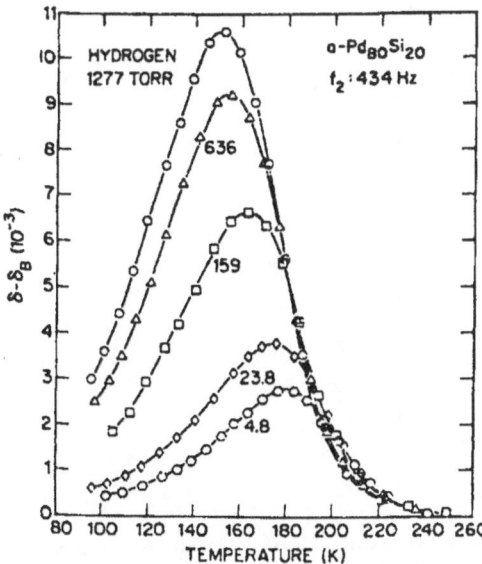

Fig. 3.16. The internal friction peak exhibited by glassy $Pd_{80}Si_{20}$ after equilibration against various pressures of hydrogen gas at room temperature (293 K–295 K) [3.18]

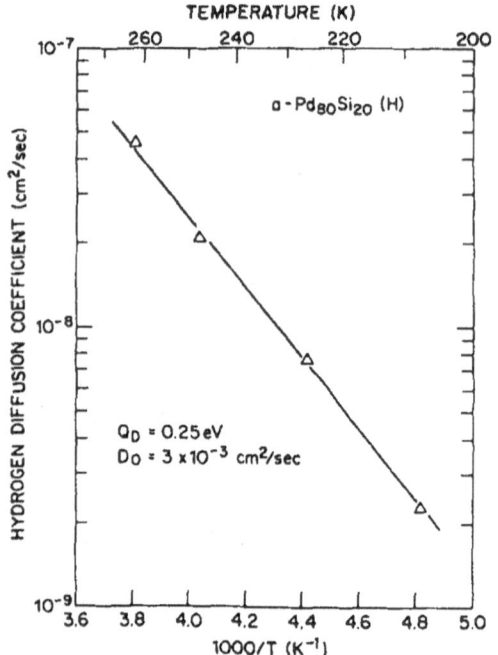

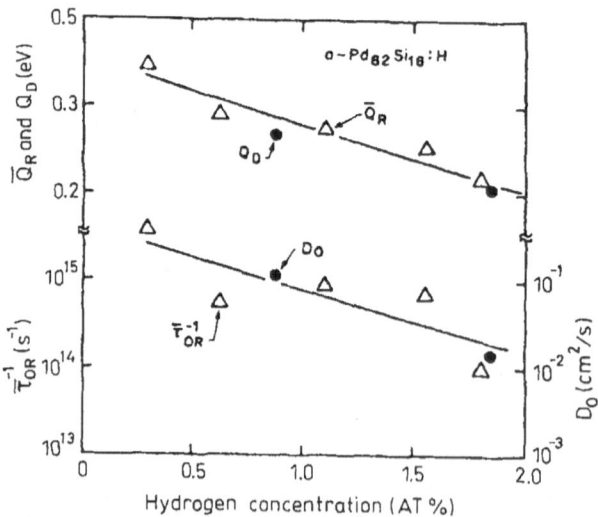

Fig. 3.18. Comparison of the rate parameters for the Snoek and Gorsky relaxations produced by hydrogen in glassy $Pd_{82}Si_{18}$. All measurements were made on the same sample (from [3.18])

dependence and—within experimental error—an equal activation energy. From this *Berry* and *Pritchet* [3.18, 136] conclude that reorientation and diffusion occur through the same family of jumps. However, this conclusion should not be taken too literally, since different jump rates contribute differently to the signal in the Snoek and Gorsky effect experiments (via the shape factor in the former case and via the size factor in the latter, see Sect. 3.2.2). In addition, when an internal friction peak is recorded by changing the sample temperature, the distribution of jump rates is not only traced out, but also heavily manipulated: in an amorphous metal with a certain distribution of site energies, the distribution of the H atoms between the available sites changes dramatically with temperature. Therefore the measured distribution of Snoek relaxation times only mirrors very crudely the distribution of H jump times. Furthermore, since the activation energy of the Snoek peak corresponds to the most probable jump rate, whilst the Gorsky effect measures a thermally weighted average diffusional jump rate, the agreement between $\bar{E}_R$ and E_0 does not prove that both measurements are governed by the same jump rates, and the field may still be open for surprises. These may come from measurements using other techniques.

There have been a few NMR studies of hydrogen diffusion in metallic glasses when the hydrogen content has been sufficiently large [3.45, 46, 3.126]. When the temperature dependence of the τ_c^{-1} parameters derived from BPP analyses [3.45] of the proton $T_{1\rho}$ data for the glass a-TiCuH$_{1.41}$ is compared in Fig. 3.19 with the results for the crystalline hydrides TiCuH$_{0.94}$ and TiH$_{1.90}$, an enhanced hydrogen mobility with reduced activation energies is clearly evident, as is the distinct non-Arrhenius character for the amorphous hydride. Very similar behaviour has also been observed [3.46] for crystalline and amorphous Zr$_2$PdH$_x$. Although the protons preferentially occupy tetrahedral sites in both TiCuH$_x$ and Zr$_2$PdH$_x$, variations in hydrogen site occupancies or structure-sensitive changes in the allowed diffusion jump paths (e.g. the presence of octahedral intermediate sites) are assumed to be responsible for this behaviour. The smaller U values derived from the lower temperature $T_{1\rho}$ data may be due to protons in less stable sites, or represent specific variations in the diffusion saddlepoint configurations for the presumably dominant tetrahedral sites. As more information on the distributions of occupied proton sites is obtained for these amorphous hydrides, it should be possible to determine which cause is responsible.

Particularly interesting newer results stem from NMR measurements on amorphous Zr$_3$RhH$_{3.5}$ by *Markert* et al. [3.138] and amorphous ZrV$_2$(H, D)$_x$ by *Rychkova* et al. [3.139] and the permeation studies on amorphous and crystalline Ti–Ni alloys [3.140]. Various workers [3.127, 140, 141] have observed that when diffusion coefficients are compared between crystalline and amorphous phases with nominally identical compositions the hydrogen mobility in glasses is substantially enhanced (i.e. by factors of two-to-four orders of magnitude). However, the NMR data of *Rychkova* et al. [3.139] indicated that these observations are not universally valid since a *reduced* rate of hydrogen diffusion occurs in amorphous ZrV$_2$(H, D)$_x$ relative to that seen in the corresponding crystalline hydrides with the C15 Laves crystal structure. On the

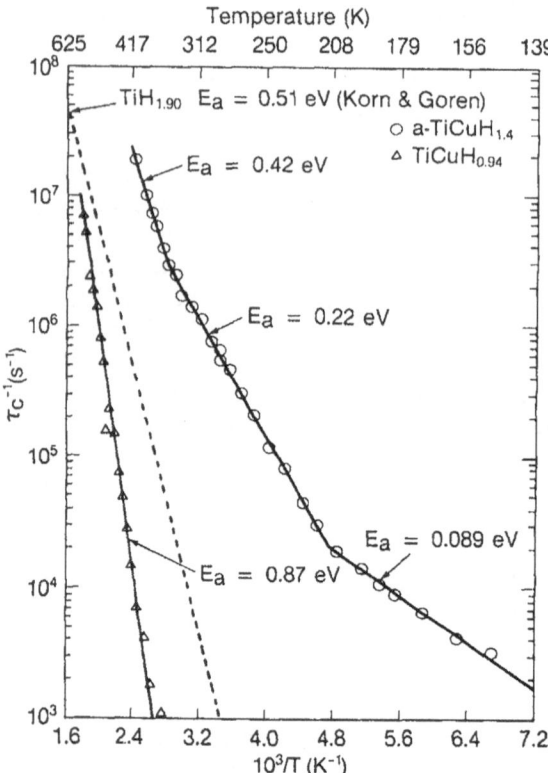

Fig. 3.19. Comparison of the proton jump rates τ_c^{-1} for crystalline and amorphous TiCuH$_x$ derived from the T_1 relaxation times. Also shown are the results for crystalline TiH$_{1.90}$. (From the $T_{1\rho}$ data of [3.137] and the TiCuH$_x$ data of [3.45])

other hand, hydrides possessing the Laves structures generally exhibit some of the highest hydrogen diffusion rates [3.47, 106–110, 142, 143] of all known intermetallic compounds and alloys. Whereas disorder in hydrogen site occupancies and local geometries could permit more rapid motion in those crystalline phases which restrict easy jump paths [3.126], similar disorder in structures where the diffusion barriers are originally quite low (i.e. the C15 and C14 Laves and hexagonal AB$_5$H$_x$ phases) could have the opposite effect so that diffusion becomes more difficult, as appears to be the case [3.139] for amorphous ZrV$_2$(H, D)$_x$. It would be most illuminating to investigate other amorphous alloy hydrides where the hydrogen mobilities in their corresponding crystalline phases are very high.

When hydrogen diffusion parameters for intermetallic or amorphous alloys are determined over limited temperature ranges (i.e. ≤ 50–100 K), the data are usually well represented [3.126, 127, 140] by the Arrhenius relation with a specific activation energy E_a. However, techniques that can monitor motion over wider temperature ranges (i.e. the NMR relaxation times) often exhibit very anomalous temperature dependencies [3.143, 126], which cannot be described by simple Arrhenius models. During the past few years, two alternative concepts have been proposed to account for such deviations. One view is that the microscopic diffusion processes can consist of two basic kinds of jump: (1) relatively rapid motion—usually with smaller energies—within localized regions of the lattice that do not directly lead to long distance transport and may involve only a portion of the hydrogen atoms and a subset of the allowed interstitial sites; and (2) jumps that permit movement of hydrogen between these regions which will determine the long distance diffusion properties. Examples of this descriptive scheme are presented in this chapter and elsewhere [3.126, 128]. However, an alternative premise is to assume that various types of lattice disorder can lead to distributions of activation energies about the mean value E_a, which would represent distinct jumps among equivalent sites in a defect-free host lattice. *Kirchheim* and co-workers [3.144] have applied variations of a Gaussian distribution model for the E_a values to explain hydrogen diffusion characteristics in amorphous alloys. Similar Gaussian distributions were also used to analyze the NMR relaxation times obtained from crystalline $V_xNb_{1-x}H_{0.2}$ [3.145], $Zr_{1-y}Hf_yV_2H_x$ [3.143] and $ZrCr_2H_{4.2}$ [3.142] as well as some amorphous hydride samples [3.126]. It was possible in most cases to generate satisfactory fits to the experimental data with reasonable values for E_a and for the widths of the distribution energy. However, *Markert* et al. [3.138] required an asymmetric distribution function to improve fits of their proton T_1 and $T_{1\rho}$ relaxation times from glassy a-$Zr_3RhH_{3.5}$, but even with this version they could not remove all disparities. Furthermore, *Bowman* et al. [3.126, 146] have described other inconsistencies that arise when distribution models are used to represent more extensive sets of NMR relaxation times from crystalline and glassy hydrides. The recent PAC results [3.141] for amorphous Zr_2NiH_x do not reveal the presence of a broad distribution of activation energies. However, the strongest evidence for the inadequacy of Gaussian distribution models comes from neutron scattering measurements [3.128, 129] and Monte Carlo simulations [3.130] on hydrogen diffusion in amorphous $Pd_{1-y}Si_yH_x$ alloys. The QNS results dictate a bimodial distribution of widely separated H-site energies that correspond to very different regimes for the jump rates. This behavior arises quite naturally from the simultaneous H-site occupancies in distorted Pd_6 octahedral and various tetrahedral sites as deduced from neutron vibrational spectra [3.129]. Consequently, prior applications [3.132] of Gaussian distributions to represent the microscopic diffusion mechanisms in this system as well as other disordered hydrides must be treated with caution. Resolution of the specific contributions of jump rate distributions versus multi-step processes is a crucial issue in hydrogen dynamics that warrants sustained and comprehensive investigations.

3.4 Hydrogen Vibrations

3.4.1 Overview

This chapter is—as mentioned in Sect. 3.2.4.6—restricted to optical or localized hydrogen vibrations, which can be uniquely accessed by neutron vibrational spectroscopy (NVS). Essentially four types of information about metal—hydrogen systems can be obtained:

i) The frequencies of the hydrogen vibrations are related to the hydrogen potential. Considerable progress has been achieved recently for binary metal hydrides [3.63], particularly by comparing the fundamental vibrations of the hydrogen isotopes with their higher harmonics, which have been measured up to fifth order at neutron spallation sources [3.147]. The deviations of the hydrogen potential from a harmonic (parabolic) shape have important implications for the detailed understanding of, for example, thermodynamic properties. Unfortunately this type of INS investigation has not yet been performed on intermetallic hydrides; therefore in Sect. 3.4.2 we shall describe the relation between localized mode frequencies and hydrogen potential for the example of α-PdH$_{0.01}$.

ii) The hydrogen vibrations in concentrated hydrides can be understood in terms of optical phonons. Their dependence on the reduced wave vector q results from H–H interactions. Intermetallic hydrides have not yet been investigated in this respect because the necessary single crystalline samples are not available. For reviews on binary metal–hydrogen systems we refer to the literature [3.63, 149, 150]. The q dependence of the "optical phonons" is absent in dilute systems (α-phases) and there the hydrogen vibrations can be assumed to be localized and to originate from three-dimensional Einstein oscillators.

iii) The degeneracy of the vibrational modes depends on the point symmetry of the corresponding interstitial hydrogen site. In simple cases this qualitative argument already enables statements to be made about the type of interstitial site occupied by the hydrogen atoms. In a bcc metal, for instance, both the tetrahedral and the octahedral interstices are tetragonally distorted, i.e. one expects the vibrational spectrum to consist of two modes with the intensity ratio 1:2, but for the tetrahedral site the mode at higher frequency is doubly degenerate, whereas for the octahedral site this is the case for the mode at lower frequency. Additionally, owing to its larger size, an octahedral site generally gives rise to lower vibrational frequencies. Most of the localized mode measurements on intermetallic hydrides, on amorphous metal hydrides and on metal–oxide hydrides (hydrogen bronzes), have been interpreted in these terms. These will be reviewed in Sect. 3.4.3, together with a recent more quantitative attempt to assign the different vibrational peaks.

iv) The fourth type of information that can be obtained is the hydrogen occupancy of a certain site which is—apart from in spectroscopic methods based on electromagnetic radiation—directly evident from the intensity of the corresponding vibrational modes. The fraction of hydrogen at a certain site characterizes the geometric structure of the hydride, of course, but it also gives an insight into the energy structure. In a system with two different sites, for instance, the changes in occupation with temperature enable a determination of the energy difference between these two sites, as will be outlined in Sect. 3.4.4.

3.4.2 Hydrogen Potential

Using NVS, the hydrogen potential has been investigated in some bcc metals [3.69, 70, 150–152], and quite recently in α-PdH$_x$ [3.153]. An important breakthrough has been achieved in the theoretical understanding of hydrogen potentials. Ho et al. [3.154], using frozen phonon calculations, were able to calculate the hydrogen potential in β-NbH from first principles. The agreement with the measured anharmonicity of the H potential [3.69, 155] is remarkable. Because of its simplicity (cubic symmetry, few anharmonicity parameters) and its importance as a prototype metal–hydrogen system, the results on α-PdH$_x$ will be described in the following.

We consider a hydrogen atom at the octahedral site. If we allow for small deviations from harmonicity, using only the lowest order nonvanishing anharmonic terms, the hydrogen potential can be described phenomenologically as

$$V(x, y, z) = c_2(x^2 + y^2 + z^2) + c_4(x^4 + y^4 + z^4) + c_{22}(x^2y^2 + y^2z^2 + z^2x^2),$$
$$(3.53)$$

where the second term only influences the shape of the potential, whereas the third term also brings about a coupling between the vibrations in the $x, y,$ and z directions (for symmetry reasons cubic terms do not appear). If the corrections to the harmonic shape of the potential are sufficiently small, the energy eigenvalues for a particle of mass m oscillating in this potential are obtained from first order perturbation theory as

$$E_{nml} = \sum_{\substack{j=n \\ m \\ l}} \{ \hbar\omega_0(j + \tfrac{1}{2}) + \beta(j^2 + j + \tfrac{1}{2}) + \gamma[(2n + 1)(2m + 1)$$

$$+ (2m + 1)(2l + 1) + (2l + 1)(2n + 1)] \}, \qquad (3.54)$$

where

$$\omega_0 = \sqrt{\frac{2c_2}{m}}, \quad \beta = \frac{3\hbar^2 c_4}{4mc_2}, \quad \gamma = \frac{\hbar^2 c_{22}}{8c_2 m},$$

and where n, m and l are the quantum numbers for the vibrations in $x, y,$ and z directions respectively. Hence the lowest excitation energies $\varepsilon_{nml} = E_{nml} - E_{000}$ are:

$$\varepsilon_{100} = \varepsilon_{010} = \varepsilon_{001} = \hbar\omega_0 + 2\beta + 4\gamma \quad \text{(fundamental vibration)}$$
$$\varepsilon_{200} = \varepsilon_{020} = \varepsilon_{002} = 2\hbar\omega_0 + 6\beta + 8\gamma \quad \text{(first overtone)}$$
$$\varepsilon_{110} = \varepsilon_{011} = \varepsilon_{101} = 2\hbar\omega_0 + 4\beta + 12\gamma \quad \text{(combination vibration)}. \qquad (3.55)$$

Experimentally, vibrational modes of H in Pd H$_{0.005}$ were observed at 69 meV, 138 meV and 156 meV (Fig. 3.20). From a comparison with (3.55), the parameters

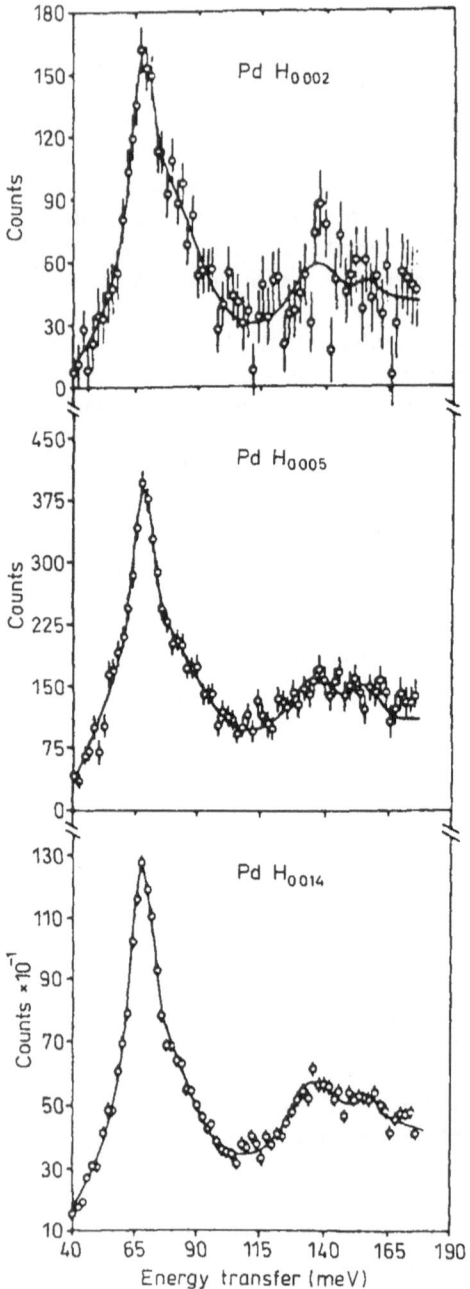

Fig. 3.20. Vibrational spectra of H in α-PdH$_x$ at 298 K. Note that for all concentrations there is a peak at 69 ± 0.5 meV and a second broad maximum around 138 meV with a shoulder at higher energy (clearly shown in PdH$_{0.014}$). The lines are a result of a fit to the data, which gives values of 137 and 156 meV for the excited state levels [3.153]

$\hbar\omega_0$, β and γ could be determined for the Pd–H α-phase, and resulted in $\hbar\omega_0 = 50\,\text{meV}$, $\beta = 9.5\,\text{meV}$, and $\gamma = 0$. A vanishing γ value indicates a negligible coupling of the vibrations in different directions in Pd–H. A positive β value and thus a positive c_4 in (3.53) means that in the energy range of the levels involved in the neutron scattering experiment, the hydrogen potential is steeper than a harmonic potential. The harmonic frequency ω_0 and the anharmonicity parameter β, both corrected for the isotope mass ratio, also satisfactorily describe the fundamental frequency of α-PdD$_x$. Therefore there is no evidence for different electronic forces for the two hydrogen isotopes.

These results have important consequences for understanding the super-conductivity of the Pd–H(D) system. It has been shown [3.156] that the existence of relatively high T_c's is in this system ($T_c = 9\,\text{K}$ for PdH$_{1.0}$ and $T_c = 11\,\text{K}$ for PdD$_{1.0}$) is the result of a strong electron–phonon coupling to the optic modes. As the origin of the anomalous isotope effect in T_c, differences in the electronic band structure (as proposed in [3.157]) can now be ruled out. This is in agreement with *Wicke* et al. [3.158] who have recently measured the magnetic susceptibility of PdH$_x$ and PdD$_x$ above the critical point for the concentration range $0 \leqq x \leqq 0.55$, and who do not observe any electronic difference between the Pd–H and Pd–D system. Instead the NVS experiment confirms the view of *Ganguly* [3.159, 160] that the anharmonicity of the optical hydrogen vibrations brings about the anomalous T_c.

Since the energy levels of the anharmonic oscillator H in Pd differ from those of an harmonic oscillator, the oscillatory partition function also differs

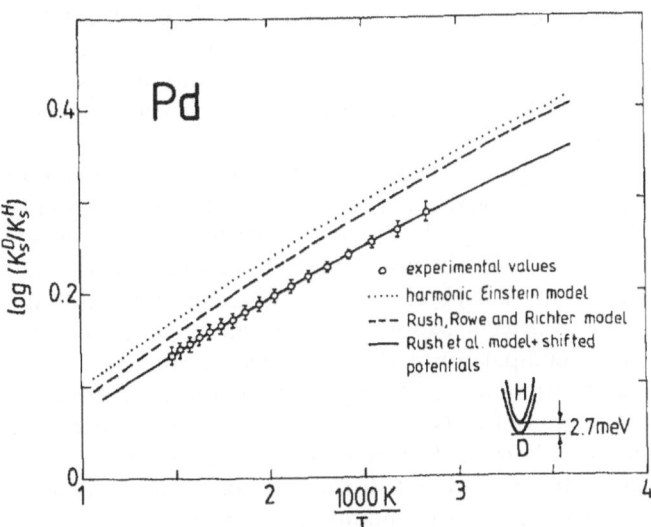

Fig. 3.21. Comparison of experimental values of the ratio of the Sieverts' constants for D and H, as obtained from p–c–T data, with thermodynamic calculations using different models for the H/D potential [3.162]

from the harmonic value; it can be calculated numerically as

$$Z = \left[\sum_n \exp(-\varepsilon_n/kT) \right]^3 \qquad (3.56)$$

with

$$\varepsilon_n = \hbar\omega_0(n + \tfrac{1}{2}) + \beta(n^2 + n + \tfrac{1}{2}). \qquad (3.57)$$

For the oscillatory part of the entropy at 400 K, for example, this gives the result $1.25\,k_B$, instead of the harmonic value $1.37\,k_B$. The total entropy of solution, as calculated for instance by *Magerl* et al. [3.161], diminishes by about 5%. (This effect increases with decreasing temperature.) Thus it is evident that for a quantitative microscopic understanding of thermodynamic properties of metal–hydrogen systems, the anharmonicity of the hydrogen potential must be taken into account. Recent thermodynamic (p–c–T) measurements by *Lässer* [3.162] (Fig. 3.21) have indeed shown that the agreement between theory (statistical thermodynamics) and experiment is improved when the anharmonicity of the potential is taken into account.

3.4.3 Hydrogen Site Determination

a) Hydrogen in Intermetallic Compounds

A considerable number of intermetallic hydrides have been investigated using NVS; examples are $LaNi_5H_x$ [3.163–166, 71], $TeFeH_x$ [3.167, 168], TiCuH [3.169], $ZrCoH_x$ [3.170], $ZrNiH_x$ [3.171, 172], $Ti_{1.2}Mn_{1.8}H_3$ [3.173], $ZrBe_2H$ [3.174], Zr_2NiH_x [3.175]. We will not describe all the results in detail, but instead, for a few examples, discuss characteristic and (we believe) general features of NVS on intermetallic hydrides.

The localized modes of H in $LaNi_5H_6$ have been measured by several authors [3.71, 163–166] and are shown in Fig. 3.22. The neutron scattering intensity is rather broad and separate peaks are not visible. Similar large widths for the localized modes are observed in practically all intermetallic hydrides and may be partially due to mutual H–H interactions (dispersion), although irreversible activation-induced defects also make a substantial contribution to the linewidth in $LaNi_5H_6$. A comparison of the two spectra in Fig. 3.22 clearly shows that absorption–desorption cycles tend to eliminate the structure in the spectra. Thus the hydrogenation of $LaNi_5$—and certainly also of other intermetallic compounds—not only destroys the macroscopic morphology (pulverization) but also introduces strain and severe distortions at an atomic level in the bulk material which are locally probed by the homogeneously dissolved hydrogen atoms.

The localized hydrogen modes in $FeTiH_x$ have been investigated by *Eckert* et al. [3.167] and—more extensively—by *Shapiro* et al. [3.168]. Figure 3.23

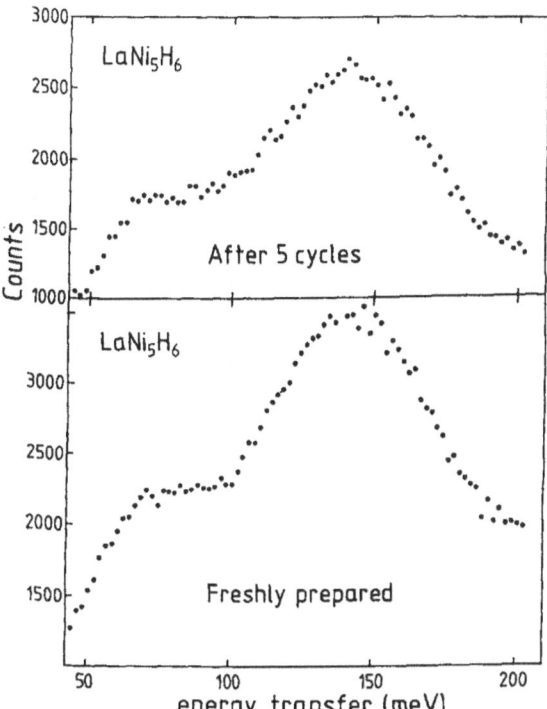

Fig. 3.22 Localized hydrogen vibrations in β-LaNi$_5$H$_6$: *top* activated sample, *bottom* hydrogenated once without cycling [3.71]

shows some of Shapiro's spectra, recorded at 10 K: spectra of FeTiH$_{0.06}$ (what they believe to be an α-phase), and of β_1-FeTiH$_{0.94}$ and β_2-FeTiH$_{1.4}$. The main peak in the γ-phase appears at about the same energy is in the "α-phase" and is of similar width. These spectra vary considerably for the different FeTi–hydride phases, and were qualitatively discussed in terms of the well-known structures of FeTiH$_x$ [3.176]. A conclusive interpretation of the γ-phase results was not possible. For thermodynamic reasons, the existence of FeTiH$_x$ α-phase at 10 K is doubtful. A similarly broad spectrum was observed by *Hempelmann* et al. for activated α-LaNi$_5$H$_{0.36}$ [3.71] and attributed to hydrogen atoms trapped at activation-induced defects. If, however, an α-phase sample is prepared without ever being transformed to any hydride phase, then well-defined vibrational peaks are observed, e.g. in ZrNiH$_{0.6}$ [3.172] and in LaNi$_5$H$_{0.15}$ [3.71] (Fig. 3.24). In the latter case an initial quantitative investigation has been undertaken in order to assign the different vibrational peaks to certain hydrogen sites. The frequencies of the localized vibrations of hydrogen in metals are obtained by solving the eigenvalue problem

$$\underline{D}\hat{u} = \omega^2 \hat{u}, \tag{3.58}$$

where the eigenvectors $\hat{u}$ indicate the directions of the vibrational displacements.

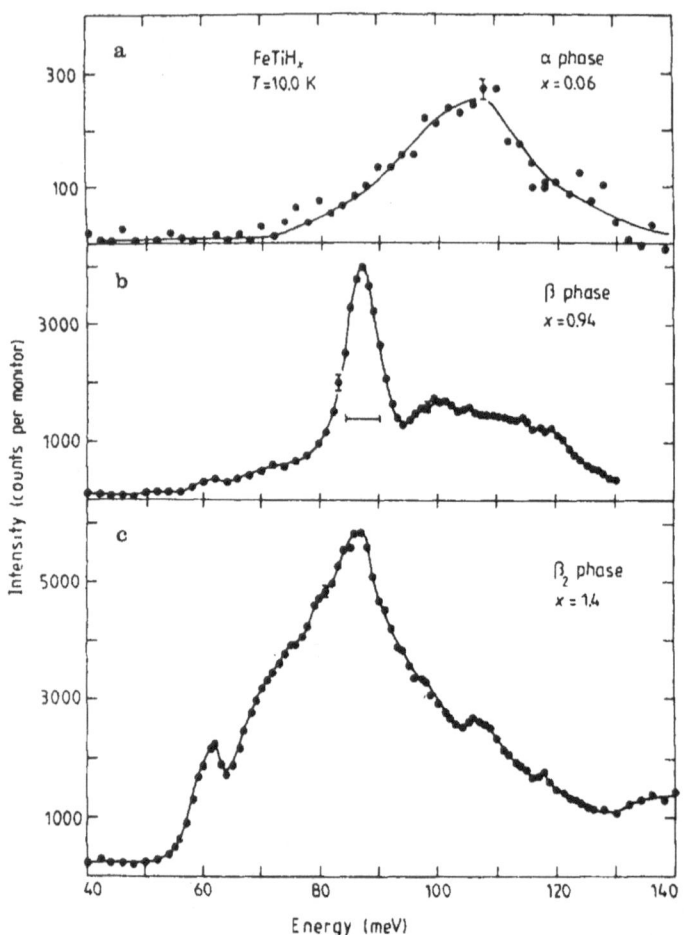

Fig. 3.23. Hydrogen vibration in FeTiH$_x$: (**a**) Vibrational spectrum of α-phase (FeTi$_{0.06}$). An energy-dependent background has been subtracted (**b**) Vibrational spectrum of β_1-phase (FeTiH$_{0.54}$) prepared by desorption of hydrogen (**c**) Vibrational spectrum of β_2-phase (FeTiH$_{1.4}$) prepared by absorption. The horizontal line indicates the size of the resolution (after [3.168])

To a good approximation we can regard the metal atoms as completely immobile; then D is a 3×3 dynamic matrix. The elements D_{ij} of this matrix represent the force acting on the hydrogen atom in the direction i if it is displaced in the direction j. The force constants are generally expressed by appropriate second derivatives of the potential. The restoring force $\boldsymbol{F}$ can be approximately modelled by Z longitudinal springs with $s^n = f^n \hat{s}^n$, pointing towards the Z metal atoms of the first coordination shell (e.g. an octahedron):

$$\boldsymbol{F} = \sum_{n=1}^{Z} f^n(\hat{u} \cdot \hat{s}^n) \cdot \hat{s}^n, \tag{3.59}$$

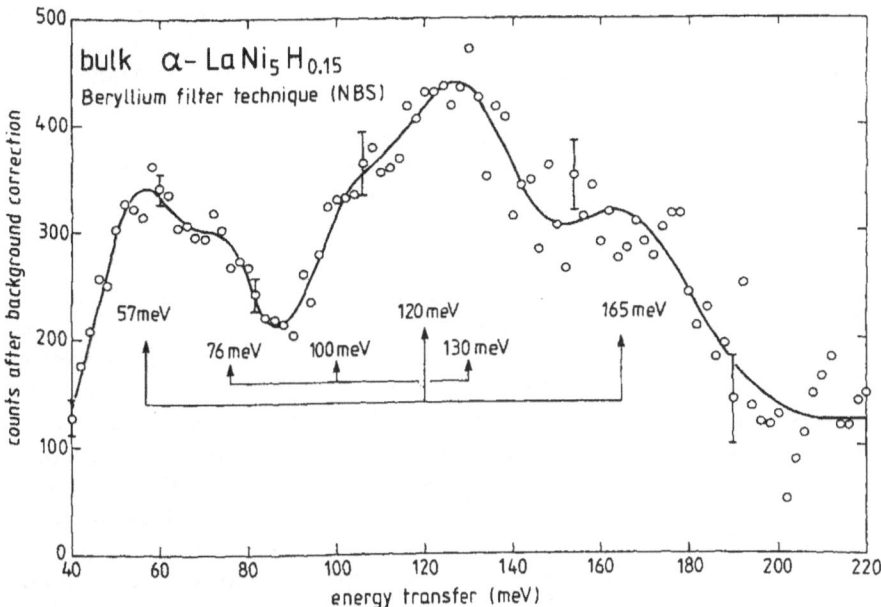

Fig. 3.24. Localized hydrogen vibrations in virgin bulk α-LaNi$_5$H$_{0.15}$ [3.71]

The vectors $\hat{s}^n$ indicate the directions of the Z springs; their magnitude is given by the distance-dependent force constants

$$f^n = f^M(r^M) = \left(\frac{c^M}{r^M}\right)^2,\tag{3.60}$$

where r^M is the distance between the hydrogen atom and the metal atom M, and the values c^M characterize the hydrogen–metal interaction and are adjustable parameters. The $1/r^2$ dependence of f was chosen in order to reproduce the $1/r$ dependence of the hydrogen frequencies. This dependence has been observed experimentally in a large variety of binary metal hydrides with the CaF$_2$ structure [3.177], and has also been proposed on the basis of theoretical considerations by *Sugimoto* and *Fukai* [3.179], who treated the hydrogen vibrations in group Vb metals quantum mechanically using a semi-empirical double Born–Mayer potential. Simple algebra transforms (3.59) into the form of (3.58) with

$$D_{ij} = \sum_{n=1}^{Z} \hat{s}_i^n \hat{s}_j^n f^n \quad i,j = 1,2,3.\tag{3.61}$$

For α-LaNi$_5$H$_{0.15}$ the atomic coordinates and the hydrogen–metal distances for all hypothetical sites were obtained from the well-known structure of the host lattice; thus with two adjustable parameters, c^{La} and c^{Ni}, a frequency pattern could be calculated and compared with the experimental spectrum (Fig. 3.24), which consisted of at least five vibrational modes. The result of the comparison was that hydrogen in α-LaNi$_5$H$_{0.15}$ occupies the octahedral $3f$ and the tetrahedral $6\,m$ sites.

The same formalism was applied to neutron vibrational spectra taken on ZrV$_2$H$_x$, $0.5 \leqq x \leqq 4.5$ [3.178]. It was concluded that, in addition to Zr$_2$V$_2$ and ZrV$_3$ tetrahedral sites, which are occupied at low and intermediate hydrogen concentrations, V$_4$ tetrahedral sites are accessible to hydrogen at the highest possible hydrogen contents.

An important aid in the assignment of the different vibrational modes to certain hydrogen sites is a consideration of the intensities of the modes. However, in addition to the problems mentioned in Sect. 3.2.4b, for strongly anisotropic sites anharmonic effects can also strongly influence the vibrational intensities. This was pointed out by *Benham* et al. [3.172], who studied ZrNiH$_x$ with x values between 0.6 and 2.7. In the α-phase with $x \leqq 0.6$, hydrogen occupies a site whose geometry is that of a distorted square pyramid, with the hydrogen situated close to the base and with an adjacent site on the other side of the basal plane. Thus the authors expect a wide and very anharmonic potential well, and consider this a plausible explanation for the factor of five attenuation in intensity that they observed for the corresponding vibrational mode. In the β-phase of the ZrNi-H system, the hydrogen atoms occupy two other sites which are only slightly anisotropic; in these cases the intensities are much less problematic.

b) Hydrogen in Amorphous Metals

The investigation of the structure and interatomic bonding of metal glasses has attracted increasing scientific interest. Structural investigations are mostly performed with X-rays, and recently also by neutron scattering [3.180], and can be interpreted only in terms of pair correlation functions. However, many questions, in particular on the local topology in the glassy state, remain open. Since the H vibrations are sensitive essentially to the immediate surroundings of the proton, they are well suited local probes for investigating this problem.

The first such measurements were reported by *Rush* et al. [3.169], who compared the vibrational spectra of glassy and crystalline TiCuH. Their results are shown in Fig. 3.25. Large differences between the density of states distributions for the two samples are evident. The crystalline sample shows a narrow frequency distribution with features at 142 and 157 meV and is very similar to that of γ-TiH$_2$. Since in both structures H occupies regular tetrahedral sites with four Ti next-nearest neighbours, this similarity is not surprising. In the amorphous metal the peak occurs at the same frequency, but with a strongly increased width (75 meV). Thus, on the average, the H sites occupied in the amorphous and crystalline structures are very similar. The large width indicates that the tetrahedron may be heavily distorted and fluctuates in its chemical composition. The wing towards frequencies below 100 meV may also show that

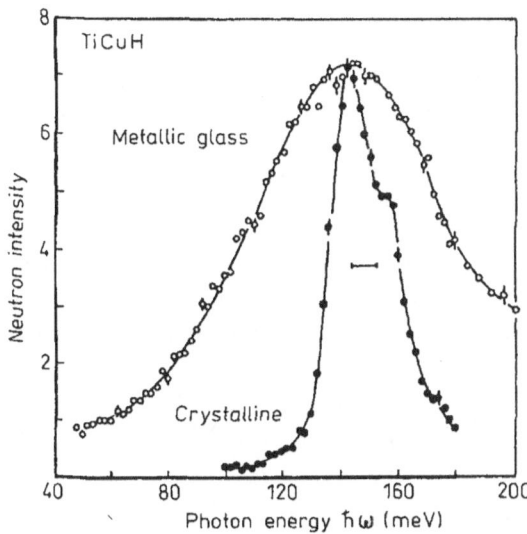

Fig. 3.25. Neutron spectra measured at 78 K for crystalline $TiCuH_{0.93}$ and amorphous $TiCuH_{1.3}$. The energy resolution (FWHM) near the peak is indicated by the horizontal bar [3.169]

octahedral site occupation occurs as well. The observation of a broader density of states disagrees with the microcrystalline or microcluster model of a metglass, in which the local environment would not differ from that of the crystal.

Later experiments on other glasses such as ZrNiH [3.171, 181–183], Ti_2NiH [3.182], and Zr_2Pd [3.184], qualitatively all revealed the same result: the local H spectra in the amorphous substance centre around similar frequencies to those in the crystalline state, but the frequency distributions are generally much broader and washed out. Thus, on the basis of these measurements, we conclude that in amorphous metals the H atoms remain essentially at the same polyhedral sites as in the crystalline reference substance; changes in topology are therefore rather restricted.

At very low H content, however, reasonably well-resolved spectra were obtained for $Pd_{85}Si_{15}H_x$ with x down to 0.13 [3.129], which demonstrates the existence of a complex distribution of site energies with at least two classes of occupied sites. From the vibrational frequencies these were identified as octahedral and tetrahedral sites.

c) Hydrogen at Metal Surfaces and in Metal Oxides

In order to be dissolved in a metal, hydrogen has first to be adsorbed and chemisorbed at the metal surface, and thereafter to penetrate through a surface layer, which in most cases consists of a metal (sub)oxide. For the storage material FeTi, for example, the ordered suboxides $FeTiO_x$, $Fe_7Ti_{10}O_3$ [3.185], and Ti_4Fe_2O [3.113, 114], were observed and discussed as being involved in the dissociation process of the H_2 molecule at the surface.

In comparison with other surface techniques, the most valuable feature of NVS is its ability to highlight the vibration of hydrogen atoms. It is therefore a suitable technique for studying hydrogen chemisorbed on metals [3.186, 187], although it is still necessary to use high surface area samples to have sufficient hydrogen in the neutron beam. An example is hydrogen on Raney nickel [3.188, 189]. This is a material of high specific area, made by the reaction of a nickel/aluminium alloy with potassium hydroxide. It contains about 4% aluminium and is widely used as a catalyst for hydrogenation. The neutron spectrum of hydrogen on Raney nickel, shown in Fig. 3.26, has a strong doublet at energies of 120 and 140 meV, and a broad weak band at 240–280 meV not observed by *Cavenagh* et al. [3.189]. The intensity of the latter feature is slightly greater than one would expect for the second harmonic of the main doublet, and in addition the q dependence of the scattering intensity is nearly quadratic, whereas

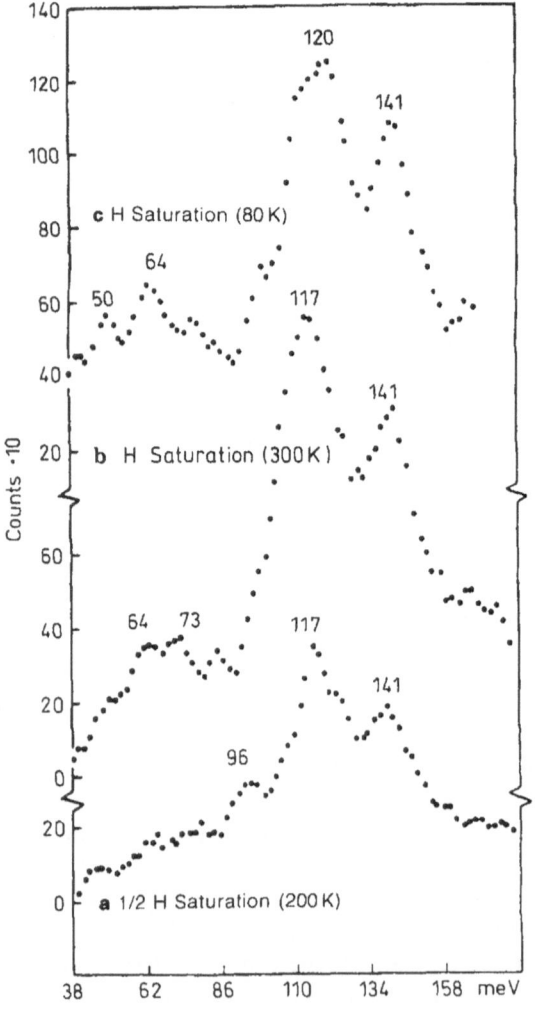

Fig. 3.26. Neutron vibrational spectra due to hydrogen on Raney nickel, recorded at 80 K: **(a)** 0.5 of saturation hydrogen coverage, sample annealed at 200 K prior to recording spectrum, **(b)** saturation hydrogen coverage, sample annealed at 300 K prior to recording spectrum, **(c)** saturation hydrogen exposure, with sample held at 80 K (after [3.189])

a two-quantum process should vary as q^4. Therefore, near 280 meV it is suspected that there is another weak fundamental.

A force field calculation with a simple empirical model of the Ni–H stretching force constants shows that large differences are to be expected between the frequencies of hydrogen terminally bound to a single nickel atom, and those of multiply bound hydrogen. For hydrogen bound to three nickel atoms in a trigonal-pyramidal configuration, two stretching frequencies of symmetry A and E are expected in the region 120–140 meV, whereas a terminally-bound hydrogen would have a frequency of about 275 meV. The dominant contribution to the spectrum is therefore from multiply bound hydrogen; the observed two to one intensity ratio for vibrational modes is in accordance with the trigonal–pyramidal site and inconsistent with a tetragonal–pyramidal site, which can also be ruled out for other reasons [3.189]. A small fraction of the hydrogen on Raney nickel is terminally bound, accounting for the weak band at 275 meV. Hydrogen chemisorbed by platinum and palladium has also been investigated [3.190, 191], but in much less detail than that chemisorbed by Raney nickel, and will not be discussed here. To the knowledge of the authors, hydrogen chemisorbed on intermetallic compounds has not yet been studied by NVS. It is, however, anticipated that part of the scattering intensity of activated α-phase samples is due to surface hydrogen.

The study of hydrogen in metal (sub)oxides is at a very early stage [3.192] and this holds particularly for NVS experiments. Some stoichiometric metal oxides, such as WO_3 or MoO_3, which themselves are semiconductors, exhibit an Anderson-type non-metal–metal transition upon hydrogenation [3.193]. Inelastic neutron scattering spectra of metallic $H_{0.4}WO_3$ [3.194] and H_xMoO_3 $(0.3 \leqq x \leqq 2.0)$ [3.195] show only vibrations that would be expected from a metal hydroxide. We therefore restrict ourselves here to metal suboxides, in which hydrogen enters interstitial sites with a purely metallic surrounding. An illustrative example is the recent work of *Hirabayashi* and co-workers who have investigated $ZrO_{0.4}H_{0.1}$ [3.196] and $TiO_{0.3}H_{0.1}$ [3.196, 197]. The oxygen atoms in these metal–oxygen solid solutions are located at the octahedral sites, and tend to form ordered structures below about 500 °C [3.198, 199]. It is known [3.200] that oxygen increases the solubility of hydrogen in α-Zr, but decreases it in β-Zr. Figure 3.27 shows the vibrational spectrum of hydrogen in $ZrO_{0.4}H_{0.1}$. Clearly a fundamental mode is observed at about 135 meV, whereas the corresponding mode in $TiO_{0.3}H_{0.1}$ appears near 86 meV. These results indicate that hydrogen in the α-phase of $ZrO_{0.4}$ occupies tetrahedral sites, whereas in $TiO_{0.4}$ it occupies octahedral ones. The same behaviour has been found by *Hempelmann* et al. [3.201] for oxygen-free Zr and Ti. For the quantitative evaluation, the vibrational mode in $ZrO_{0.4}H_{0.1}$ at about 135 meV was fitted by a superposition of two Lorentzians at 129 and 136 meV with the intensity ratio 1:2. This result is not surprising because two excitation energies (one of which is doubly degenerate), are expected for the tetrahedral site in $ZrO_{0.4}H_{0.1}$; the observed axial ratio of $c/a = 1.605$ is less than that of the ideal hcp lattice.

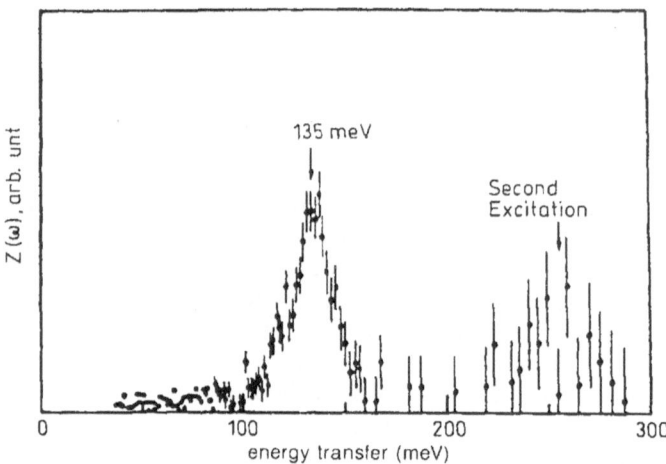

Fig. 3.27. Frequency distribution (density of states) curve $Z(\omega)$ of $ZrO_{0.4}H_{0.1}$ [3.196]

A further hardly investigated class of materials are oxygen-stabilized intermetallic compounds, the best known examples being η-carbides, e.g. Ti_4Ni_2O [3.202] or $Zr_3V_3OD_x$ [3.203]; surely in the future NVS can and will contribute considerably to the microscopic understanding of hydrogen in these systems.

3.4.4 Hydrogen Occupancy

An instructive example of the capability of NVS to detect quantitatively the occupancy of a certain interstitial site is the work by *Richter* et al. [3.204] on hydrogen trapping in $Nb_{0.99}Cr_{0.01}H_{0.009}$. Figure 3.28 presents the outcome of this experiment and shows the gradual intensity transfer from a high-temperature peak around 106 meV to a higher energy peak at lower temperatures. A second peak develops near 123 meV and grows with decreasing temperature. At 210 K the separate peaks are no longer distinguishable. Below 210 K a new centre of vibrational intensity develops at around 177 meV, which dominates the intensity at low T.

A careful analysis of the data showed that in NbCrH both trapping and precipitation of the hydride phase takes place; i.e. the binding energy ΔE at the trap is smaller than the enthalpy of the precipitate ΔH. A simple thermodynamical approach was used to model this situation: each impurity was associated with one trapping site [3.115], the potential energy of which is lowered by ΔE with respect to the undisturbed lattice. In thermal equilibrium the chemical potential of trapped and dilute protons must be equal

$$kT\ln(c_d/6) = -\Delta E + kT\ln\left(\frac{c_t}{c_i - c_t}\right). \tag{3.62}$$

c_d and c_t are the H concentrations in the dilute phase and the trapped state respectively, and c_i and is the impurity concentration. The factor 1/6 enters

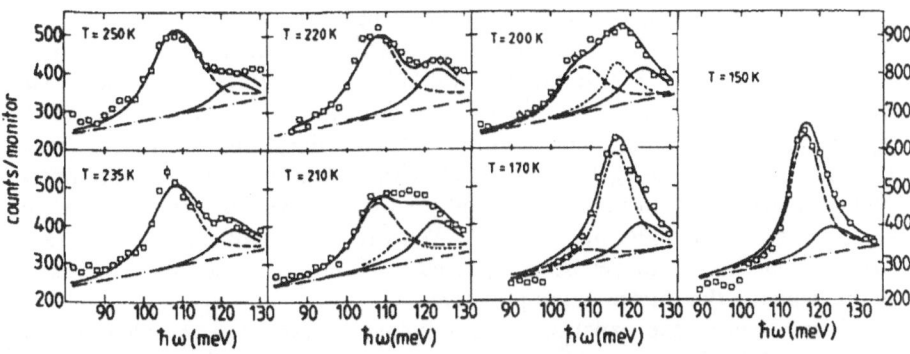

Fig. 3.28. Temperature dependence of the intensity distribution in the region of the lower fundamental vibration in $Nb_{0.99}Cr_{0.01}H_{0.009}$. The upper line represents the result of a fit with the combined trapping and precipitation model (see text). The broken line represents the α-phase contribution, the dashed line the intensity from protons in the hydride phase, while the lower solid line displays the intensity originating from protons in the trap. Finally, the dot-dashed line shows the background level [3.204]

because the number of tetrahedral sites in Nb is 6 times higher than the number of Nb atoms. The concentration of trapped protons follows from this as

$$c_t = \tfrac{1}{2}(c_H + c_i + 6e^{-\Delta E/kT}) - \tfrac{1}{2}[(c_H + c_i + 6e^{-\Delta E/kT})^2 - 4c_Hc_i]^{1/2} \qquad (3.63)$$

If the solubility in the α-phase, c_α, becomes smaller than c_d, then c_d in (3.62) has to be replaced by c_α, and we get instead

$$c_t = \frac{c_\alpha c_i}{c_\alpha + 6\exp(-\Delta E/kT)}. \qquad (3.64)$$

The result of a fit with this model is indicated by the various lines in Fig. 3.28. It is evident that the combined trapping and precipitation model allows a good description of the experimental data. In particular, it accounts for the gradual growth of the trapping peak above 210 K, as well as for the sudden intensity increase of the central β-phase peak below 210 K. For the binding energy a value of $\Delta E = 105 \pm 10$ meV, smaller than the enthalpy of the hydride phase ($\Delta H = 120$ meV), was determined. The vibrational energy of the trapping peak amounts to $\hbar\omega_t = 123$ meV. Its shift from the α-phase value of 14% is the largest observed so far in the context of H trapping in Nb [3.152] and indicates a severely stiffened potential.

3.5 Conclusion and Outlook

We hope this review has demonstrated that from a scientific point of view, as well as from the standpoint of applications, there is a widespread interest in the dynamical behaviour of H in intermetallics and amorphous metals. It also shows that the investigation of H vibrations and transport phenomena is a very lively field. However, in spite of all these efforts, a detailed microscopic picture of H diffusion in these materials has not yet evolved. NMR experiments suffer inherently from the multi-jump nature of diffusion, and also because of their hypersensitivity to paramagnetic impurities, with the result that nuclear relaxation rates cannot easily be associated with details of the diffusion process. Neutron scattering, in principle, could give the answer, but this task requires single crystals which are not available. With respect to amorphous substances, QNS studies have not yet been reported, but they will encounter similar difficulties as for the intermetallic powders.

Nevertheless, the experiments conducted so far have revealed motional characteristics of hydrogen that appear to be common to all intermetallic compounds. We must always deal with a distribution of jump rates, including jumps between energetically different sites. Thus H-trapping phenomena must be taken into account in dealing with diffusion. Another common feature are the fast local H-jump phenomena that exist, but which do not contribute to long-range diffusion. They may originate from structural pecularities, as well as from H-correlation effects. As a consequence of the various H jumps present, long-range diffusion must be measured over distances which sample all these different motions. Experiments that are sensitive to local jumps only, such as standard NMR techniques, Mössbauer effect, or QNS at large momentum transfers, are bound to fail in predicting the correct macroscopic diffusion coefficient. FGSE-NMR and QNS at small q appear to be the only microscopic methods which have the potential to reveal the correct D value, though the discrepancies in the case of $LaNi_5H_6$ are disturbing. Most of the macroscopic methods are surface sensitive, and in the case of intermetallic powders they cannot be expected to yield reliable values for D. For met glasses, however, which exist as ribbons, the Gorsky effect presumably provides the easiest and most reliable method for determining D.

For the near future, QNS experiments on some prototype α-phase single crystals are anticipated to shed more light on the details of the diffusive process, though the effects of H concentration (viz. site blocking, trap saturation and correlation effects) cannot be approached in this way. These must first be understood in simpler systems. The development and more widespread use of NMR-FGSE techniques will also help obtain more information on long-range H diffusion.

With respect to vibrational spectroscopy, investigations of intermetallics and metglasses are only just beginning. Systematic studies of the H potential, such as have been conducted on fcc and bcc binary metal-H systems, have not yet been reported for intermetallics. Here, experiments on virgin α-phase samples are expected to yield major insights. Attempts to obtain structural information on the basis of point symmetry considerations for vibrational modes have been published for α-$LaNi_5H_x$, and further experiments with similar objectives are anticipated. These also could be used as a diagnostic tool to reveal precipitation and segregation effects. Investigations of the thermal repopulation of H sites are not available so far for intermetallics and metglasses. They could reveal information on the energy differences between interstitial sites and would be most useful in combination with diffusion studies. Due to the short range of the metal–H interaction, the H vibrations are very sensitive to distortions of the neighbouring host atoms. First experiments on $LaNi_5H_6$ showed a considerable broadening of the H spectrum due to cycling. In the future protons at low concentrations might be used as local probes to monitor dislocations, grain boundaries, and lattice strains, using spectroscopic methods.

References

3.1   J. Völkl, G. Alefeld: In *Diffusion in Solids: Recent Developments*, ed. by A.S. Nowick, J.J. Burton (Academic, New York 1975) p. 231
3.2   J. Völkl, G. Alefeld: In *Hydrogen in Metalls I—Basic Properties*, ed. by G. Alefeld, J. Völkl, Topics Appl. Phys. Vol. 28 (Springer, Berlin, Heidelberg 1978) p. 321
3.3   A. Zeilinger, W.A. Pochmann: J. Phys. F **7**, 575 (1977)
3.4   E. Brauer, R. Gruner, F. Rauch: Ber. Bunsenges. Phys. Chem. **87**, 341 (1983)
3.5   G. Sicking, M. Glugla, B. Huber: Ber. Bunsenges. Phys. Chem. **87**, 418 (1983); B. Jungblut, G. Sicking: J. Less-Common Met. **101**, 373–382 (1984)
3.6   R.-Y. Lin, H.H. Johnson: J. Non-Crystalline Solids **51**, 45 (1982); Y. Sakamoto, K. Boba, W. Kurahashi, K. Takao, S. Takayama: J. Non-Crystalline Solids **61&62**, 691 (1984)
3.7   N. Boes, H. Züchner: Phys. Status Solidi (A) **17**, K111 (1973); N. Boes, H. Züchner: Z. Naturforsch. A **31a**, 754, 760 (1976)
3.8   G. Arnold, J.M. Welter: Metall. Trans. **14A**, 1573 (1983)
3.9   A. Heidemann, G. Kaindl, D. Salomon, H. Wipf, G. Wortmann: Phys. Rev. Lett. **36**, 213 (1976)
3.10   H. Wipf, A. Heidemann: J. Phys. C **13**, 5757 (1980)
3.11   P.J. West, D. Salomon: J. de Phys. (Paris) Coll. C2, **40**, C2-616 (1979)
3.12   U. Gonser, N. Blaes, R. Preston: Z. Metallkd. **73**, 605 (1982)
3.13   E.F. Wagner, R. Wordel, M. Zelger: J. Phys. F **14**, 535 (1984)
3.14   Y.S. Hwang, D.R. Torgeson, R.G. Barnes: Solid State Commun. **24**, 773 (1977)
3.15   M. Peretz, J. Barak, D. Zamir, J. Shinar: Phys. Rev. B **23**, 1031 (1981)
3.16   M.Yu. Belyaev, A.V. Skripov, V.N. Kozhanov, A.P. Stepanov, E.V. Galoshina: Solid State Commun. **48**, 1049 (1983)
3.17   A.S. Nowick, B.S. Berry: *Anelastic Relaxation in Crystalline Solids* (Academic, London 1972)
3.18   B.S. Berry, W.C. Pritchet: In *Nontraditional Methods in Diffusion*, ed. by G.E. Murch, H.K. Birnbaum, J.R. Cost (The Metallurgical Society of AIME 1983) p. 83
3.19   N. Boes, H. Züchner: J. Less-Common Met **49**, 223 (1976)
3.20   D.J. Pine, R.M. Cotts: Phys Rev. B **28**, 641 (1983); Rev. Sci. Instrum. **55**, 614 (1984)
3.21   H.G. Schöneich, H. Züchner: Ber. Bunsenges. Phys. Chem. **87**, 566 (1983)
3.22   R. Kirchheim, R.B. McLellan: J. Electrochem. Soc. **127**, 2439 (1980)
3.23   J.L. Snoek: Physica **6**, 591 (1939)
3.24   W.S. Gorsky: Z. Phys. Supplment **8**, 457 (1935)
3.25   G. Schaumann, J. Völkl, G. Alefeld: Phys. Rev. Lett. **21**, 891 (1968)
3.26   R.M. Cotts: Ber. Bunsenges. Phys. Chem. **76**, 760 (1972)
3.27   R.M. Cotts: In *Hydrogen in Metals I—Basic Properties*, ed. by G. Alefeld, J. Völkl, Topics Appl. Phys. Vol. 28 (Springer, Berlin, Heidelberg 1978) p. 227
3.28   R.C. Bowman, Jr.: In *Metal Hydrides*, ed. by G. Bambakidis (Plenum, New York 1981) p. 109
3.29   R.G. Barnes: In *Nuclear and Electron Spectroscopics Applied to Material Sciences*, ed. by E.N. Kaufmann, G. Shenoy (Elsevier, Amsterdam 1981) p. 189
3.30   E.F.W. Seymour: J. Less-Common Met. **88**, 323 (1982)
3.31   R.M. Cotts: In Proc. International Symposium on Electronic Structure and Properties of Hydrogen in Metals, ed. by P. Jena, C.B. Satterthwaite (Plenum, New York 1983) p. 451
3.32   H.T. Weaver: Phys. Lett. A **35**, 417 (1971)
3.33   R.C. Bowman, Jr., A. Attalla, B.D. Craft: Scripta Metall. **17**, 937 (1983)

3.34 T.T. Phua, R.G. Barnes, D.R. Torgeson, D.T. Peterson, M. Belhoul, G.A. Styles: In *Electronic Structures and Properties of Hydrogen in Metals*, ed. by P. Jena, C.B. Satterthwaite (Plenum, New York 1983) p. 467; M. Belhoul, G.A. Styles, E.F.W. Seymour, T.T. Phua, R.G. Barnes, D.R. Torgeson, D.T. Peterson: J. Phys. F: **12**, 2455 (1982); T.T. Phua, B.J. Beaudry, D.T. Peterson, D.R. Torgeson, R.G. Barnes, M. Belhoul, G.A. Styles, E.F.W. Seymour: Phys. Rev. B **28**, 6227 (1983)

3.35 R.C. Bowman, Jr., E.L. Venturini, B.D. Craft, A. Attalla, D.B. Sullenger: Phys. Rev. B **27**, 1474 (1983)

3.36 W.D. Williams, E.F.W. Seymour, R.M. Cotts: J. Mag Reson **31**, 271 (1978)

3.37 R.F. Karlicek, Jr., I.J. Lowe: J. Mag. Reson. **37**, 75 (1980)

3.38 L.D. Bustard, R.M. Cotts, E.F.W. Seymour: Phys. Rev. B **22**, 15 (1980)

3.39 A. Abragam: *The Principles of Nuclear Magnetism* (Oxford University Press, London 1961)

3.40 N. Bloembergen, E.M. Purcell, R.U. Pound: Phys. Rev. **73**, 679 (1948)

3.41 C.A. Sholl: J. Phys. C: **14**, 447 (1981)

3.42 Y. Fukai, S. Kazama: Acta Metall. **25**, 59 (1977)

3.43 R.F. Karlicek, Jr., I.J. Lowe: J. Less-Common Met. **73**, 219 (1980); H. Chang, I.J. Lowe, R.J. Karlicek, Jr.: In *Nuclear and Electron Resonance Spectroscopics Applied to Material Science*, ed. by E.N. Kaufmann, G.K. Shenoy (Elsevier, Amsterdam 1981), p. 331

3.44 T.C. Jones, T.K. Halstead: J. Less-Common Met. **73**, 209, 6362 (1982)

3.45 R.C. Bowman, Jr., A.J. Maeland, W.K. Rhim: Phys. Rev. B **26**, 6362 (1982)

3.46 R.C. Bowman, Jr., A. Attalla, A.J. Maeland, W.L. Johnson: Solid State Commun. **47**, 779 (1983)

3.47 R.C. Bowman, Jr., B.D. Craft, A. Attalla, J.R. Johnson: Int. J. Hydrogen Energy **8**, 801 (1983)

3.48 M. Yu, Belyaev, A.V. Skripov, V.N. Kozhanov and A.P. Stepanov: Sov. Phys. Solid State **26**, 1285 (1981)

3.49 P.A. Fedders: Phys. Rev. B **18**, 1055 (1978)

3.50 T.K. Halstead, K. Metcalfe, T.C. Jones: J. Mag. Reson. **47**, 292 (1982)

3.51 J. Shinar: J. Less-Common Met. **104**, 87 (1984); J. Shinar, D. Davidov, D. Shaltiel: Phys. Rev. B **30**, 6331 (1984)

3.52 L. van Hove: Phys. Rev. **95**, 249 (1954)

3.53 D. Richter, B. Alefeld, A. Heidemann, N. Wakabayashi: J. Phys. F **7**, 569 (1977)

3.54 See for example: *Topics in Current Physics*, Vol. 3, ed. by S.W. Lovesey, T. Springer (Springer, Berlin, Heidelberg 1977); C.G. Windsor: *Pulsed Neutron Scattering*, (Taylor and Francis, London 1981); R. Scherm, H. Stiller (eds.): Proc. *Workshop on Neutron Scattering Instrumentation for SNQ*, Jülich Rept. Jül–1954 (1984), KFA Jülich, D-5170 Jülich

3.55 C.T. Chudley, R.J. Elliott: Proc. Phys. Soc. **77**, 353 (1961)

3.56 J.M. Rowe, J.J. Rush, L.A. de Graaf, G.A. Ferguson: Phys. Rev. Lett. **29**, 1250 (1972)

3.57 K.W. Kehr, D. Richter: Solid State Commun. **20**, 477 (1976)

3.58 D.K. Ross, D.T. Wilson: IAEA Report SM-219/80, Vienna (1978); K.W. Kehr, R. Kutner, K. Binder: Phys. Rev. B **23**, 4931 (1981)

3.59 K.W. Kehr, K. Binder: In *Topics in Current Physics*, Vol. 36, (Springer, Berlin, Heidelberg 1984) p. 181

3.60 A.D. LeClaire: In *Physical Chemistry*, Vol. 10, ed. by H. Eyring, D. Henderson, W. Jost (Academic, New York 1970) p. 261

3.61 D. Richter, R. Hempelmann, O. Hartmann, E. Karlsson, L.O. Norlin, S.F.J. Cox, R. Kutner: J. Chem. Phys. **79**, 4564 (1983)

3.62 U. Potzel, J. Völkl, H. Wipf, A. Magerl: J. Less-Common Met. **103**, 182 (1984)

3.63 T. Springer, D. Richter: In *Methods in Experimental Physics*, ed. by K. Sköld and D. Price (Academic, New York 1985)

3.64 S.M. Shapiro, D. Richter, Y. Noda, H. Birnbaum, Phys. Rev. B **23**, 1594 (1981)

3.65 W. Bührer, A. Furrer, W. Hälg, L. Schlapbach: J. Phys. F **9** L141 (1979)

3.66 V. Lottner, H.R. Schober, W.J. Fitzgerald: Phys. Rev. Lett. **42**, 1162 (1979)

3.67 S.W. Lovesey: *Theory of neutron scattering from condensed matter*, (Clarendon, Oxford 1984)

3.68 B. Dorner, A. Griffin: J. Chem. Phys. **78**, 890 (1983)

3.69 D. Richter, S.M. Shapiro: Phys. Rev. B **22**, 599 (1980)

3.70 R. Hempelmann, D. Richter, A. Kollmar: Z. Phys. B **44**, 159 (1981)

3.71 R. Hempelmann, D. Richter, G. Eckold, J.J. Rush, J.M. Rowe, M. Montoya: J. Less-Common Met. **104**, 1 (1984)

3.72 R. Hempelmann, D. Richter: unpublished data

3.73 J. Tomkinson, M. Warner, A.D. Taylor: Molec. Phys. **51**, 381 (1984)

3.74 J. Howard, J.M. Nicol, B.C. Boland, J. Tomkinson, J. Eckert, J.A. Goldstone, A.D. Taylor: Mol. Phys. **53**, 323 (1984)
3.75 D. Richter: In *Neutron scattering and muon spin rotation*, Springer Tracts in Modern Physics **101**, 85 (1983)
3.76 R. Hempelmann: J. Less-Common Met. **101**, 69 (1984); Z. Phys. Chemie NF **154**, 221 (1987)
3.77 R.C. Bowman, Jr., A. Attalla, G.C. Carter, Y. Chabre: In *Hydrides for energy storage*, ed. by A.F. Andressen, A.J. Maeland (Pergamon, Oxford 1978) p. 67
3.78 D.L. Lindner: Inorg. Chem. **17**, 3721 (1978)
3.79 E. Lebsanft, D. Richter, J. Töpler: J. Phys. F **9**, 1057 (1979)
3.80 R.C. Bowman, Jr., W.E. Tadlock: Solid State Commun. **32**, 313 (1979)
3.81 A.S. Pedersen, P.J. Møller, O.T. Sørenson: Ber. Bunsenges. Phys. Chem. **87**, 104 (1983)
3.82 J. Töpler, H. Buchner, H. Säufferer, K. Knorr, W. Prandl: J. Less-Common Met. **88**, 397 (1982)
3.83 S. Hayashi, K. Hayamizu, O. Yamamoto: J. Phys. Chem. Solids **45**, 555 (1984)
3.84 S.D. Goren, C. Korn, M.H. Mintz, Z. Gavra, Z. Hadari: J. Chem. Phys. **73**, 4758 (1980)
3.85 M. Yamaguchi, I. Yamamoto, T. Ohta: Phys. Lett. A **66**, 147 (1978)
3.86 J. Senegas, M. Pezat, J.P. Darnaudery, B. Darriet: J. Phys. Chem. Solids **42**, 29 (1981)
3.87 S. Hayashi, K. Hayamizu, O. Yamamoto: J. Chem. Phys. **79**, 2308 (1983)
3.88 J. Senegas, A. Mikou, M. Pezat, B. Darriet: J. Solid State Chem. **52**, 1 (1984)
3.89 D. Noréus, L.G. Olsson: J. Chem. Phys. **78**, 2419 (1983)
3.90 S. Hayashi, K. Hayamizu, O. Yamamoto: J. Chem. Phys. **79**, 5572 (1983)
3.91 T.K. Halstead: J. Solid State Chem. **11**, 114 (1974)
3.92 T.K. Halstead, N.A. Abood, K.H.J. Buschow: Solid State Commun. **19**, 425 (1976)
3.93 S. Tanaka, J.D. Clewley, T.B. Flanagan: J. Phys. Chem. **81**, 1684 (1977)
3.94 R.F. Karlicek, Jr., I.J. Lowe: Solid State Commun. **31**, 163 (1979)
3.95 R.C. Bowman, Jr., D.M. Gruen, M.H. Mendelsohn: Solid State Commun. **32**, 501 (1979)
3.96 E.F. Khodosov, A.I. Linnik, G.F. Kobzenko, V.G. Ivanchenko: Phys. Met. Metall. **44**, 178 (1977)
3.97 M. Rubinstein, L.J. Swartzendruber, L.H. Bennett: J. Appl. Phys. **50**, 2046 (1979)
3.98 D. Richter, R. Hempelmann, L.A. Vinhas: J. Less-Common Met. **88**, 353 (1982)
3.99 E. Lebsanft, D. Richter, J. Töpler: Z. Phys. Chem. N.F. **116**, 175 (1979)
3.100 P. Fischer, A. Furrer, G. Busch, L. Schlapbach: Helv. Phys. Acta **50**, 421 (1977)
3.101 D. Noréus, L.G. Olsson, U. Dahlborg: Chem. Phys. Lett. **67**, 432 (1979)
3.102 R.C. Bowman, Jr., B.D. Craft, A. Attalla, M.H. Mendelsohn, D.M. Gruen: J. Less-Common Met. **73**, 227 (1980)
3.103 J. Shinar, A. Malchin, D. Davidov, N. Kaplan: J. Less-Common Met. **73**, 255 (1980)
3.104 F.E. Spada, H. Oesterreicher, R.C. Bowman, Jr., M.P. Guse: Phys. Rev. B **30**, 4909 (1984)
3.105 C. Lartigue, A. Percheron-Guégan, J.C. Achard, M. Bée, A.J. Dianoux: J. Less-Common Met. **101**, 391 (1984); A. Percheron-Guégan, C. Lortigue: Mat. Sci. Forum **31**, 125 (1988)
3.106 J. Shinar, D. Davidov, D. Shaltiel, N. Kaplan: Z. Phys. Chem. NF **117**, 69 (1979)
3.107 M. Arita, N. Takashima, Y. Ichinose, M. Someno: Z. Metallkde. **72**, 238 (1981)
3.108 K. Hiebl: Mat. Res. Bull. **17**, 757 (1982)
3.109 R. Hempelmann, D. Richter, R. Pugliesi, L.A. Vinhas: J. Phys. F **13**, 59 (1983); O. Bernauer, J. Töpler, D. Noréus, R. Hempelmann, D. Richter: Int. J. Hydrogen Energy **14**, 187 (1989)
3.110 R. Hempelmann, D. Richter, A. Heidemann: J. Less-Common Met. **88**, 343 (1982)
3.111 R. Hempelmann, D. Richter, O. Hartmann, E. Karlsson, R. Wäppling: J. Chem. Phys. **90**, 1935 (1989)
3.112 J. Töpler, E. Lebsanft, R. Schätzler: J. Phys. F **8**, L25 (1978)
3.113 K. Hiebl, E. Tuscher, H. Bittner: Monat. Chem. **110**, 9 (1979)
3.114 K. Hiebl, E. Tuscher, H. Bittner: Monat. Chem. **110**, 869 (1979)
3.115 D. Richter, T. Springer: Phys. Rev. B **18**, 126 (1978)
3.116 V. Shitikov, G. Hilscher, H. Stampfl, H. Kirchmayr: J. Less-Common Met. **102**, 29 (1984)
3.117 J.J. Didisheim, K. Yvon, D. Shaltiel, P. Fischer: Solid State Commun. **31**, 47 (1979)
3.118 J.C. Achard, C. Lartigue, A. Percheron-Guégan, A.J. Dianoux, F. Tasset: J. Less-Common Met. **88**, 89 (1982)
3.119 A.L. Bowman, J.L. Anderson, N.G. Nereson: In C.J. Kevane and Th. Moeller, *Proc. 10th Rare Earth Research Conf.*, Carefree AZ, April 1973 in Rep. CONF 730402-PI, P2, 1873 (National Technical Information Service, Springfield, VA)
3.120 D. Noréus, L.G. Olsson, P.E. Werner: J. Phys. F **13**, 715 (1983); A.F. Andresen: In *Proc. Int. Symp. on Hydrides for Energy Storage*, Geilo, August 14–19 (1977), ed. by A.E. Andresen, A.J. Maeland (Pergamon, Oxford 1978) p. 61

3.121 A. Percheron-Guégan, C. Lartigue, J.C. Achard, P. Germi, F. Tasset: J. Less-Common Met. **74**, 1 (1980);
J.C. Achard, F. Givord, A. Percheron-Guégan, J.L. Soubeyroux, F. Tassett: J. de Phys. Coll. C5, **40**, 218 (1979)

3.122 P. Thompson, J.J. Reilly, L.M. Corliss, J.M. Hastings, R. Hempelmann: J. Phys. F **16**, 675 (1986)

3.123 C. Lartigue, A. Percheron-Guégan, J.C. Achard, J.L. Soubeyroux: J. Less-Common Met. **113**, 127 (1985)

3.124 C. Lartigue, A. LeBait, A. Percheron-Guégan: J. Less-Common Met. **129**, 65 (1987)

3.125 F.E. Spada, R.C. Bowman, Jr., J.S. Cantrell: J. Less-Common Met. **129**, 261 (1987)

3.126 R.C. Bowman, Jr.: Mater, Sci. Forum **31**, 197 (1988)

3.127 Y.S. Lee, D.A. Stevenson: J. Non-Cryst. Solids **72**, 249 (1988)

3.128 D. Richter, G. Driesen, R. Hempelmann, I.S. Anderson: Phys. Rev. Lett. **57**, 731 (1986)

3.129 J.J. Rush, T.J. Udovic, R. Hempelmann, D. Richter, G. Driesen: J. Phys. Condens. Matter **1**, 1061 (1989)

3.130 G. Driesen, K.W. Kehr: Phys. Rev. B **39**, 8132 (1989)

3.131 H.J. Frost: Acta Metall. **30**, 889 (1982)

3.132 R. Kirchheim: Acta Metall. **30**, 1069 (1982);
R. Kirchheim, F. Sommer, G. Schluckebier: Acta Metall. **30**, 1059 (1982)

3.133 P.M. Richards: Phys. Rev. B **27**, 2059 (1983)

3.134 B.S. Berry, W.C. Pritchet: Scripta Metall. **15**, 637 (1981)

3.135 H.U. Künzi, E. Armbruster, H.J. Güntherodt: Proc. 4th International Conference on Rapidly Quenched Materials (Sendai 1981) p. 1653

3.136 B.S. Berry, W.C. Pritchet: Phys. Rev. B **24**, 2299 (1981)

3.137 C. Korn, S.D. Goren: Phys. Rev. B **22**, 2727 (1980)

3.138 T.T. Markert, E.J. Cotts, R.M. Cotts: Phys. Rev. B **37**, 6446 (1988)

3.139 S.V. Rychkova, M.Y. Belyaev, A.V. Skripov, A.P. Stepanov: Sov. Phys. Solid State **30**, 1285 (1988); Solid State Commun. **71**, 1119 (1989)

3.140 J.J. Kimm, D.A. Stevenson: J. Non-Cryst. Solids **101**, 187 (1988)

3.141 P. Boyer, A. Baudry: J. Less-Common Met. **129**, 213 (1987); A Chikdene, A. Baudry, P. Boyer: J. Phys. F **18**, L187 (1988); A. Chikdene, A. Baudry, P. Boyer: Z. Phys. Chem. NF **163**, 443 (1989)

3.142 K. Morimoto, M. Saga, H. Fujii, T. Okamoto, T. Hihara: J. Phys. Soc. Japan **57**, 647 (1988)

3.143 J. Shinar: Mater. Sci. Forum **31**, 143 (1988)

3.144 R. Kirchheim: Progress in Materials Science **32**, 261 (1988);
R. Kirchheim, U. Stolz: Acta Metall. **35**, 281 (1987) and references therein

3.145 L. Lichty, J. Shinar, R.G. Barnes, D.R. Torgeson, D.T. Peterson: Phys. Rev. Lett. **55**, 2895 (1985)

3.146 R.C. Bowman, Jr., D.R. Torgeson, R.G. Barnes, A.J. Maeland, J.J. Rush: Z. Phys. Chem. NF **163**, 425 (1989)

3.147 S. Ikeda, N. Watanabe, K. Kai: Physica B **120**, 131 (1983)

3.148 J.M. Rowe, J.J. Rush: Proc. IAEA Symposium on Neutron Scattering, Vienna 1977

3.149 T. Springer: In *Hydrogen in Metals I*, ed. by G. Alefeld, J. Völkl, Topics Appl. Phys. Vol. 28 (Springer, Berlin, Heidelberg 1978) p. 75

3.150 J.J. Rush, A. Magerl, J.M. Rowe, J.M. Harris and J.L. Provo: Phys. Rev. B **24**, 4903 (1981)

3.151 J. Eckert, J.A. Goldstone, D. Tonks, D. Richter: Phys. Rev. B **27**, 1980 (1983)

3.152 D. Richter: J. Less-Common Met. **89**, 293 (1983)

3.153 J.J. Rush, J.M. Rowe, D. Richter: Z. Phys. B **55**, 283 (1984)

3.154 K.M. Ho, H.J. Tao, X.Y. Zhu: Phys. Rev. Lett. **53**, 1586 (1984)

3.155 J. Eckert, J.A. Goldstone, D. Tonks, D. Richter: Phys. Rev. B **27**, 1980 (1983)

3.156 see references in the review by B. Stritzker and H. Wühl: In *Hydrogen in Metals II*, ed. by G. Alefeld, J. Völkl, Topics Appl. Phys. Vol. 29 (Springer, Berlin, Heidelberg 1978) p. 243

3.157 R.J. Miller, C.B. Satterthwaite: Phys. Rev. Lett. **34**, 144 (1975)

3.158 E. Wicke: J. Less-Common Met. **101**, 17 (1984); E. Wicke, J. Blaurock: J. Less-Common Met. **130**, 351 (1987)

3.159 B.N. Ganguly: Z. Phys. **265**, 433 (1973); Phys. Rev. Lett. B **14**, 3848 (1976)

3.160 D.A. Papaconstantopoulos, B.M. Klein: Phys. Rev. Lett. **35**, 110 (1975);
D.A. Papaconstantopoulos, B.M. Klein, E.N. Economu, L.L. Boyer: Phys. Rev. B **17**, 141 (1978)

3.161 A. Magerl, N. Stump, H. Wipf, G. Alefeld: J. Phys. Chem. Solids **38**, 683 (1977)
3.162 R. Lässer: Proc. International Meeting Hydrogen in Metals, Belfast, UK 1985, Z. Phys. Chem., Neue Folge **143**, 23 (1985)
3.163 W. Kley, W. Drechsel: Rep. EUR 5615e (Commission of the European Communities, Luxembourg 1977)
3.164 D.K. Ross: Internal Progress Rep. 80-04 (University of Birmingham 1980); private communication, 1982
3.165 J.C. Achard, C. Lartigue, A. Percheron-Guégan, A.J. Dianoux, F. Tasset: J. Less-Common Met. **88**, 251 (1982)
3.166 M.J. Benham, D.K. Ross: Proc. International Meeting Hydrogen in Metals, Belfast (UK), 1985; C. Lartigue, A. Percheron-Guégan: Z. Phys. Chem. Neue Folge: **147**, 219 (1986)
3.167 J. Eckert, J.A. Goldstone, D. Richter: J. Phys. F **11** L101 (1981)
3.168 S.M. Shapiro, F. Reidinger, J.F. Lynch: J. Phys. F **12**, 1869 (1982)
3.169 J.J. Rush, J.M. Rowe, A.J. Maeland: J. Phys. F **10**, L283 (1980)
3.170 S.J.C. Irvine, D.K. Ross, I.R. Harris, J.D. Browne: J. Phys. F **14**, 2881 (1984)
3.171 T. Kajitani, H. Kaneko, M. Hirabayashi: Sci. Rep. Res. Inst., Tohoku Univ., Ser. A **29**, 210 (1981)
3.172 M.J. Benham, J.D. Browne, D.K. Ross: J. Less-Common Met. **103**, 71 (1984)
3.173 G. Sicking, E. Magomedbekov, R. Hempelmann: Ber. Bunsenges. Phys. Chem. **85**, 686 (1981)
3.174 A.F. Andresen, K. Otnes, A.J. Maeland: J. Less-Common Met. **89**, 201 (1983)
3.175 H. Kaneko, T. Kajitani, M. Hirabayashi, M. Ueno, K. Suzuki: J. Less-Common Met. **89**, 237 (1983)
3.176 for a review see A.F. Andresen: J. Less-Common Met. **88**, 1 (1982)
3.177 D.K. Ross, P.F. Martin, W.A. Oates, R. Khoda Bakksh: Z. Phys. Chem., Neue Folge **144**, 221 (1979)
3.178 R. Hempelmann, D. Richter, O. Hartmann, E. Karlsson, R. Wäppling: J. Chem. Phys. **90**, 1935 (1988)
3.179 H. Sugimoto, Y. Fukai: J. Phys. Soc. Jpn. **51**, 2554 (1982); J. Phys. F **11**, L137 (1981)
3.180 K. Suzuki: J. Less-Common Met. **89**, 183 (1983)
3.181 H. Kaneko, T. Kajitani, M. Hirabayashi, M. Ueno, K. Suzuki, Proc. 4th Int. Conf. on Rapidly Quenched Materials (Sendai 1981) p. 1605
3.182 K. Kai, N. Hayashi, Y. Tomizaki, S. Ikeda, N. Watanabe, K. Suzuki: KENS Report IV (Isukuba 1983)
3.183 M. Hirabayashi, H. Kaneko, T. Kajitani, H. Suzuki, M. Ueno: AIP Conf. Proc. **89**, 87 (1982)
3.184 A. Williams, J. Eckert, X.L. Yeh, M. Atzmon, K. Samwer: J. Non-Crystalline Solids **61&62**, 643 (1984)
3.185 T. Schober: J. Less-Common Met. **89**, 63 (1983)
3.186 R. Thomas: In *Emission and Scattering Techniques*, ed. by P. Day (D. Reidel, Dordrecht 1981) p. 251
3.187 C.J. Wright, C.M. Sayers: Rep. Prog. Phys. **46**, 773 (1983)
3.188 A.J. Renouprez, P. Fouilloux, G. Coudurier, D. Tocchetti, R. Stockmeyer: J. Chem. Soc. Faraday I, **73**, 1 (1977)
3.189 R.R. Cavanagh, R.D. Kelley, J.J. Rush: J. Chem. Phys. **77**, 1540 (1982)
3.190 J. Howard, T.C. Waddington, C.J. Wright: J. Chem. Phys. **64**, 3897 (1976)
3.191 I.J. Braid, J. Howard, J. Tomkinson: J. Chem. Soc. Faraday II **79**, 253 (1983)
3.192 R. Schöllhorn: Angew. Chem. **92**, 1015 (1980)
3.193 G. Hollinger, P. Pertosa: Chem. Phys. Lett. **74**, 341 (1980)
3.194 C.J. Wright: J. Solid State Chem. **20**, 89 (1977)
3.195 P.G. Dickens, J.J. Birtill, C.J. Wright: J. Solid State Chem. **28**, 185 (1979)
3.196 S. Mukawa, T. Kajitani, M. Hirabayashi, J. Less-Common Met. **103**, 19 (1984)
3.197 S. Yamanaka, H. Ogawa, M. Miyake: J. Less-Common Met. **172–174**, 85 (1991)
3.198 M. Hirabayashi, S. Yamaguchi, H. Asano, K. Hiraga: In *Order-Disorder Transformations in Alloys*, ed. by H. Warlimont (Springer, Berlin, Heidelberg 1974) p. 266
3.199 T. Arai, M. Hirabayashi: J. Less-Common Met. **44**, 291 (1976)
3.200 A. Brown and D. Hardie: J. Nucl. Mat. **4**, 110 (1961)
3.201 R. Hempelmann, D. Richter, B. Stritzker: J. Phys. F **12**, 79 (1982)
3.202 M.H. Mueller, H.W. Knott: Trans. AIME **227**, 674 (1963)
3.203 F.J. Rotella, H.E. Flotow, D.M. Gruen, J.D. Jorgensen: J. Chem. Phys. **79**, 4522 (1983); D.G. Westlake: J. Chem. Phys. **79**, 4532 (1983)
3.204 D. Richter, J.J. Rush, J.M. Rowe: Phys. Rev. B **27**, 6227 (1983)

4. Hydride Formation and Decomposition Kinetics

Norbert Gérard and Shuichiro Ono

With 12 Figures and 2 Tables

This chapter summarizes results obtained for the kinetics of hydride formation and decomposition for the most important compounds, i.e. the AB_5 family (LaNi$_5$ type), magnesium alloys, FeTi and the Laves phases.

Heterogeneous kinetics is the result of a sequence of many steps; this chapter is centered on the intrinsic kinetics and its goal is the identification of the rate-determining step. It is essentially concerned with the comparison of the many available experimental data, and their interpretation in terms of theoretical models.

An effort is made here to distinguish between reliable and less reliable results. With the same aim of clarification, reference is made to heat and mass transport and to surface processes.

4.1 Overview

Most reports in the literature dealing with solid–gas heterogeneous kinetics, in contrast to homogeneous kinetics, begin by considering the complexity of the process which transforms a molecule in the gas phase into a bound atom in the lattice of the final solid product. This process is classically divided into a sequence of elementary steps consisting of gas transport, surface adsorption and dissociation, transition from surface to bulk, diffusion, nucleation and growth. The main objective of intrinsic kinetics is the identification of the *slowest* mechanism, or in other words of the *rate-determining step*. It is assumed that only one of the above steps has a rate slow enough, compared to the others, to control the whole reaction. It is also assumed that heat and mass transport are minimal and therefore not regulating steps. In addition, it has been emphasized that, in kinetics, the measured quantity is the overall reaction rate [4.1] and that the conclusions drawn from theoretical fits to the experimental results often lead to contradictory interpretation for the mechanisms because a given curve can be fitted equally well using equations stemming from different models [4.2].

Hydriding reaction kinetics can be treated with the same models as developed for oxygen or nitrogen solid–gas reactions [4.2–6]. On the other hand, special characteristics of the metal/hydrogen/hydride systems contribute significantly to make the interpretation of the kinetics very complicated. These characteristics are exothermic formation, fast reaction rates even near ambient temperature, poor thermal conductivity of the hydrided phase and embrittlement of the products. In addition, the system appears to be very sensitive to small changes both in the surface or bulk morphology and in the purity of the solid and gaseous phases.

The result is an abundance of data which can differ significantly even when experiments have been carried out under apparently the same conditions. Morphological parameters of the samples and heat transport are so dependent on the experimental approach that, according to *Mintz* et al. [4.1], the determination of the intrinsic mechanism can be performed only for massive samples and not for powders. Since the experimental data are influenced by a complex mixture of factors, the most probable mechanism will emerge only after artefacts and true parameters have been identified.

This chapter aims neither to give an exhaustive survey of the whole field of hydride kinetic studies nor is it a work on heterogeneous kinetic theory (however references are given to basic books and papers). It should be considered instead as an attempt to clarify the reasonably established facts and hypotheses which need further investigations in the kinetics of AB_5(LaNi$_5$), FeTi, Laves phases and Mg-based alloys. It is to be noted that special attention has been paid to experiments consisting of a kinetic analysis of the response of systems undergoing a well-defined perturbation.

References to additional reviews are given at the end of this chapter. The overall kinetics of a metal hydride bed are also discussed in Chap. 5.

The importance of kinetic studies on hydride/H$_2$ systems lies in the fact that kinetics appears to be one of the fundamental factors determining whether a storing system is of interest in practical applications. The major aims of kinetic studies are to identify the regulating steps and to find ways to speed up the kinetics. According to *Boldyrev* [4.3], "In solid gas reactions, the most successful chemical control can be achieved by an appropriate pretreatment". Of course, if an experiment carried out on samples of only a few milligrams shows slow kinetics, the same will undoubtedly be true for masses of a hundred kilograms. In contrast, direct transposition of satisfactory kinetic results from the laboratory scale to large containers (upscaling) may be affected by the need to consider heat and mass transfer [4.4].

In addition, one needs kinetic equations for studies that simulate metal hydride beds. These studies are essential in understanding and designing various hydride application systems. As an example of the simplest one-dimensional model, the heat balance equation can be represented as

$$C_p \rho \, \frac{T}{t} = \lambda \frac{\partial^2 T}{\partial L^2} - r \rho \Delta H, \tag{4.1}$$

where C_p is the specific heat capacity, ρ the density, T the absolute temperature, t is time, λ is thermal conductivity, L is the distance, ΔH is the reaction enthalpy and r is the rate constant as a function of temperature, pressure and reacted fraction. During the calculation, the simulation results are fitted to the experimental results. Successive improvements can then be made to the rate equation and the model used to describe the metal hydride bed. An approximate rate equation, however, is essential for the simulation study.

4.2 Methods

4.2.1 General Kinetic Characteristics

We will start with a short description of the experimental kinetic curves and a discussion of their meaning. The first step concerns the curve of transformed fraction ξ versus time for both formation and decomposition. The second step is the investigation of the P–T dependence of the reaction; this is established by determining the reaction rates at various constant pressures and temperatures.

Theoretically, one would deduce the mechanism appropriate to the system investigated by determining the elementary rate-limiting steps from:

a) A comparison of the equation $f(\xi) = kt$, which fits the experimental curves, with the models of reaction progress as functions of nucleation and growth (NG) and the geometry of the solid. Such models have been developed and compiled in basic books [4.3–7].
b) The value of the activation energy obtained from an Arrhenius plot: $\ln K = f(1/T)$.
c) The P dependence of the reaction rate which discriminates between different growth modes [4.2].

In fact, the results drawn from experiments are rarely pure in form. Such is the case, for example, for activation energy [4.8], whose value often results from a superposition of elementary phenomena. The same is true for the P and T parameters which are not independent since the true P parameter is $(P - P_e)$ where P_e is the thermodynamic pressure, whose value is a function of T.

The kinetic curves $f(\xi) = KT$, which *Boldyrev* classes into two limiting types [4.3], contain two types of information: firstly qualitative information from the shape and secondly quantitative information from the slope (i.e. the rate constant).

The first type of curve is shown in Fig. 4.1 and occurs if the nucleation process is slow. It exhibits a sigmoidal shape, and the rate constant K_e is obtained from the slope of the quasi linear part.

The second type (Fig. 4.2) has a slope which decreases as time increases. According to *Delmon* [4.5], in the first type, K_e is an intricate function of velocities of nucleation and of interface progress. In the second type, if the reaction proceeds over the entire surface, the initial slope is a product of the rate constant at the interface with the initial surface area. The attempt to make a kinetic analysis supposes the two necessary pathways of all solid–gas reaction. One is located on the "skin" of the solid and corresponds to the passage of the gas from the gas phase to the solid phase; it can be represented by a sequence of elementary

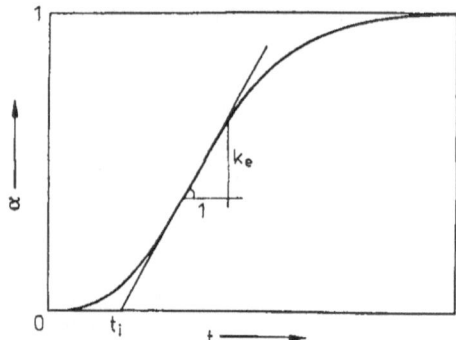

Fig. 4.1. Typical kinetic curve expressing the degree of transformation as a function of time for reactions involving nucleation and growth of nuclei [4.3]

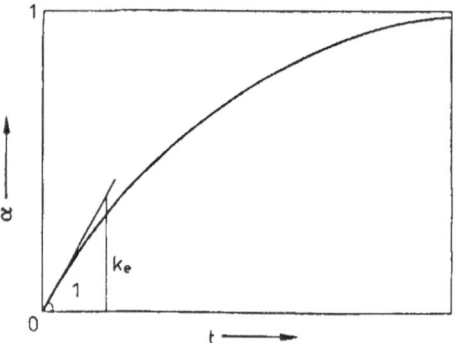

Fig. 4.2. Typical kinetic curve for the case where the reaction proceeds evenly on the entire surface of all reactant particles [4.3]

steps as given below [4.9]:

$$\tfrac{1}{2}H_2(gas) \xrightarrow{1} \tfrac{1}{2}H_2(surface) \xrightarrow{2} H \xrightarrow{3} H(entering\ site)$$

$$\xrightarrow{4} [H](adjacent\ to\ surface) \xrightarrow{5} [H](bulk),$$

where step 1 represents transport up to the surface, step 2 chemisorption onto the surface, step 3 surface migration, step 4 the surface hydride formation and step 5 solid diffusion.

The second pathway consists of two parts [4.3]. At first there is a nucleation and then (or simultaneously) the growth of nuclei involving the formation of "reaction borders" caused by transport phenomena and chemical reactivity. If there is no generation of new centers after the reaction has started, the reactivity of the surface is uniform. However, the reactivity can be modified in the vicinity of the nuclei in two ways: either the nucleus growth rate is greater than in other places, and nucleation during the reaction is slow in other places or new nuclei are formed preferentially around existing ones. According to *Boldyrev* [4.3], these first stages are "mostly influenced by crystal defects". To conclude the discussion of these general characteristics, one has to underline the fact that, in pure kinetics, heat accumulation, which depends on heat transfer and mass transport, does not interfere with the chemical process. In general, authors assume that on the basis of qualitative considerations of the geometry and mass of samples, the above artefacts are negligible. In fact, heat is produced (or consumed) at the microscopic scale of the nucleus. However, it is extremely difficult to measure a temperature perturbation on a surface or to analyse all the consequences of this heat perturbation in terms of reactivity.

4.2.2 Experimental Methods and Devices

A device designed for studying heterogeneous kinetics has to fulfil two main conditions: it must deliver a signal, as a function of time, which is proportional to the reaction transformation (for example the volume or mass of gas fixed or released); secondly, it must allow the reaction to proceed as quickly as possible, when one parameter is modified, in order to go from one equilibrium state to a new one. As a consequence, special attention has to be paid to the following points:

1) The means chosen to push the system out of equilibrium must introduce the smallest possible perturbation to the other parameters of the reaction and reduce the transition period (the time for establishing the new $P-T$ conditions) to a value that is negligible compared to the whole reaction time (less than 10%).

2) The device should not have a blind period at the beginning of the recording. The initial part of the curve $\xi = f(t)$ is of fundamental interest for its interpretation. It should be noted that when the reaction is started by a fast change in pressure, it includes an adiabatic expansion and recompression which create successive fast changes in temperature.

The devices and methods can be divided into three classes:

—volumetric devices,
—thermogravimetric devices,
—other methods.

a) Volumetric Devices

Such devices were first developed in hydride kinetic studies as an extension of the Sieverts apparatus built for the investigation of $P-c-T$. In the simplest version, the recorded parameter is the change in pressure resulting from the reaction progress [4.10–13]. This pressure change depends on the volume of the reacting vessel, the mass of the sample and, of course, on the sensitivity of the detector; with an electronic transducer the initial pressure change can be as low as 5% [4.9]. Decomposition kinetics can be followed by direct measurements of the released gas quantity with burets [4.14]. Some kinetic experiments have

been carried out in isothermal conditions but under pressure sweep. The authors
assume that one can mathematically separate the pressure from reacted fraction
dependences [4.15]. *Josephy* et al. [4.16] proposed an access to kinetics by
measuring the hydrogen flow between a reactor maintained under isothermal
conditions by a Ni ballast and a volumetric reservoir. For *Suda* et al. [4.17] the
"step-wise method" which consists in giving step-wise changes in the hydride
composition, as in an ordinary determination of a $P–c–T$ equilibrium, and
measuring the pressure change, is considered by the authors as "an excellent
technique for detailed studies of reactions covering both, the α and the β phase".

In more sophisticated volumetric devices, the problem of maintaining
constant pressure during the experiment has been solved either by successive
regulated adjustments through electrovalves or, as we have described [4.18],
by modifying the length of a bellow which is part of the reaction vessel; the
changes are piloted through a servo-motor by a zero transducer which works
between the reaction vessel and a reference pressure. This device has been
developed for samples of mass not exceeding a few milligrams. It is true that
volumetric devices offer an easier means of building reactors with either a high
heat transfer coefficient [4.19–21] or with special powder-bed geometry [4.20].
However, their main disadvantages are a sensitivity which varies with the
absolute pressure and a blind period during the pressure change which initiates
the reaction.

Reactor Design

As hydriding/dehydriding kinetics depends strongly on the temperature change
of the metal hydride bed, any reactor or reaction cell designed without considering the thermal effects of those reactions should be avoided regardless of the
sample amounts used and the temperature level as well. There are mainly three
categories of reaction cells as summarized in the following.

Autoclave Reactor. The autoclave type reactors have been employed in many
experiments. They undoubtedly have poor heat-transfer characteristics and no
consistent results should be expected from such experimental set-ups because
of their comparatively large sample space and small contact surface for the
rapid heat transmission. In this regard, the kinetic data obtained from those
reactors should be understood as being subject to thermal effects caused by the
hydriding/dehydriding reactions, for example, data has been taken under non-
isothermal conditions.

Thin Disc Reactor. In such reaction cell, the sample can be located within a
narrow space and over a large flat area. Thus, in a finely powdered state, it
might be expected to exhibit good heat transfer characteristics. *Goodell* and
Sandrock [4.19] have used a reaction cell of 15 mm thickness and 43 mm
diameter, in which copper has been used as the reactor body with a time constant
of $7 s^{-1}$. However, this type of reaction cell facilitates thermal flow in only one
direction.

Double Wall Reactor. In this type of reactor, a thin annular space forms the sample holder which is divided by a filter with a mean filter rating of 1–2 μm. This shape provides improved heat-transfer characteristics without adding extra penalties such as hydrogen flow resistance. The heat transfer has been greatly improved in comparison with the single-wall reactor, because of the thinner sample space with a comparatively large heat transfer area. *Suda* and coworkers [4.17] have designed a double-wall reactor. The sample layer between outer and inner tube is less than 0.5 mm thick. Cooling and heating fluid at a given temperature is pumped through the inner tube located at the center of the reactor. This shape gives the largest heat transfer area with a thin and small sample space.

In a modification of the double-wall reactor the sample is contained between two coaxial porous steel tubes. The H_2 gas is applied from the inside and flows through the sample in a radial direction.

b) Thermogravimetry

High-pressure temperature-controlled microthermobalances are the ideal tools for hydride kinetics studies. Both commercial designs [4.22, 23] and specially designed vessels containing a commercial modulus [4.24] can be used. Sometimes vacuum microbalances are used, but usually only for decomposition measurements to check volumetric measurements [4.25, 26]. In general, microthermobalances have a mass resolution of a few micrograms, which enables one to work with very small samples (10–200 mg). The sensitivity does not depend upon the pressure and therefore the measured P–c–T curves are better defined than with volumetric methods. Temperature measurements are generally made outside the sample and require heat transfer through the gas phase.

c) Other Methods

To end this section we briefly mention some other methods which are based on the effects of H insertion in the solid phase:

a) change in electrical resistivity with hydrogen absorption in thin films of $LaNi_5$ [4.27],

b) acoustic emission during hydride formation [4.28],

c) X-ray diffraction studies of the structural changes accompanying the formation/decomposition reactions [4.29]. Systems have been developed which allow simultaneous diffraction and thermogravimetric analysis [4.30].

d) morphological observations carried out on samples quenched at a defined degree of reaction. These give quasi-quantitative kinetic results and serve as precise tools for determining mechanisms [4.31].

4.3 Results of LaNi$_5$ and Substituted Compounds of the Same Family

Pure LaNi$_5$ shows an absorption plateau pressure of a few bars at ambient temperature. There is complete reversibility between formation and decomposition with a hysteresis between the corresponding equilibrium pressure. An α-phase solid solution precedes the β-LaNi$_5$H$_6$ hydride. The system requires an activation stage which involves decrepitation into small particles.

4.3.1 Kinetics in the Primary Solubility Range (α-Phase)

Experimental results of *Flanagan* [4.9] on both unactivated and activated samples and of *Belkbir* [4.32] on an activated LaNi$_{4.9}$Cu$_{0.1}$ sample show reasonable agreement. The diffusion coefficients are 3×10^{-7} cm^2/s at 434 K (extrapolated value at 291 K: 2×10^{-9} cm^2/s [4.91]) and 1.1×10^{-9} cm^2/s measured at 291 K [4.32]. These values are one order of magnitude smaller than those derived from NMR and neutron scattering (Chap. 3). The conclusions differ only in their details: *Flanagan* states "the time dependence which is $t^{1/2}$ agrees with diffusion control and the $P_{H_2}^{1/2}$ dependence suggests that the rate is controlled by hydrogen dissolved in layers adjacent to the surface". According to *Belkbir*, the experimental curves agree better with the following equation:

$$-\ln(1 - \xi) = K_1 t \tag{4.2}$$

(ξ is the transformation ratio), which corresponds to diffusion in the bulk, with variable conditions throughout the reaction.

As far as the $P_{H_2}^{1/2}$ dependence is concerned, *Belkbir* states that as in the primary solubility domain, the composition (c_∞) depends upon $P_{H_2}^{1/2}$ according to Sievert's law; the instantaneous velocity is correctly expressed by the equation.

$$\left(\frac{dc}{dt}\right)_{t \to 0} = K_{1c\infty} \tag{4.3}$$

with K_1 the rate constant of (4.2) and $c_\infty = B(T)P_{H_2}^{1/2}$. Under these conditions,

$$\left(\frac{dc}{dt}\right)_{t \to 0} = G(P_{H_2}). \tag{4.4}$$

This rate dependence on P_{H_2} is interpreted as being the result of a dissociative slow step in which only one hydrogen atom of the adsorbed molecule diffuses from its dissociation site into the bulk [4.33]. *Suda* et al. [4.34] obtained, in α-phase formation of LaNi$_{4.7}$Al$_{0.3}$, a reaction order with respect to hydrogen pressure equal to 1.5. They suggested that chemisorption could be the rate-controlling step even though this implies a reaction order close to 1.

4.3.2 Kinetics in the Two-Phase Region (α–β)

The identification of the rate-determining step in this domain is probably the most debated point in hydride kinetics.

If one refers to Table 4.1 which summarizes the data obtained by twelve authors under the same conditions (25 °C, absolute pressure 5 atm.), it appears that the reaction rate determined for $H/M = 0.5$ varies from 0.01 to 35 min^{-1}. A number of schemes have been developed to explain this dispersion. They can be described under three headings: surface activity, heat effect, and mixed processes controlled by several intrinsic steps [4.35, 36].

a) Surface Activity

Flanagan [4.9] first noticed that LaNi$_5$ reaction rates were "markedly inhibited by small pressures of helium" (1 torr of He decreased the rate of absorption by a factor of 3). The results of *Goodell* and *Rudman* [4.35] shown in Fig. 4.3 point to the same phenomenon: The observation that the reaction rates depend linearly on P_{H_2} after surface damage by CO, CO_2 and NH_3, are presented as evidence of a rate-limiting surface process. Interesting attempts have been made to understand the role of the surface in relation to hydrogenation kinetics. For example:

—Isotope scrambling experiments in the reaction $H_2 + D_2 \rightleftharpoons 2HD$ which give a measure of the dissociative activity could be related to kinetics [4.37].
—Kinetic studies of the catalytic properties of LnNi$_5$ (Ln = La, Ce, Pr, Nd, Sm, Gd) in hydrogenation of propylene have been carried out and compared to hydrogenation kinetics of the same intermetallic compounds [4.38].

b) Heat Effect

All authors have emphasized the fact that LaNi$_5$ kinetics are extremely fast and that the thermal conductivity of both the hydride and LaNi$_5$ is poor [4.37].

Table 4.1. Comparison of LaNi$_5$–H kinetics data

Reference	Absorption to 5 atm at 25 °C		Desorption to 1 atm at 25 °C	
	$R_{0.5}$ (min^{-1})	$t_{0.5}$ (min)	$R_{0.5}$ (min^{-1})	$t_{0.5}$ (min)
4.35a	—	—	0.09	4.9
4.35b	—	—	1.2	0.3
4.35c	—	—	0.5	0.5
4.12	< 1.7	> 0.2	< 1.7	> 0.2
4.35d	1.0	0.3	0.2	1.4
4.35e·	2.0	0.13	0.6	0.5
4.35f	0.083	8.0	0.06	6.0
4.17	1.2	0.3	0.2	1.7
4.19	20.0	0.024	3.0	0.1
4.35g	< 0.01	> 10.0	—	—
4.35h	2.5	0.16	—	—
4.35	35.0	0.02	4.0	0.06

$R_{0.5} = d([H]/[M])/dt$ at $[H]/[M] = 0.5$; $t_{0.5}$ is the time to reach $[H]/[M] = 0.5$

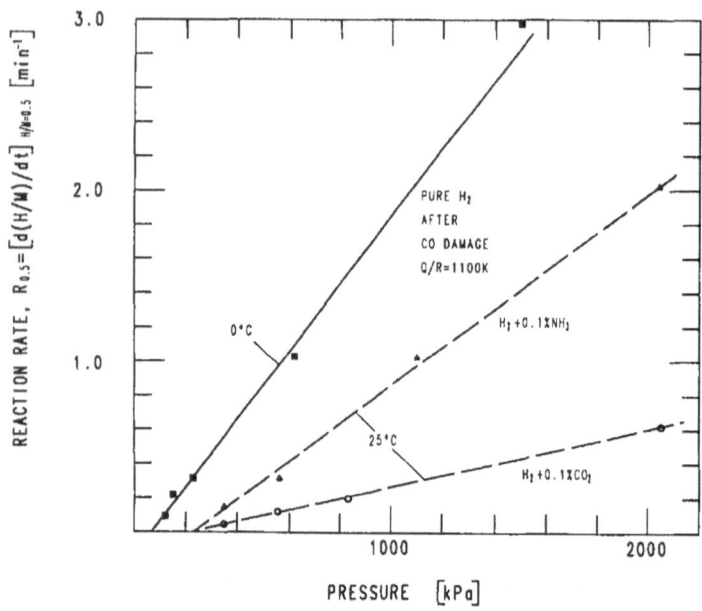

Fig. 4.3. Test results showing the linear pressure dependence of the reaction rate for hydriding LaNi$_5$ when the surface process is retarded and hence becomes rate limiting [4.35]

As a consequence, heat accumulation and heat flow become the rate-limiting step in many experiments [4.37–39]. For *Dantzer* et al. [4.40] who analyzed the contribution of heat transfer to the reaction rate "All studies of hydriding kinetics reported in the literature should include a contribution due to heat transfer"!

They made a simulation study of the effect of the heat transfer characteristics on the overall reaction rate. They used the following heat balance equation of the system:

$$[\Delta H][n_H]\frac{dc(t)}{dt} = \frac{S}{d_1 k_{\text{eff}}}[T(t) - T_0] + C\frac{dT(t)}{dt}, \tag{4.5}$$

where n_H is the amount of hydrogen reacted at the completion of the reaction, $T(t)$ represents the instantaneous temperature of the reacting sample, k_{eff} is the overall thermal conductivity, S and d_1 represent the heat exchange surface area and length over which heat is transferred, and C is the total heat capacity of the sample and reactor. The left-hand term represents the heat evolved, the first term on the right is the heat dissipated through the wall by heat conduction and the second term is the heat accumulated in the system.

Indeed this formulation is supported by the increase in reaction rates observed both in reactors with "high heat-transfer factor" [4.16, 17] and with

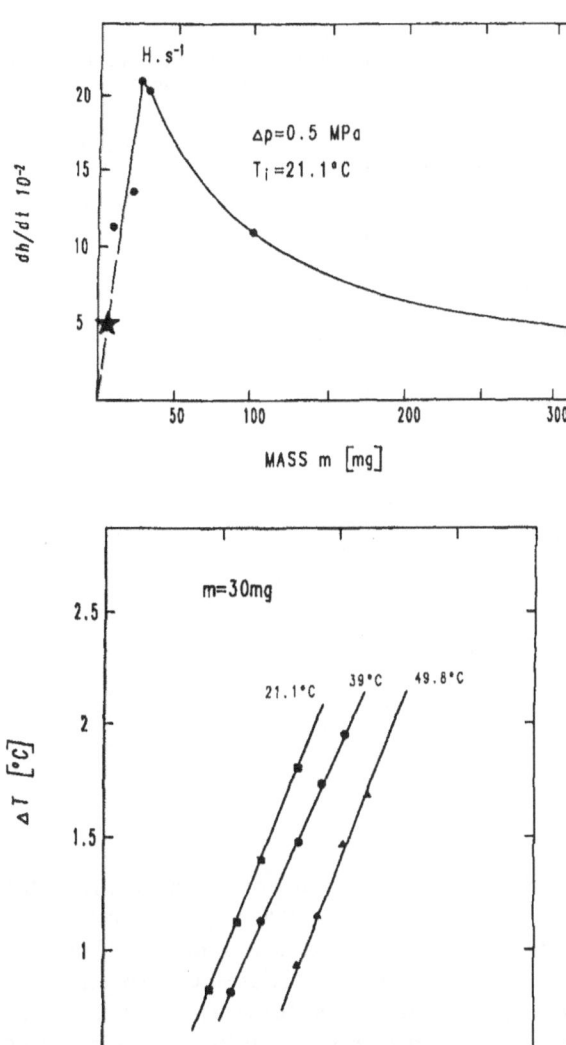

Fig. 4.4. Reaction rate as a function of sample mass. The value denoted by a star has been obtained with a mass of 5 mg [4.32]

Fig. 4.5. Difference ΔT between the actual temperature T_a and the imposed temperature T_i as a function of the hydrogen pressure for three different imposed temperatures

metal hydride powder mixed with Ni, Cu [4.16, 17, 19] or Al [4.14] powders as heat ballast and pressed into pellets.

Heat effects are not restricted to a slowing down of the rate due to the rise of equilibrium pressure, but can also play a more complex role especially in nucleation and surface reactivity. This can be deduced from a series of experiments by *Bayane* et al. [4.41], carried out by thermogravimetry on LaNi$_5$ activated samples as a function of the mass of the samples ($T = 25\,°C$, $P_{H_2} =$

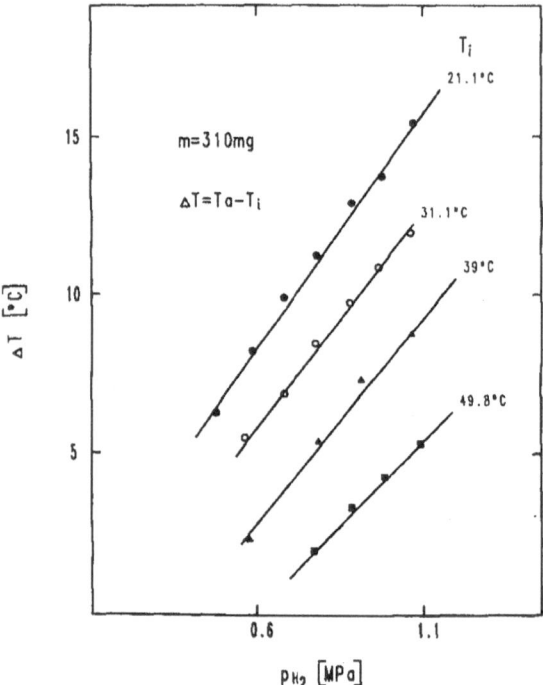

Fig. 4.6. Difference between the actual temperature inside the sample and the imposed temperature

0.5 MPa). The mass ranged from 10 to 300 mg. The surprising result is reported in Fig. 4.4, in which the reaction rate $R_{0.5}$ ($H/M = 0.5$) is plotted versus the mass of sample. The reaction rate shows a maximum near a mass of 30 mg and decreases for both higher and lower masses. In this series of experiments, and particularly to test a possible "heat effect", the temperature was directly recorded inside the sample. The results are reported in Figs. 4.5, 6 for masses of 30 mg and 300 mg, respectively. The difference, ΔT, between the actual temperature and the imposed temperature is plotted versus the hydrogen pressure P_{H_2}. This shows firstly that the "heat effect" exist only for masses larger than 30 mg. Secondly, the heat effect is enhanced when both the imposed temperature and the pressure are decreased. None of the calculated corrections, which take into account the change in temperature and the change in the equilibrium pressure, restore a reaction rate that is independent of the mass. In addition, and illustrating the complex role of the heat effect, it has been observed in the same series of experiments that the reaction rate obtained with a sample of 10 mg is increased by 12% simply by replacing the 10 mm diameter flat stainless steel sample holder with a silica sample holder of the same shape (the only difference being in the thermal conductivity of the two sample holders).

c) Intrinsic Limiting Steps: Surface Reactions, Hydrogen Diffusion, and Phase Transitions, i.e. Nucleation and Growth

If one compares the results obtained from experiments carried out under conditions where special attention has been paid to heat transfer and gas purity, one finds that any of the above three intrinsic steps can appear as the possible rate-determining step.

The experiments of *Tanaka* et al. [4.13] were carried out at a temperature of 195 K under such conditions that the evolution of hydrogen was small enough for self-heating to be neglected (195 K, $P = 5$ torr, half time for reaction 120 s). Initial hydriding/dehydring reaction rates were independent of time and dependent on P_{H_2} indicating that "mass transport through cracks is the slow step". This result is interpreted by *Park* and *Lee* [4.42] in terms of an initial rate which is controlled by chemical adsorption of hydrogen molecules on the nickel surface.

Goody et al. [4.14] measured desorption rates for pellets of LaNi$_5$ hydride: The LaNi$_5$ was first poisoned with SO$_2$ to prevent immediate hydride decomposition, mixed with Al powder and pressed. The pellets, which contain 95% Al, were outgassed overnight under vacuum at 100 °C. The half-time for reaction is about 40 s (25 °C, 1 atm). As the time dependence of the decomposition is $\ln(c_0/c) = kt$ (c_0 is the hydrogen concentration in the metal at $t = 0$), the authors consider that the rate-determining step is the diffusion of atomic hydrogen along the La$_2$O$_3$/Ni interface.

The experiments of *Goodell* et al. [4.19] were performed with thermal ballasting samples consisting of 2.5 wt% LaNi$_5$ mixed with Ni powder pressed into pellets and cleaned by several outgassing cycles at 140 °C under vacuum. These authors infer a mixed control by surface dissociative adsorption and bulk diffusion through the β phase.

Belkbir's experiments [4.32] were carried out on small samples (5–10 mg) during the activation stage and also on activated samples, both on pure LaNi$_5$ and on series of substituted compounds LaNi$_{4.9}$M$_{0.1}$ (M = Cu, Ti, Al). The pressure dependence of the reaction rate, both in formation and decomposition, is found to be proportional to ΔP_{H_2}. All the curves $\xi = f(t)$ in the slow formations of *Belkbir* (half-time equals several minutes) have a sigmoidal shape and fit Johnson and Mehl's equation:

$$- \ln(1 - \xi) = Kt^2. \tag{4.6}$$

These observations are consistent with a rate controlled by nucleation and growth (NG) process in the bulk of the solid. The only remaining problem is that of relating the P_{H_2} dependence of the reaction rate to the NG limiting step. This can be done by assuming that nucleation is proportional to the supersaturation of the precursory α-phase. This supersaturation can be reasonably considered to be proportional to P_{H_2} by extrapolation of the part of the isotherm which deviates from *Sieverts*' law [4.43]. Decomposition curves differ from formation curves; they are represented by curves $\xi = f(t)$ with continuously decreasing slopes. During the first few cycles, one observes a transition from:

$$1 - (1 - \xi)^{1/3} = K_i t \tag{4.7}$$

to

$$\ln(1 - \xi) = - K t. \tag{4.8}$$

Equation (4.7) represents a geometrical regression in the spherical symmetry with a nucleation at a constant rate on the surface of the solid. According to *Delmon* [4.5], Eq. (4.8) represents a reaction in which nuclei appear with a constant probability but not further growth occurs. Undoubtedly, in these experiments, the interface reactions are the regulating step. A formation reaction controlled by the chemical process at the α/β interface is also obtained by *Myamoto* et al [4.44] in fast formation kinetics (half-time = 20 s). Their results fit (4.7), which corresponds to a spherical regression. The pressure dependence is $\ln(P_0/P_e)$ which agrees with a slow interfacial reaction.

The inventory of factors that play a part in surface activity is developed in Chap. 2 on surfaces. For the terms concerned with nucleation, the number and nature of defects are undoubtedly important factors, although their quantitative study is made difficult by activation. An attempt to indirectly measure their influence can be found in studies of the reaction rate as a function of the frequency of formation/decomposition cycles [4.43]. The reaction rate decreases when a term with the periodicity t_1 or t_2 (t_1 = formation time + time in the hydrided state; t_2 = decomposition time + time in the decomposed state) is increased. This result is interpreted as a consequence of the annealing out of the defects which are potential sites of nucleation.

4.4 Results on Mg, Mg Alloys and Mg Intermetallic Compounds

Hydride formation/decomposition kinetics of Mg and Mg alloys are still under investigation. Magnesium alloys are attractive materials for hydrogen storage, but activation (i.e. first hydride formation) and decomposition kinetics and thermodynamic equilibrium parameters all need to be improved. An enormous improvement in the kinetics of Mg-H_2 was achieved by the preparation of fine powder of MgH_2 from metallo-organic solutions [4.45].

4.4.1 Formation/Decomposition Kinetics of MgH$_2$ from Pure Mg

The results on hydriding/dehydriding of unalloyed magnesium are described in many papers and consist of a series of contradictory experimental results and of diametrically opposed interpretations. For example, according to *Genossar* and *Rudman* [4.46], hydriding/dehydriding kinetics "can be very simply described: they are terrible!" In contrast, *Vigehom* and co-workers [4.23] claim that unalloyed magnesium is well suited for hydrogen storage. In [4.45] MgO is considered to be the dominant contaminant inhibiting hydriding/dehydriding, whereas [4.46] states that "broken-up oxides act as a barrier to sintering and contribute to fast reactions". Nevertheless, there is general agreement that the first hydriding of unalloyed magnesium needs pressures up to 3 MPa, temperatures up to 340 °C and a time ranging from 6 h [4.47] to two weeks [4.44].

According to *Isler* [4.24], the first formation from Mg powders (Prolabo 97.5% purity, medium size 300 µm) at a temperature of 340 °C and a hydrogen pressure of 4 MPa is represented by a sigmoidal curve of transformed fraction (ξ) versus time, with an inflexion point near 0.4. It includes an induction period ranging from 5 to 10 h, depending on the sample, and after 25 h reaches a value near $\xi = 0.96$. This undoubtedly corresponds to a reaction controlled by nucleation and growth as claimed earlier by *Douglass* [4.25]. An inflexion point was also observed by *Vigehom* et al. [4.23] in the formation curves. According to *Ivanov* et al. [4.48, 49] the nucleation is determined by the cracking of an oxide layer, and as a majority of Mg particles react through the formation of only one nucleus, a stress concentration in the oxide layer should favor nucleation. Even though they insist on the formation of a uniform chemisorbed hydrogen layer as the first step of the hydriding process [4.49], they reach the same conclusion as *Stander* [4.50]: a rate-controlling process of surface nucleation. Morphological observations on cross sections [4.24] confirm that the hydride appears as isolated nuclei, which develop towards the bulk and have a hemispherical shape. Identical observations were made by *Douglass* [4.25] on Mg–Al alloys. *Schober's* study, by electron microscopy on Mg single crystals [4.51], is clearly in favor of nucleation, growth and coalescence of three-dimensional MgH_2 clusters. The above observations are in agreement with the conclusions of *Genossar* and *Rudman* [4.52] that the H^- anion is the diffusing species (from Kirkendall marker experiments) as opposed to theoretical arguments which favor Mg^{++} diffusion [4.50, 53]. One of the many open questions concerns the observation by *Douglass* [4.25] on Mg–Al alloy, of MgH_2 nuclei in the bulk. This result has been confirmed by *Chene* [4.54] in a high purity Mg sample free of cracks. The above observations raise two questions: firstly, should hydrogen diffusion in Mg be considered as a slow [4.54] or a fast [4.55] phenomenon? Secondly, does a hydrogen-solid solution in the Mg lattice really exist? Such a solid solution has not been detected by high sensitivity thermogravimetry measurements, [4.23, 56] but a solid solution does appear in isotherms [4.57]. The formation of internal hydrides in Mg alloys as investigated by *Renner* and *Grabke* [4.58] also suggests its existence.

Let us return to the highly debated point of determining the rate-limiting factor in MgH_2 formation. According to the three-dimensional MgH_2 growth model, formation is controlled by H diffusion along the $Mg-MgH_2$ interface. Diffusion could be favored by flaking and cracking along the $Mg-MgH_2$ interface [4.47]. On the other hand, if one supposes that a thin MgH_2 layer is formed, H diffusion through this layer would be rate limiting [4.46–59]. In fact, both mechanisms can be observed successively [4.24–50]. The first mechanism corresponds to the first hydridation and to hydridations carried out under H pressures near the equilibrium pressure P_e. *Isler* has shown [4.24] that the number of MgH_2 nuclei is a function of $(P-P_e)$. Under rapid formation conditions, a first growth of numerous MgH_2 nuclei forms a practically continuous thick layer near $\xi = 0.3$. The reaction rate then becomes progressively controlled by transport of material through this layer [4.24, 47]. In the last stage, the reaction rate, which does not depend upon P, becomes extremely slow and finally stops, according to the grain size, between $\xi = 0.6$ and 0.80 [4.24, 59]. Each grain contains an unreacted Mg core [4.24]. A point which is far from being clearly understood is the role played by surface steps in MgH_2 formation kinetics.

Ball-milling [4.47, 60, 61] and additional vibrations on samples cleaned by ion bombardment increase the H absorption rate. The addition of indium, according to *Mintz* et al. [4.62], is only supposed to lower the activation energy of H diffusion and has no influence on the decomposition rate. Contrary to this, *Genossar* and *Rudman* [4.52] suggest that the addition of Cu causes Mg_2Cu precipitates which provide an external surface free of oxide that is supposed to adsorb (or desorb) and transfer H into (or out of) the hydriding (dehydriding) phase. However, *Eisenberg* et al. [4.63] found that Ni plated on Mg produced no effect on hydriding kinetics but, on the contrary, that decomposition rates are increased. They conclude that decomposition involves a different mechanism.

MgH_2 Decomposition. The observation by *Boldyrev* et al. [4.64] of MgH_2 whisker decomposition induced by an electron beam shows that the first stage is the formation of a superlattice with hydrogen vacancy ordering followed by the formation of a lamellar structure of Mg/MgH_2. Morphological observations of MgH_2 decompositions after partial dehydriding clearly show the reformed metal at the exterior of the grains. In addition, *Schober* [4.51] observed the formation of H_2 bubbles in the hydride.

The curves of transformed fraction ξ versus time, have a sigmoidal shape. They become linear both in the *Prout–Tompkins* Eq. [4.65]: $\ln \xi/(1 - \xi) = kt$ [4.24] and in the *Avrami–Erofeev* Eq. [4.66]: $-\ln(1 - \xi)^{1/n} = kt$ [4.42]. In these two kinetic laws, slow nucleation plays a key role in the reaction rate. But it again appears to be very difficult to separate diffusion control from phase boundary control. *Douglass* [4.25] found that in Mg/Y alloys the regulating step seems to be migration of Mg through the hydride phase but that in Mg/Ni/Y alloy, both diffusion and interfacial migration of H through the hydride phase are regulating steps. The above authors pointed out the linear dependence of the decomposition rate on (P_e-P). The activation energy, whose value ranges from $77.45\,kJ\,mole^{-1}$ [4.68], through $80\,kJ\,mole^{-1}$ [4.24] to $113\,kJ\,mole^{-1}$ [4.44], is frequently a poor discriminative tool for determining the rate-limiting step.

4.4.2 Magnesium Alloys and Intermetallic Compounds

The aim of numerous studies is the search for "catalysts" of the MgH_2 formation/ decomposition. The word catalyst is intended here to describe operations as diverse as the following: mechanical grinding of Mg [4.47, 48], co-grinding of Mg with metals [4.67–69] or compounds [4.68–70], surface chemical deposition of Ni [4.63], and true formation of Mg alloys. The alloys can be divided into two categories according to whether they consist of:

a) Mg-rich solid solutions containing small amounts of other elements (In [4.44], Th, Zr, Li, Al [4.58, 71], Ag, Al, Cd, In, Pb, Y, Zn [4.25]) or of,
b) defined compounds such as Mg_2Cu [4.72], Mg_2Ni [4.73], Mg_2Ca [4.26], Mg_2Al_3 [4.25, 26, 74], $ReMg_{12}$ (Re=Ce, La) [4.75] and La_2Mg_{17} [4.76].

The true intermetallic compounds can be further divided into three classes according to their behavior with hydrogen:

i) Compounds which form a ternary hydride such as $Mg_2Ni + 2H_2 \rightleftharpoons Mg_2NiH_4$ [4.73];
ii) compounds which undergo a disproportionation reaction to form two hydrides, for example: $2CeMg_{12} + 27H_2 \rightarrow 2CeH_3 + 24MgH_2$ [4.77];
iii) compounds which give, after disproportionation, only Mg hydride: $2Mg_2Cu + 3H_2 \rightleftharpoons 3MgH_2 + MgCu_2$ [4.72].

To end this description, we mention some of the many compounds obtained either by partial substitution of Mg in Mg–rare-earth alloys, such as $CeMg_{11}M$ with M=V, Cr, Mn, Fe, Ni, Cu, Zn [4.78–79], $La_{1-x}Mg_xNi_2$ [4.80], $La_{2-x}Ca_xMg_{17}$ [4.76], or by substituting Ni in Mg_2Ni-type alloys such as $Mg_2Ni_{1-x}M_x$ with M=Be [4.81], Co [4.26], Fe [4.26, 81], Mn [4.82]. In addition, the binary system Mg/Ni gives many alloys ranging from pure Mg to Mg_2Ni with variable amounts of the eutectic $Mg–Mg_2Ni$ [4.20].

Kinetics of Mg Alloy Hydrides. The kinetic studies carried out on many of these compounds generally consist of a comparison of the substituted compound with pure Mg which is taken as the reference. The reaction time required to give a defined transformation ratio is measured.

The kinetics of Mg_2Ni and $Mg–Mg_2Ni$ systems have been investigated in great detail [4.20, 68, 82, 83, 85] both in formation and in decomposition reactions [4.86, 89, 90]. A particular difficulty that one has to keep in mind is that Mg_2Ni is never the only phase; the samples very often include $Mg–Mg_2Ni$ eutectic in grain boundaries and variable amounts of crystalline Mg (cf. Chap. 2 of Vol. 1). The problem is also complicated by kinetic behavior that is influenced by Mg/Ni phase transitions [4.84]. The hydride formation curves are sigmoidal in shape [4.69, 84] with the fast formation of an α solid solution [4.20, 69] at the beginning. The sigmoidal shape is referred to as a nucleation and growth process, although the curves are deformed into continuous rate-decreasing curves when the imposed pressure P_i is increased. The pressure dependence is linear but, according to Song [4.69] is also a function of P_0^2. This fact suggests a limiting step which could be the forced flow of H_2 through pores and cracks. Again a possible limiting step by heat transfer is evoked [4.69, 84]. *Song* et al. [4.90, 91] suggest that surface-segregated Ni provides active dissociation sites even in the presence of oxygen, and that the associated exothermic effect can be "a rate-delaying step". But, according to *Mattsof* and *Noreus* [4.88], when experiments are carried out on a time-scale of several hours, the reaction kinetics can be interpreted as a diffusion-controlled growth of Mg_2NiH_4 in the α-$Mg_2NiH_{0.3}$ in agreement with a $P_{H_2}^{1/2}$ dependence. Dehydriding kinetics of Mg_2Ni–$H_2(D_2)$ carried out by *Lupu* et al. [4.92] showed a reaction which progresses by nucleation and growth. The $H_2(D_2)$ diffusion in the α-phase is the rate-limiting process at high thermodynamic driving forces.

Studies on Mg—10 wt % Ni and $CeMg_{12}$ [4.83] show faster kinetics than for pure Mg, but comparable mechanisms. Morphological investigations by *Douglass* [4.85] and *Boulet* and *Gérard* [4.83] confirm that in Mg_2Ni alloys the reaction starts in the Mg crystals of the eutectic phase. Mg_2Ni primary crystals are transformed into hydride beginning at their boundaries; the interface moves homothetically towards the inside from the initial boundaries. At the same time, cracks develop between the eutectic and the primary crystals and also inside the Mg_2Ni crystals [4.83].

Coming back to the many compounds referred to above, the philosophy of investigation consists either in exploring a family of substituting elements or in searching for substitution-induced modifications of a supposed limiting step in MgH_2 formation/decomposition kinetics. Interest is focussed on the following areas:

—Surface effects [4.94–96] which lead, e.g., to the formation of three-metal clusters [4.44, 46, 63, 71, 93],
—the role of electronic factors [4.80, 81],
—enhancement of the number of nucleation sites (by eutectic regions which can also play a role in H diffusion [4.25, 83] or by finely dispersed precipitates in Mg crystals [4.83],
—changes in the mechanical properties of the Mg matrix (by the addition of elements that lead to embrittlement) [4.83, 97, 98].

According to certain authors, some of the preceding alloys or compounds are considered satisfactory with regard to their kinetics or thermodynamics; such is the case for: Mg–5Ni–5Y [4.25], Mg–Ni–Fe and Mg–Ni–Al [4.81], $CeMg_{12}$ and $CeMg_{11}M$ [4.78].

A key point is the duration of the improvements achieved with regard to formation/decomposition cycles. Superficial oxidation develops progressively; strains and stress fields disappear by plastic deformation and sintering; dispersed phases coalesce and foreign atoms migrate. Generally, the initial properties are gradually lost [4.68, 83, 86], though they can sometimes be partially restored [4.97].

4.5 Results on FeTi and Its Family

Although FeTi is one of the most extensively studied materials, little attention has been paid to its kinetics, probably because it shows fairly complicated features in its $P-c-T$ isotherms. It has two plateaux corresponding to the $\alpha \rightarrow \beta$ and $\beta \rightarrow \gamma$ transitions. The plateau of the $\beta \rightarrow \gamma$ transition is not as flat as that of the $\alpha \rightarrow \beta$ or as those of Mg_2Ni or $LaNi_5$. Uncertainty in the plateau pressure makes it difficult to analyze the experimental data accurately because of the inaccuracy in the driving force of pressure difference.

4.5.1 Activation of the Alloy

Activation, i.e. the first hydriding, is much more of a concern for FeTi than for other hydrides, because the alloy is relatively hard to activate and requires a rather involved process. *Sandrock* et al. [4.99] proposed to distinguish between "first-stage activation" which is the first hydriding of the alloys and "second-stage activation" which refers to the first few cycles. The alloy is generally handled in air and thus covered by a surface oxide which inhibits hydriding. The first-stage activation corresponds to the breach of this deactivated surface and the decrepitation of the bulk sample into powder. In the subsequent cycles, the kinetics speed up, probably due to a reduction in particle size.

First-Stage Activation. Several parameters have been identified which affect the ease of first-stage activation. The main parameters are (a) the FeTi alloy composition, (b) the oxygen content, (c) the addition of ternary elements, (d) the formation temperature–pressure conditions and (e) the FeTi particle size. The first three parameters have been studied extensively by *Sandrock* et al. [4.100],

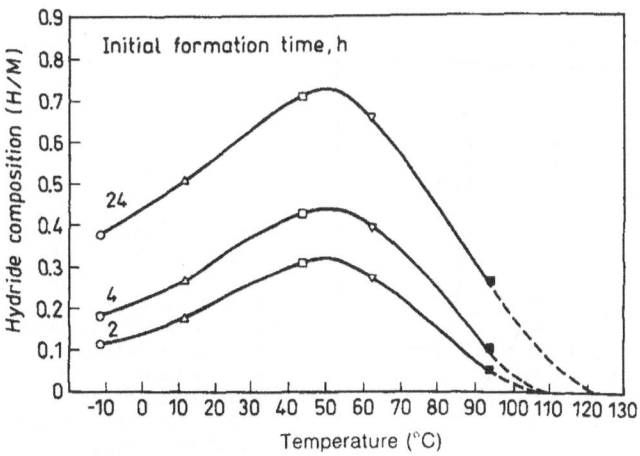

Fig. 4.7. The effect of formation temperature on the initial formation at 69 atm. The sample is FeTi with 4 wt.% mischmetal

and the last two have been studied by *Johnson* and *Pangborn* [4.101]. The results on the effect of formation temperature for FeTi with 4 wt% mischmetal are shown in Fig. 4.7. In the test conducted with the optimum formation temperature of 52 °C at a formation pressure of 71 atm, the particle size was found to be the most significant parameter affecting the initial formation rate; for particle sizes less than 100 mesh (US series sieve designation), the initial rate seems to be directly proportional to the surface area.

Second-Stage Activation. According to Sandrock et al. [4.100], the second stage of activation is very dependent on the alloy microstructure. Improvement of the kinetics during this stage appears to be due to fragmentation as was determined from micrographs. Second-phase precipitates such as $Fe_7Ti_{10}O_3$, Fe_2Ti or excess Ti seem to promote this fragmentation process.

4.5.2 Kinetics of the Fully Activated FeTi

The kinetics of activated FeTi is very fast, and in most experimental systems limited by heat transfer. The kinetics is nearly an order of magnitude slower than that of $LaNi_5$. *Goodell* et al. [4.102] confirmed this heat transfer limitation by using a reactor with very good heat transfer. The thermal time constant is about

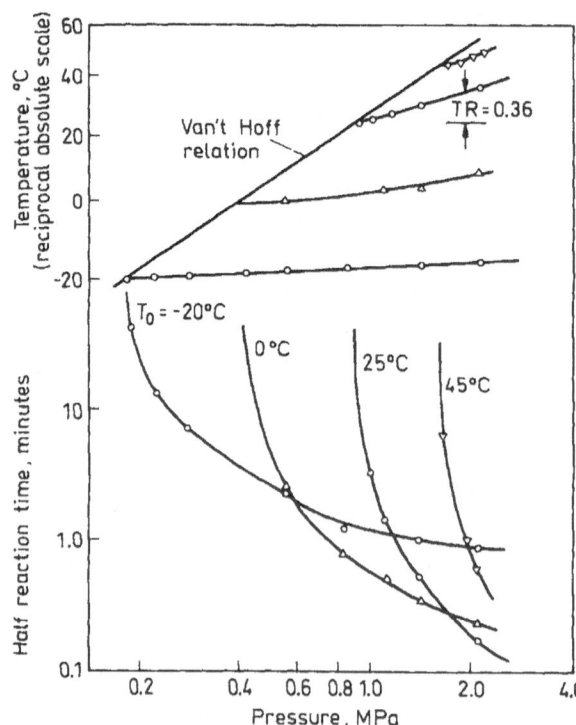

Fig. 4.8. Comparison of isobaric absorption half reaction times for several alloys at a reactor test temperature of 25 °C. Sample identification includes final hydrogen transfer, the plateau pressure at 25 °C and the thermal effect ratio. Improved reactor conductivity results and estimated isothermal response are also included [4.102]

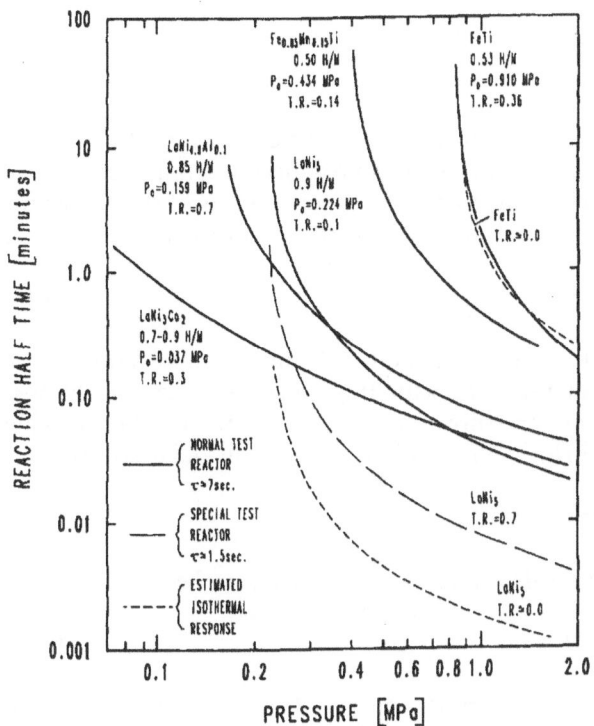

Fig. 4.9. Examples of isobaric absorption kinetic test data for FeTi showing the time when half the reaction is completed ($H/M = 0.3$). The maximum temperature excursions are also shown and compared with the 0.3 H/M dynamic van't Hoff relation [4.102]

7 s and that of the thermocouple used is 0.3 s. Some of their isobaric absorption kinetic data are shown in Fig. 4.8. They introduced a thermal effect ratio $TR = (T^0 - T_0)/(T_H - T_0)$ to identify the heat transfer contribution quantitatively. Here T^0 is the peak sample temperature, T_0 is the nominal reactor temperature and T_H is the temperature at which the equilibrium pressure is equal to the applied hydrogen pressure. If TR equals zero, true isothermal conditions are attained and intrinsic kinetics are represented. If TR equals unity, the kinetics are thermally controlled. The authors made comparisons between tests at $T_0 = 25\,°C$ and at a pressure equal to twice the 25 °C absorption plateau pressure. The results are displayed in Fig. 4.9, in which the reaction data for FeTi are only partially affected by heat transfer ($TR = 0.36$), while those of LaNi$_5$ are controlled by heat transfer ($TR = 1.0$). As a result, the data for FeTi more closely reflect the intrinsic reaction kinetics.

Sandrock and *Goodell* analyzed the data on the monohydride formation and decomposition reactions on non-isothermal bases [4.103]. The guiding principle of their kinetic model is as follows: The relations between reaction fraction ξ, reaction rate r, pressure P, temperature T, and time t are represented as

$$f(\xi) = Ag(P)h(T)t, \tag{4.9}$$

and

$$r = Bg(P)h(T)[f(\xi)] - 1, \tag{4.10}$$

where A and B are constants. The reaction rates were determined at a fixed hydrogen transfer level corresponding to a fixed reaction fraction of about 0.3. In addition, it was assumed that the temperature function was of the Arrhenius type. Therefore the analysis was based on an expression of the form.

$$r = Cg(P)\exp(-Q/RT). \tag{4.11}$$

The activation energy Q, pressure function $g(P)$ and constant term C were determined by correlation of the data with this equation. The best fit was obtained using the logarithmic pressure ratio function as the reaction driving force. The same expression can be applied for both absorption and desorption:

$$r = 19 \times 10^6 \ln(P/P_0)\exp(-9550/RT) \tag{4.12}$$

where the plateau pressure at the fixed reaction fraction, P_0 can be expressed as

$$\ln P_0 = -\Delta H/RT + \Delta S/R. \tag{4.13}$$

For absorption $\Delta H = -5664.1$ cal/mole, $\Delta S = -23.43$ cal/mol. °C and for desorption $\Delta H = 7124.4$ cal/mol., $\Delta S = 26.54$ cal/mol. °C.

Concerning the logarithmic pressure function, the authors indicated that this form can result form the inclusion in the rate equation of an expression for the free energy change which accompanies the phase transformation, and form the concentration dependence of diffusion processes, as *Flanagan* pointed out in his review [4.9]. They argue that the logarithmic pressure function should not occur for processes controlled by surface or external mass transfer, but is appropriate for bulk phase reactions controlled by processes such as diffusion, interface movement or nucleation and growth.

The last function to be determined is the reaction extent function $f(\xi)$. In other kinetic studies [4.104, 105] reaction behavior of the "first-order" type has been reported. *Goodell* and *Sandrock* [4.103] showed, however, that the thermal effect tends to give all rate equations a superficial first-order appearance. It is probably impossible to clearly discriminate the reaction extent function. The available data, on the other hand, seem to indicate that the kinetics is controlled by nucleation and crystal growth processes. In this case the Johnson–Mehl–Avrami equation can be reasonably applied in the form

$$f(\xi) = -[\ln(1-\xi)]^{1/n}, \tag{4.14}$$

where n originally represents the growth dimensionality. The first part of the reaction is well described by this expression with $n = 2$. For the later part, the

Table 4.2. Kinetic model for iron–titanium hydrogenation/dehydrogenation

Reactions:

$$1.89\text{FeTiH}_{0.08} + \text{H}_2 \rightarrow 1.89\text{FeTiH}_{1.14}$$
$$2.13\text{FeTiH}_{1.02} \rightarrow \text{H}_2 + 2.13\text{FeTiH}_{0.08}$$

Kinetic Model:

A. Basic Form—applicable to isothermal, isobaric reaction conditions.

$$f(\alpha) = A f(P) f(T) t$$

B. Open Form—applicable to variable temperature and pressure reaction conditions.

$$t = \int_0^\alpha \mathcal{R}(\alpha) d\alpha$$

where

$$\mathcal{R} = A \psi f(P) f(T) [f'(\alpha)]^{-1}$$

with

$$T = T_0 + g(\mathcal{R}) \qquad \text{(from heat balance)}$$
$$P = P_1 + g(\mathcal{R}) \qquad \text{(flow restriction, if any)}$$

Parameter Functions:

$$f(\alpha) = \xi [-\ln(1-\alpha)]^{1/n}$$

where $n = 1.7 - 0.6\alpha$

$$f'(\alpha) = f(\alpha) \left\{ \frac{-1}{(1.7 - 0.6\alpha)(1-\alpha)\ln(1-\alpha)} + \frac{0.6\ln(-\ln(1-\alpha))}{(1.7 - 0.6\alpha)^2} \right\}$$

$$f(P) = \ln(P/P_0)$$

where $\pm \ln P_0 = -\Delta H^*/RT + \Delta S^*/R$

$$f(T) = \exp(-Q/RT)$$

Definitions:

α = fraction of reaction completed
P = hydrogen partial pressure
T = sample temperature; t = time
$\mathcal{R}$ = reaction rate; A = constant
ψ = maximum hydrogen transfer
ξ = regime control parameter $\equiv 1$ for $T > 20\,^\circ\text{C}$ or for $0.7 < |\ln P/P_0| < 1.4$ if $T < 20\,^\circ\text{C}$

Dimensions and Constants:
For t in minutes; T in Kelvin; P in atm.; $\mathcal{R}$ in H/M/min.:

Term	Absorption	Desorption
ψ (H/M)	0.53	0.45
A (H/M/min.)	43×10^6	51×10^6
Q (cal/mole)	9550	9550
ΔH^* (cal/mole)	-5664	7124
ΔS^* (cal/mole·K)	-23.43	26.54

appropriate value of n tends to decrease to about unity. This variation of n is in fact expected from the nucleation and growth model [4.106]. A value n in the range 1.7–0.6 was derived [4.102]. A complete description of the rate equation for FeTi is presented in Table 4.2. This can be applied to model simulations for any practical system design, although applicability is limited to the monohydride region.

Park and *Lee* [4.107] investigated the kinetics of FeTi. There is some doubt as to whether their data actually reflect the intrinsic kinetics, because the reactor used seems to have had a rather poor heat transfer characteristic. The temperature increase upon reaction as measured by inserting a thermocouple into the reactor was below 2 °C. The pressure sweep technique was used. The reacted fraction is obtained from the pressure change. Since the driving force $(P-P_{H_2})$ changes with time, the data cannot be analyzed theoretically. They introduced a mathematical approach to obtain a set of kinetic data at constant pressure from the pressure-sweep data in which the reacted fraction F is defined as:

$$F = -[n(P_c) - n(P_i)]/nf, \qquad (4.15)$$

where nf is the equilibrium number of moles of hydrogen in the FeTi–H system calculated from the $P-c$ isotherm ($nf = 3.3 \times 10^{-2}$ mol at 294 K), $n(P_i)$ is the total number of moles of hydrogen gas present in the system, and $n(p_0)$ is the number of moles of hydrogen gas remaining after the pressure has changed from P_i to P_0; $n(P)$ can be obtained using Van der Waal's equation.

Figure 4.10 shows F versus t plots converted from the pressure-sweep data. The reaction rates are a little lower than those obtained by *Goodell* et al. [4.102]. The following rate equations were proposed for FeTi:

For the initial stage ($F < 0.25$)

$$F = 3.5 \times 10^{-2} \exp(-800/RT) \frac{P_0 - P_{H_2}}{T^{1/2}} t \qquad (4.16)$$

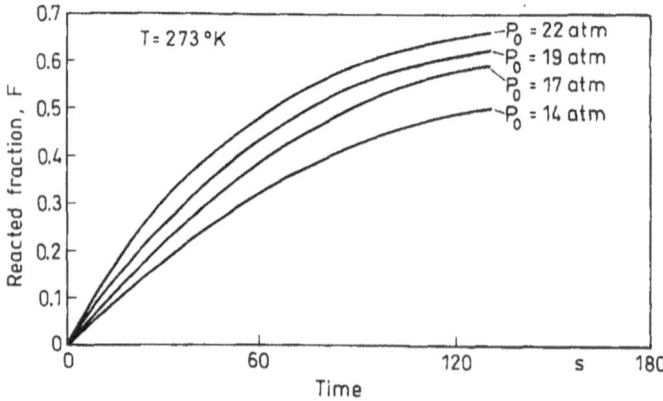

Fig. 4.10. The hydriding kinetics of FeTi at 273 K and various constant pressures [4.106]

and for the later stage $(0.35 < F < 0.55)$

$$(1 - F^*)^{1/3} - (1 - F)^{1/3} = 5.6 \times 10^{-3} \exp(-1000/RT)(P_0^{1/2} - P_H^{1/2})(t - t^*)$$
(4.17)

where F^* and t^* denote the reacted fraction and the time of the initial stage. It was proposed that the rate-controlling step of the initial stage is chemisorption, and that of the later stage is the hydriding reaction. The rate increases with increasing pressure at constant temperature and with decreasing temperature at constant pressure.

Park and *Lee* [4.107] reported an investigation into the effect of the evacuation time (desorption time) on the absorption rate in the subsequent hydrogen uptake. They found that the initial hydrogen absorption rate of the following cycle increased as the evacuation time decreased until a saturation value was attained. The change in the rate with the evacuation time seems to be due to the remaining hydrogen in the sample which promotes hydride nucleation in the subsequent reaction. This result indicates that the reaction in the activated state is controlled by the nucleation and crystal growth as emphasized by *Rudman* in his review [4.105]. This nucleation and the growth controlling mechanism is also obtained by *Chung* and *Lee* [4.108] for FeTi and $Fe_{0.85}Mn_{0.1}Ti$ but is not observed at all for $Fe_{0.9}Ni_{0.1}Ti$.

4.6 Results on AB_2 Laves Phase Alloys

AB_2-type alloys of Ti and Mn are the most interesting materials in this group. They have been developed by Matsushita Electric Industrial Co. [4.109]. They have hexagonal (C14) structure whose stoichiometric composition is AB_2. Non-stoichiometric alloys of low Mn composition were shown to have better hydrogen sorption properties. According to *Bernauer* et al. [4.110], the optimized composition $Ti_{0.98}Zr_{0.02}V_{0.43}Fe_{0.09}Cr_{0.05}Mn_{0.15}$ has such fast absorption kinetics that heat transfer is the rate-limiting parameter. The fast kinetics results in a high bulk diffusion constant

$$D = (1.8 \pm 0.8) \times 10^{-4}(cm^2 s^{-1}) \exp\left(\frac{-190 \pm 30\,meV}{k_B T}\right)$$

measured with quasi-elastic neutron spectroscopy.

When the representative composition is $TiMn_{1.5}$, the excess Ti atoms replace Mn atoms and the more appropriate formula is $Ti(Ti_xMn_{1-x})_2$ [4.111]. The initial activation is very easy for this alloy and the reaction rate for the fully activated state is extremely fast. Some rather qualitative results by *Gamo* et al. [4.109] are shown in Fig. 4.11. For the absorption measurements using the pressure-sweep method, the experimental conditions were selected to give the following relations: $\log(P_i/P_e) = 0.2$ and $\log(P_f/P_e) = 0.1$, where P_i is the

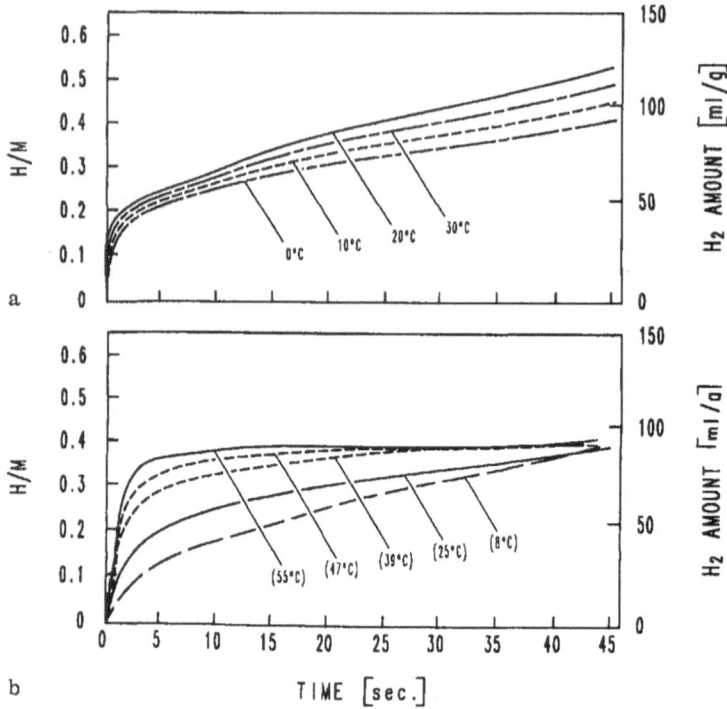

Fig. 4.11 a, b. Hydrogen absorption speed (a) and desorption speed (b) [4.109]

initial hydrogen pressure applied. P_e is the equilibrium pressure at the experimental temperature and P_f is the pressure at the completion of absorption. For the desorption, after the sample had reached equilibrium with the applied pressure and temperature at $H/M = 0.6$, the system was immediately opened to air. Isothermic conditions can hardly be maintained and the later stage is completely controlled by heat transfer. It is well understood that the initial rate within the first several seconds is extremely fast. It should be noted for absorption, as the authors pointed out, that the maximum rate was obtained at 20 °C, under the same driving force of $\log(P_i/P_e)$.

Wallace and his group studied $ZrMn_2$-based hexagonal Laves-phase hydrides [4.112]. They presented the kinetic data on non-stoichiometric Zr-poor $ZrMn_{3.8}$, $(Zr_{0.63}Mn_{0.37})Mn_2$ and $Zr_{0.7}Ti_{0.3}Mn_2$. Their results also showed that the initial rate of absorption is very rapid and the curves obtained are similar to those of $TiMn_{1.5}$. At 30 °C, the material absorbs 92% of the saturation value in 30 s. The release of hydrogen is 90% complete after 2 min at 80 °C, which is slower than the uptake. They indicate that the thermodynamic properties of this Zr-poor Laves-phase hydride may be reflected in the fast reaction rate. The enthalpy of hydride formation is about -17 kJ per mole of H_2 and much smaller than that for LaNi$_5$ and FeTi. This means that the heat-transfer effect is much less severe.

Perewesenzew et al. [4.113] assume that the hydrogen sorption characteristics can be changed by changing the composition of the intermetallic compound: $Zr(Cr_{1-x}V_x)_2$; ($x = 0$; $x = 0.2$; $x = 0.4$).

Smith et al. [4.114] studied hydriding kinetics and surface compositions of Laves-phase compounds of rare earth ErT$_2$ (T = Mn, Fe, Co, Ni). The rate of absorption follows the trend ErMn$_2$ > ErFe$_2$ > ErCo$_2$ > ErNi$_2$ with the rates differing by about a factor of 500 from ErMn$_2$ to ErNi$_2$. The kinetic results are correlated with the observed surface compositions determined by Auger electron spectroscopy. The surface enrichment by the *d*-band transition metal is greatest for ErMn$_2$ and least for ErNi$_2$. It seems that the sorption kinetics are more rapid for Mn- and Fe-containing systems than for similar systems containing Co or Ni.

Shaltiel and his group have applied thermal desorption spectroscopy (TDS) to investigate the kinetics of desorption as well as the energetics of hydrogen in hydrides, for the cubic Laves phases ZrV$_2$ and HfV$_2$ [4.115], see also Chap. 7.

The spectrum of HfV$_2$H$_x$ is shown in Fig. 4.12 and the desorption rate is plotted against that of ZrV$_2$H$_x$ in Fig. 4.11a. The overall structures of the spectra and their development with increasing initial concentration are similar in both figures and both have three peaks I, II and III. By integrating the spectrum with respect to time, the positions of the extrema can be determined as a function of x. For example, in the case of ZrV$_2$H$_x$, the maxima I, II and III correspond to $x = 0.7$, 2.5 and 4.3, respectively. These results can be correlated with the structural data obtained from neutron diffraction. The hydrogen atoms are located in tetrahedral interstitial sites. There are 17 tetrahedra per AB_2 formula

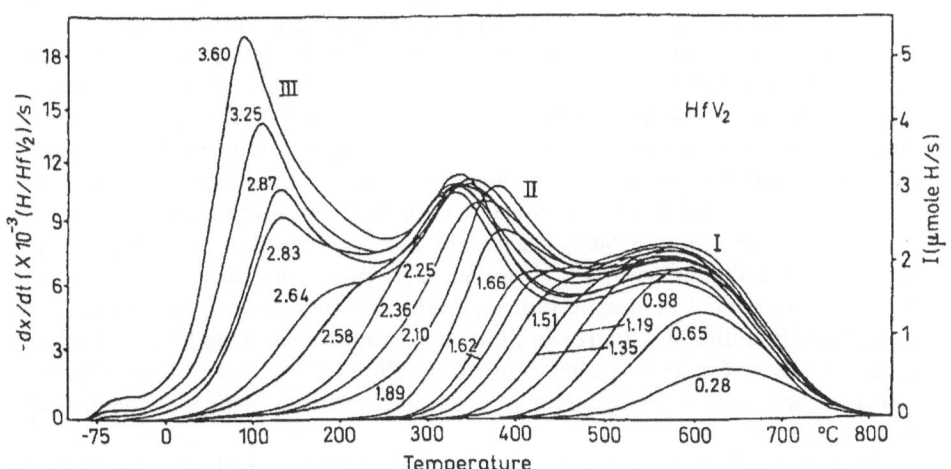

Fig. 4.12. The thermal desorption spectra of HfV$_2$H$_x$. The numbers on the curves denote the initial hydrogen concentration [4.114]

unit of which 12 are composed of two A atoms and two B atoms (A_2B_2), four are of three B atoms and one A atom (AB_3), and one is composed of four B atoms (B_4). The neutron diffraction data for ZrV_2D_x [4.116] show that at room temperature only the A_2B_2 site is occupied at concentrations up to $x = 2.5$. At $x = 2.5$, the occupancy of the AB_3 sites starts to increase while that of the A_2B_2 site increases more slowly for $x > 2.5$. These results indicate that the A and B lines are associated with hydrogen atoms desorbing from the A_2B_2 while the C line is associated with hydrogen atoms desorbing from the AB_3. For the occurrence of the I and II peaks, the following explanation has been proposed: In the A_2B_2 sites, the main contribution to the bond energy comes from the Zr–H bond. For $x > 1$, the excess hydrogen atoms have to share a Zr atom with another hydrogen atom [4.117]. At the beginning of the spectrum, i.e. at the lowest temperature at which desorption occurs, the desorption rate shows an exponential dependence on $1/T$:

$$ -\frac{dN}{dt} = I = I_0 \exp\left(\frac{-E}{T}\right) $$

by plotting $\ln I$ vs $1/T$, a straight line was obtained from which the activation energy E and pre-exponential factor I_0 were extracted [4.115].

The kinetics of the Daimler–Benz alloy $Ti_{0.98}Zr_{0.02}V_{0.43}Fe_{0.09}Cr_{0.05}Mn_{0.15}$ were analyzed by *Domschke* et al. [4.21] under isochoric conditions. Kinetics comparable to those of $LaNi_5$ were obtained.

4.7 Conclusion

To summarize the kinetics of the major hydride-forming compounds, one can say that formations are very fast for activated compounds of the families $LaNi_5$, FeTi and $ZrMn_2$. In many experiments, even on the laboratory scale, heat transfer is often the rate-limiting step. Of course, it is always rate limiting in pilot plants. A promising way to increase heat transfer is with PMH (pressed metal hydrides) or "compacts" obtained by mixing the active material with Al [4.118], Ni, Cu, etc. However, this has the negative consequence of a decrease in the ratio of stored H_2 to the mass (Chap. 5). From a fundamental point of view, the rate-determining steps are usually nucleation and growth or hydrogen diffusion for activated samples. However, after poisoning, surface effects again become dominant. For magnesium formation/decomposition, the kinetics are slow and rate limited by hydrogen diffusion through a growing layer of hydride or metal. Faster kinetics can be obtained by alloying Mg, but further improvements are necessary since the advantageous characteristics of the alloy deteriorate with cycling.

For the analysis by statistical rate theory of the transport rate of hydrogen across the hydrogen–metal interface and for experiments on the H_2–Mg system, the reader is referred to [4.119] and [4.120] respectively.

Results of the analysis of the kinetics of hydrogen–metal–hydride systems, including metal hydride beds, with emphasis on the effects of non-isothermal conditions can be found in [4.121–4.123].

References

4.1   M.H. Mintz, J. Bloch: Prog. Solid State Chem. **16**, 163 (1985)
4.2   K.C. Hong, K. Sapru: Int. J. Hydrogen Energy **12**, 165 (1987)
4.3   V.V. Boldyrev, M. Bulens, B. Delmon: "The control of the reactivity of solids" in *Studies in Surface Sciences and Catalysis* (Elsevier, Amsterdam 1979)
4.4   M. Ron, M. Elembach: Proc. Int. Symp. on Hydrides for Energy Storage, Geilo Norway (1977), ed. by A.F. Andresen, A.J. Maeland (Pergamon, Oxford 1978) p. 417
4.5   B. Delmon: *Introduction à la cinétique hétérogène* (Techniq, Paris 1973)
4.6   P. Barret: *Cinétique hétérogène* (Gauthier Villars, Paris 1979)
4.7   A.Y. Rozovsky: *Topochemical Reaction Kinetics* (Kimia, Moscow 1973)
4.8   G. Bertrand, M. Lallemant, G. Watelle: Propos sur l'interprétation de l'énergie d'activation expérimentale. J. Therm. Anal. **13**, 525 (1978)
4.9   T.B. Flanagan: Proc. Int. Symp. on Hydrides for Energy Storage (1977), ed. by A.F. Andresen, A.J. Maeland (Pergamon, Oxford 1978)
4.10 C.E. Lundin, F.E. Lynch: A.R.P.A. report No. 2552 (1975)
4.11 F.A. Kuijpers: Thesis Technological Univ. Delft (1973), Philips Res. Rept. Suppl. No. 2 (1973)
4.12 O. Boser: J. Less-Common Met **46**, 91 (1976)
4.13 S. Tanaka, J.D. Clewley, T.B. Flanagan: J. Phys. Chem. **81**, 121 (1977)
4.14 A.J. Goody, D.S. Strokes, J.A. Gazzilio: J. Less-Common Met. **91**, 149 (1983)
4.15 A. Karty, J. Grunzweig-Genossar, P.S. Rudman: J. Appl. Phys. **50**, 7200 (1979)
4.16 Y. Josephy, M. Ron: Z. Phys. Chem. NF **164**, 1233 (1989)
4.17 X.L. Wang, S. Suda: Z. Phys. Chem. NF **164**, 1235 (1989); S. Suda, N. Kobayashi, K. Yoshida: J. Less-Common Met. **73**, 119 (1980)
4.18 N. Gérard, L. Belkbir, E. Joly: J. Phys. E **12**, 476 (1979)
4.19 P.D. Goodell, G.D. Sandrock, E.L. Huston: J. Less-Common Met. **73**, 135 (1980)
4.20 E. Akiba, K. Nomura, S. Ono, S. Suda: Int. J. Hydrogen Energy 7, 787 (1982)
4.21 T. Domschke, E. Schütt, I. Haas: Int. J. Hydrogen Energy **14**, 671 (1989)
4.22 H.M. Lutz, R. Schmitt, F. Steffeno: Thermochimica Acta **24**, 369 (1978)
4.23 A.S. Pedersen, J. Kjoller, B. Larsen, B. Vigehom: Int. J. Hydrogen Energy **8**, 205 (1983)
4.24 J. Isler: Thesis, 'Etude cinétique des réactions d'hydruration et de deshydruration du Mg et du mélange Mg/LaNi₅'. Dijon Univ. (1979) (in French)
4.25 D.L. Douglass: Proc. Int. Symp. on Hydrides for Energy Storage 9 (1977), ed. by A.F. Andresen, A.J. Maeland (Pergamon, Oxford 1978) p. 151
4.26 M.H. Mintz, Z. Gavara, G. Kimel, Z. Hadari: J. Less-Common Met. **74**, 263 (1980)
4.27 C.Y. Adachi, K.I. Niki, J. Shiokawa: J. Less-Common Met. **88**, 213 (1982)
4.28 C.J.M. Northrup, W.J. Kass, A.J. Beattle: Proc. Int. Symp. on Hydrides for Energy Storage. Geilo Norway (1977) (Pergamon, Oxford 1978) p. 205
4.29 T. Hirata, T. Matsumoto, M. Amarro, Y. Sasaki: J. Phys. **11**, 251 (1981)
4.30 N. Gérard: J. Phys. E **7**, 509 (1974)
4.31 J.M. Boulet: Thesis. 'Contribution à l'étude des mécanismes de formation et de décomposition des hydrures des alliages de Mg/Ce et Mg/Ni' Dijon Univ. (1982) (in French)
4.32 L. Belkbir: Thesis. 'Contribution à l'étude cinétique de l'évolution des systèmes constitués par l'hydrogène et les alliages de LaNi₅ faiblement substitués'. Dijon Univ. (1979)
4.33 W. Auer, M.J. Grabke: Ber. Bunsenges. Phys. Chem. **1**, 78 (1974)
4.34 X. Wang, S. Suda: Proc. of the MRS Int. Meeting on Advanced Materials (May 30–June 3, 1988) Tokyo, Japan, ed. by Y. Moro-oka, S. Ono, Y. Sasaki, S. Suda (Mater. Res. Soc., Pittsburgh 1989) vol. 2, p. 137
4.35 P.D. Goodell, P.S. Rudman: J. Less-Common Met. **89**, 117 (1983)
4.35a J.H.N. van Vucht, F.A. Kuijpers, H.C.A.M. Bruning: Philips Res. Rep. **25**, 133 (1970)
4.35b J.J. Reilly, R.H. Wiswall: Rep. BNL 17136, 1972 (Brookhaven National Laboratories Upton, NY)

4.35c C.E. Lundin, F.E. Lynch: Proc. 10th Intersociety Energy Conversion Engineering Conf., Newark, DE, 1975 (IEEE, New York 1975) p. 1380
4.35d M. Kitada: J. Met. Soc. Jpn. **41**, 412 (1977)
4.35e P.D. Goodell, G.D. Sandrock, E.L. Huston: Final Rep. SAND 79-7095, 1978 (Sandia National Laboratories Contract 13-0524)
4.35f L. Belkbir, N. Gerard, A. Percheron-Guégan, J.C. Achard: Int. J. Hydrogen Energy **4**, 541 (1979)
4.35g L. Mangi, Q. Zhenzhong, W. Pingsen: Acta Metall. Sin. **16**, 65 (1980)
4.35h T. Tanaka, M. Miyamoto, K. Yamamichi: Proc. 15th Fall Meet., October 13–14, 1981 (Society of Chemical Engineers, Kanazawa 1981) p. 411
4.36 J.I. Han, J.Y. Lee: Int. J. Hydrogen Energy **14**, 181 (1989)
4.37 P.S. Rudman: J. Less-Common Met. **89**, 117 (1983)
4.38 I.R. Konenko, N.M. Parfenova, K. Klabunovskii, E.M. Savistkii, V.P. Morodin, T.P. Makarochikina: Isvestiya akademii Nauk SSSR **5**, 981 (1980)
4.39 H. Bjurström, S. Suda: Int. J. Hydrogen Energy **14**, 19 (1989)
4.40 P. Dantzer, E. Orgaz: J. Less-Common Met. **147**, 27 (1989)
4.41 C. Bayane, M. El Hammioui, E. Sciora, N. Gérard: J. Less-Common Met. **107**, 213 (1985)
4.42 C.N. Park, J.Y. Lee: J. Less-Common Met. **83**, 39 (1982)
4.43 E. Sciora-Joly: Thesis. Contribution à l'étude des lois de vitesse des systèmes hétérogènes solides-gaz. Dijon Univ. (1981) (in French)
4.44 M. Miyamoto, K. Yamaji, Y. Nakata: J. Less-Common Met. **89**, 111 (1983)
4.45 B. Bogdanovic: Angew. Chem. **97**, 253 (1985); Angew. Chem. Int. Edu. Engl. **24**, 262 (1985)
4.46 J. Genossar, P.S. Rudman: Z. Phys. Chem. NF **116**, 799 (1979)
4.47 T.N. Dymova, Z.K. Sterlyadkina, V.G. Sofronov: Russ. J. Inorg. Chem. **6**, 309 (1961)
4.48 K.B. Gerasimov, E.L. Goldberg, E.Yu. Ivanov: J. Less-Common Met. **131**, 99 (1987)
4.49 K.B. Gerasimov, E.Yu. Ivanov: Mater. Lett. **3**, 497 (1985)
4.50 C.M. Stander: Z. Phys. Chem. NF **104**, 229 (1977)
4.51 T. Schober: Metall. Trans. A **12A**, 951 (1981)
4.52 J. Genossar, P.S. Rudman: J. Less-Common Met. **73**, 113 (1980)
4.53 M.H. Mintz, Z. Gavara, G. Kimmel, Z. Hadari: J. Less-Common Met. **74**, 262 (1980)
4.54 J. Chene: Private communication; A. Roustila: Thesis, Orsay Univ. (1982) (in French)
4.55 R. Fromageau, J. Hillaret, E. Ligeon, C. Mairey, G. Revel, P. Tzanetakis: J. Appl. Phys. **52**, 7191 (1981)
4.56 L. Belkbir, E. Joly, N. Gerard: Int. Hydrogen Energy **6**, 285 (1981)
4.57 J.F. Stampfer, C.E. Holley, J.F. Suttle: J. Am. Chem. Soc. **82**, 3504 (1960)
4.58 J. Renner, H.J. Grabke: Z. Metal. **63**, 289 (1972)
4.59 P.S. Rudman: J. Appl. Phys. **50**, 7195 (1979)
4.60 B. Tanguy, J.L. Soubeyroux, M. Pezat, J. Portier, P. Hagenmuller: Mater. Res. Bull. **11**, 1441 (1976)
4.61 E. Ivanov, I. Kornstanchuk, A. Stepanov, J. Boldyrev: J. Less-Common Met. **131**, 25 (1987)
4.62 M.H. Mintz, Z. Gavara, Z. Hadori: J. Inorg. Chem. **4**, 765 (1978)
4.63 E.G. Eisenberg, D.A. Zagnoli, J.J. Sheridan: J. Less-Common Met. **74**, 223 (1980)
4.64 B. Bokhonov, E. Ivanov, J. Boldyrev: Mater. Lett. **5**, 218 (1987)
4.65 P.W.M. Jacobs, F.C. Tomkins: In *Chemistry of Solid State*, ed. by W.E. Garner (Butterworths, London 1955)
4.66 B.V. Erofeev: Cr. Acad. Sci. URSS **52**, 511 (1946)
4.67 M. Krussanova, T. Terzieva, P. Peshev, E. Yu Ivanov: Mater. Res. Bull. **22**, 405 (1987)
4.68 C.M. Stander: J. Inorg. Nucl. Chem. **39**, 221 (1977)
4.69 M.Y. Song, 3rd Cycle Thesis, No. 354, Univ. Bordeaux Talence, "A study of the Hydriding Kinetics and Characteristics of the Mg_2Ni-H_2 system"
4.70 J.P. Faust, E.D. Whitney, H.D. Batma, T.L. Heying, C.E. Folge: J. Appl. Chem. **10**, 187 (1960)
4.71 J. Renner, H.J. Grabke: Ber. KFA. Jülich **6**, 205 (1972)
4.72 J.J. Reilly, R.H. Wiswall: Inorg. Chem. **83**, 2234 (1968)
4.74 E. Freund, C. Guillerm: IFP Rep. IFP **26-209**, 58 (1978)
4.75 B. Darriet, M. Pezat, A. Hbika, P. Hagenmuller: Int. J. Hydrogen Energy **5**, 173 (1980)
4.76 M. Khrussanova, M. Terzieva, P. Peshev: Int. J. Hydrogen Energy **11**, 331 (1986)
4.77 B. Darriet, A. Hbika, M. Pezat: J. Less-Common Met. **75**, 427 (1980)
4.78 M. Pezat, B. Darriet, P. Hagenmuller: J. Less-Common Met. **74**, 427 (1980)
4.79 M. Khrussanova, M. Pezat, B. Darriet, P. Hagenmuller: J. Less-Common Met. **86**, 153 (1982)
4.80 H. Oesterreicher, H. Bittner: J. Less-Common Met. **73**, 339 (1980)

4.81   D. Lupu, A. Biris, E. Indrea, N. Aldea, R.V. Bucur, M. Maraniv: Int. J. Hydrogen Energy
        8, 797 (1983)
4.82   S. Muroi, M. Saito: Mem. Fac. Liberal Arts educ. part 2. (Yamanashi Univ.) **32**, 31 (1981)
4.83   J.M. Boulet, N. Gérard: J. Less-Common Met. **89**, 151 (1983)
4.84   F. Stucki: Int. J. Hydrogen Energy **8**, 49 (1983)
4.85   D.L. Douglass: Metall. Trans. A **6A**, 2179 (1975)
4.86   K. Nomura, E. Akiba, S. Ono: Int. J. Hydrogen Energy **6**, 295 (1981)
4.87   E. Akiba, K. Nomura, S. Ono, S. Suda: Proc. 3rd. World Hydrogen Energy Conference,
        Tokyo, Japan (Pergamon, Oxford 1980) p. 76
4.88   S. Mattsof, D. Noreus: Int. J. Hydrogen Energy. **12**, 333 (1987)
4.89   W.B. Jung, K.S. Nahm, W.Y. Lee: Int. J. Hydrogen Energy **15**, 641 (1990)
4.90   M.Y. Song, B. Darriet, M. Pezat, J.Y. Lee, P. Hagenmuller: J. Less-Common Met. **118**, 235
        (1986)
4.91   M.Y. Song, M. Pezat, B. Darriet, P. Hagenmuller: J. Solid State Chem. **56**, 191 (1985); Int.
        J. Hydrogen Energy **12**, 27 (1987)
4.92   D. Lupu, A. Biris, G. Mihailescu, R. Sârbu, D. Vomica: Int. J. Hydrogen Energy **13**, 685 (1988)
4.93   M.Y. Song, M. Pezat, B. Darriet, P. Hagenmuller: J. Mater. Sci. **20**, 2958 (1985)
4.94   L. Schlapbach, D. Shaltiel: Mater. Res. Bull. **14**, 972 (1979)
4.95   H.C. Siegmann, L. Schlapbach, C.R. Brundle: Phys. Rev. Lett. **40**, 972 (1978)
4.96   L. Schlapbach, A. Seiler, F. Stucki, H.C. Siegmann: J. Less-Common Met. **73**, 145 (1980)
4.97   N. Gérard, G. Boscher, M. El Hammioui: J. Less-Common Met. **109**, 1 (1985)
4.98   E.I. Ivanov, M. Pezat, B. Darriet, P. Hagenmuller: Rev. Chim. Min. **20**, 60 (1983)
4.99   G.D. Sandrock, J.J. Reilly, J.R. Johnson: Proc. 11th. Intersociety Energy Conversion
        Engineering Conf. (1976) p. 965
4.100   G.D. Sandrock: BNL Final Report 352310S (June 1976)
4.101   D.G. Johnson, J.B. Pangborn: J. Less-Common Met. **73**, 127 (1980)
4.102   P.D. Goodell, G.D. Sandrock, E.L. Huston: J. Less-Common Met. **73**, 135 (1980)
4.103   P.D. Goodell, G.D. Sandrock: Department of Energy/Brookhaven National Lab. 51174 (1980)
4.104   J.J. Reilly: Proc. Int. Symp. Hydrides for Energy Storage, Geilo Norway 1977 (Pergamon,
        Oxford 1978) p. 301
4.105   P.S. Rudman: J. Less-Common Met. **89**, 93 (1983)
4.106   J.Y. Lee, S.M. Byun, C.N. Park, J.K. Park: J. Less-Common Met. **87**, 149 (1982)
4.107   C.N. Park, Y.J. Lee: J. Less-Common Met. **87**, 149 (1982)
4.108   H.S. Chung, J.Y. Lee: Int. J. Hydrogen Energy **11**, 335 (1986)
4.109   T. Gamo, Y. Moriwaki, N. Yanagihara, T. Iwaki: National Technical Report (Matsushita)
        29, 78 (1983)
4.110   O. Bernauer, J. Topler, D. Noreus, R. Hempelmann, D. Richter. Int. J. Hydrogen Energy
        14, 187 (1989)
4.111   D. Fruchard, J.L. Soubeyroux, R. Hempelmann: J. Less-Common Met. **99**, 307 (1984)
4.112   F. Pourarian, V.K. Sinha, W.E. Wallace, H.K. Smith: J. Less-Common Met. **88**, 451 (1982)
4.113   A. Perevesenzew, E. Lanzel, O.J. Eder, E. Tuscher, P. Weinzierl: J. Less-Common Met. **143**,
        39 (1988)
4.114   H.K. Smith, W.E. Wallace, R.S. Craig: J. Less-Common Met. **94**, 85 (1983)
4.115   A. Stern, A. Resnik, D. Shaltiel: J. Less-Common Met. **88**, 431 (1982)
4.116   J.J. Didisheim, K. Yvon, P. Fisher, P. Tissot: Solid State Commun. **38**, 637 (1981)
4.117   A. Stern, S.R. Kreitzman, A. Resnik, D. Shaltiel, V. Zevin: Solid State Commun. **40**, 837 (1981)
4.118   R.Y. Josephy, M. Ron: Z. Phys. Chem. NF **147**, 233 (1986)
4.119   C.A. Ward, B. Farahbakhsh, R.D. Venter: Z. Phys. Chem. NF **147**, 89 (1986)
4.120   A. Krozer, B. Kasemo: J. Phys. **1**, 1533 (1989); J. Vac. Sci. Technol. A **5**, 1003 (1987)
4.121   S. Suda: Int. J. Hydrogen Energy **12**, 323 (1987)
4.122   X.-L. Wang, S. Suda: Int. J. Hydrogen Energy **15**, 569 (1990)
4.123   M. Groll, W. Supper, R. Werner: Z. Phys. Chem. NF **164**, 1485 (1989); R. Werner, M. Groll:
        J. Less-Common Met. (1991) in press

5. Applications

Gary Sandrock, Seijirau Suda, and Louis Schlapbach

With 14 Figures and 9 Tables

The aim of this chapter is to provide a comprehensive review of actual and potential applications of intermetallic hydrides. We shall also explore the material properties that determine whether a certain application is feasible.

5.1 Historical Survey

Before the mid 1960s the uses of metal hydrides were rather limited. They included ZrH_2 for neutron moderators, UH_3 or $PdH_{0.6}$ for laboratory H_2 purification/storage, and miscellaneous hydrides for specialty metallurgical applications [5.1]. One of the reasons for the limited number of hydride applications centered around the fact that almost all published work up to that time was on binary (elemental) hydrides [5.1-2]. Although numerous binary hydrides are possible, their properties cannot be controlled. In particular, with very few exceptions, their thermodynamic stabilities were either very high or very low.

In 1958, *Libowitz* et al. discovered that the intermetallic compound ZrNi could be made to form a true ternary hydride ($ZrNiH_3$) which not only was reversible but had a stability (dissociation pressure) intermediate between the respective binary hydrides, i.e. the very stable ZrH_2 and the very unstable NiH [5.3]. Although this discovery was not fully appreciated for a decade, the stage had been set for the tailoring of hydride stability necessary for most of the applications we will discuss. Because thousands of intermetallic compounds are possible, the limited world of binary hydrides was suddenly and markedly enlarged to encompass ternary and higher component hydrides.

The full appreciation of the potential of intermetallic hydrides came in the 1970s when pioneering work at Philips Laboratory in the Netherlands and Brookhaven National Laboratory in the U.S. resulted in the discovery of the first examples of AB_5 (LaNi$_5$) and AB (TiFe) families of intermetallic hydrides [5.4, 5]. These hydrides could be formed readily and reversibly at room temperature and at modest hydrogen pressures. Because this period was also one of uncertain future fossil fuel supplies and suggestions of a *Hydrogen Economy* [5.6], early work tended to concentrate on applying intermetallic hydrides as alternative fuel or energy storage media. As we shall show in this chapter, applications and potential applications for intermetallic rechargeable hydrides have developed in areas well beyond the early concepts of fuel and energy storage.

We, the authors of this chapter, are representatives of organizations in our respective countries (U.S., Japan and Switzerland) dedicated to developing commercial applications for intermetallic hydrides. As such, we will present a review of potential applications from the point of view of practical hydride properties required. Before presenting the individual histories of various proposed applications, we will briefly review the properties important to hydride device design.

There are numerous good reviews of intermetallic hydrides that include discussions of applications [5.7–30]. In this review we will try to cover the broad spectrum ranging from hydride devices that are commercially available today through devices that have seen only the prototype stage to concepts that have been proposed but never realized in practice. Commercial viability is dependent on both technical and economic factors. It is beyond the scope of this chapter to discuss economic factors (e.g. markets, production costs, etc.) in more than just limited qualitative terms. In keeping with the spirit of this book, our main intent is to interrelate intermetallic hydride properties with general technical requirements of devices. In addition to the reviews cited above, the proceeding of the International Symposia on the Properties and Applications of Metal Hydrides are good sources of application-related hydride papers [5.31–35].

5.2 Application-Related Properties

The designer of hydride devices must pay special attention to a variety of hydride properties, in particular the list to be dealt with in this section. Although many of these properties are reviewed in detail in other chapters, we will briefly reexamine the list with special reference to applications. For those subjects that have not been covered in earlier chapters, such as heat transfer, gaseous impurity effects and cyclic stability, more detailed reviews will be given.

5.2.1 Pressure–Temperature Relations

The basic P–T properties of a hydride-forming intermetallic compound are the first keys to its application in hydride devices. These properties are best presented in the form of approximately straight lines on a van't Hoff plot or in terms of the van't Hoff equation (discussed in more detail by *Flanagan* and *Oates* in Chap. 3 of Vol. I), which relates the plateau H_2 pressure P to the absolute temperature T, enthalpy change ΔH, entropy change ΔS and gas constant R:

$$\ln P = \frac{\Delta H}{RT} - \frac{\Delta S}{R} \tag{5.1}$$

Alternatively, the parameters ΔH, ΔS and R can be combined to provide a simplified engineering equation:

$$\ln P = \frac{A}{T} + B. \tag{5.2}$$

Van't Hoff plots for several elemental and intermetallic hydrides are shown in Fig. 5.1. The device designer must begin to choose his hydride formers from such data. For example, for a storage system he must match hydride properties that are consistent with his available input pressure and heat sink temperature with his desired output pressure and heat source temperature. Similarly, for a hydride heat pump he must try to narrowly match two or more hydrides to achieve his intended goals. In addition, the engineering implications of the enthalpy change ΔH must be considered from a heat transfer point of view (Sects. 5.2.8–10).

Extensive experimental work on the hydriding properties of intermetallic compounds over the last two decades has resulted in a host of possibilities available for the device designer. Figure 5.1 shows a number of intermetallic hydrides that span the temperature range -20–$300\,^\circ$C and the pressure range 0.01–10 MPa. Note that only elements and binary intermetallics (ternary hydrides) are shown. By using ternary and higher order intermetallics (quaternary and higher hydrides), a large number of additional van't Hoff lines can be produced between those in Fig. 5.1. By selecting the basic intermetallic, the substitution element and its level, the P–T properties can be tuned to the

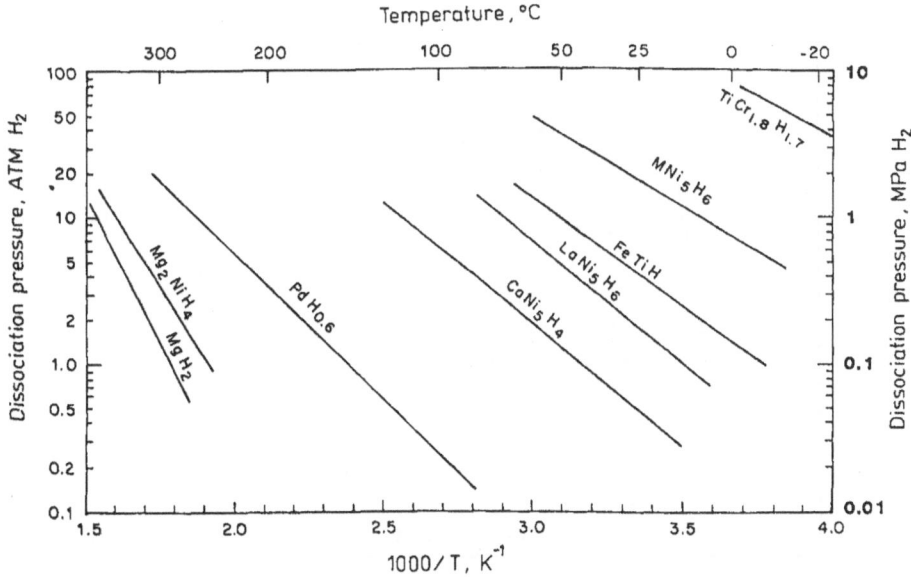

Fig. 5.1. Van't Hoff plots (dissociation) for various intermetallic and elemental hydrides. After [5.21]

particular requirements of a given application. One of the most notable examples of this versatility is shown in the system mischmetal-nickel (MmNi$_5$) where a broad range of properties can be achieved by partially substituting Ca or Ti for Mm or by partially substituting Cu, Fe, Mn, Al, Co, Cr, or Si for Ni [5.36, 37].

5.2.2 Plateau Slope

The van't Hoff plot of Fig. 5.1 represents the *P–T* behavior of a given hydride as a single line. This is generally not the actual case because of the plateau slope and hysteresis that occur with most hydride forming materials. When plateau slope and hysteresis are taken into account, van't Hoff lines are blurred into bands.

First we must consider plateau slope, shown schematically in Fig. 5.2. It can be represented in various ways, but we find the representation

$$\text{slope} = \frac{d(\ln P)}{d(H/M)} \tag{5.3}$$

most convenient. The plateau slope need not be constant all along the plateau.

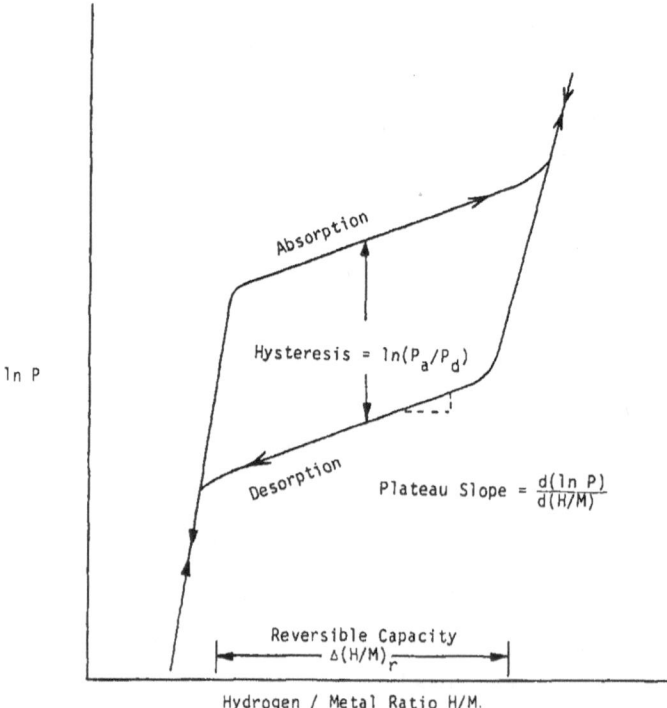

Fig. 5.2. Hypothetical presssure–composition (P–C) loop with definitions of hysteresis, plateau slope and reversible capacity

Carefully prepared binary intermetallics such as $LaNi_5$ [5.38], TiFe [5.5] and Mg_2Ni [5.39] generally show very little plateau slope. Ternary intermetallics usually exhibit plateau slope because of the metallurgical segregation that occurs naturally during solidification [5.36, 40]. Careful preparation techniques, especially annealing before use, can minimize plateau slope. For most hydride applications a near-zero plateau slope is highly desirable.

5.2.3 Hysteresis

Hysteresis is also shown in Fig. 5.2 and is usually quantified as

$$\text{hysteresis} = \ln \frac{P_a}{P_d}, \tag{5.4}$$

where P_a and P_d are absorption and desorption pressures at some value of H/M in the plateau range. The thermodynamic origins of pressure hysteresis are discussed in Chap. 3 of Vol. I. It can vary markedly from system to system. Hysteresis is a very important property in relation to hydride device design. In almost all practical cases minimum hysteresis is desired.

It is often risky to use published hysteresis data for the design of hydride devices. This is not because the data are necessarily wrong but because hysteresis is not a unique materials property and is dependent on the sample's prior history and on the test procedure used. Hysteresis is usually larger when determined by the *dynamic* (continuous) method as opposed to the *static* (point by point) method [5.41–43]. Dynamic hysteresis can be as much as 100% larger than static hysteresis [5.41, 42]. Even with the static method, the magnitude of the hysteresis is often dependent on the size of the aliquot [5.44]. While most hydride devices operate in a dynamic mode, unfortunately most published hysteresis loops are of the static type. In addition, dynamic isotherms sometimes exhibit a peak (or valley) before reaching the absorption (or desorption) plateau [5.41, 43], a nucleation phenomenon not easily detected in a static isotherm. These so-called onset overpressures and underpressures can be significant, sometimes as much as 25% of the plateau pressure itself [5.41, 42]. They are a function of testing rate and temperature.

Because of the uncertainty associated with hysteresis data, it is wise to obtain device design data using a method that closely approximates the operating cycle of the device with samples that have been appropriately cycled. Static PTC data may be fine for a storage container. It is becoming increasingly clear that dynamic data are usually more appropriate for thermodynamic devices like heat pumps or heat engines [5.45–48]. Often one desires temperature hysteresis rather than pressure hysteresis, because the two are not always precisely inter-relatable on a van't Hoff plot [5.49].

5.2.4 Hydrogen Capacity

For most applications, the higher the hydrogen capacity, the better. This is obvious, for example, in H_2 storage for vehicles. Although less obvious, it is also desirable for stationary applications, such as heat pumps where the hydride mass itself results in sensible heat efficiency losses [5.29].

Hydrogen capacity can be easily defined from P–C isotherms. In Fig. 5.2, for example, we have defined the reversible capacity $(\Delta H/M)_r$ as approximately the plateau width. For cyclic systems (compressors, heat pumps, heat engines, etc.) the important parameters are *available hydrogen content* or *transferable hydrogen*. Although usually related to the reversible capacity as defined in Fig. 5.2, the available H_2 content (or transferable H_2) really depends on the performance and efficiency of a system. In the case of heat pumps or multistage compressors, the operating conditions and properties of two or more intermetallic hydrides in tandem dictate the actual transferable H_2 [5.50].

5.2.5 Cost and Availability

Most commercial devices, sooner or later, experience cost competition. Depending on the device (e.g. a storage container), the cost of the intermetallic hydride former can be a significant fraction of the cost of the overall device. Of the common intermetallic compounds of interest today (i.e. AB, AB_2, AB_5, and A_2B) the A elements (e.g. Ti, Zr, Ca, La or other rare earth elements, and Mg) tend to be expensive because of the energy required to separate them from their ores. Thus, the raw materials for the hydriding compounds are generally quite expensive. Secondly, the cost of production can be high, especially for those compounds which require special preparation techniques or are very sensitive to stoichiometry or impurity effects (Vol. I, Chap. 2). As we shall discuss later, the moderately high cost of intermetallic hydriding compounds inhibits certain applications that require large intermetallic inventories (e.g. bulk H_2 storage) but may have only minor impact on devices that cycle rapidly with only a small intermetallic inventory (e.g. compressors or heat pumps).

Hydriding quality intermetallic compounds are commercially available. Ergenics, Inc. (USA) has produced and sold a wide variety of compounds for several years under the trade name HY-STOR [5.51]. Research quantities of HY-STOR alloys can be purchased from Aldrich Chemical Co. [5.52]. Chuo Denki Kogyo, Japan Metals and Chemicals, and Santoku Metal Industry have begun production of various intermetallic compounds in Japan [5.53]. In Europe, hydriding compounds can be purchased from MPD Technology (UK) and Gesellschaft für Elektrometallurgie (FRG) [5.54].

5.2.6 Activation

Before operating any reversible hydride device, the metal hydride former must be activated, i.e. hydrided the first time. The activation of brittle intermetallic compounds consists of hydrogen first penetrating the natural oxide film on as-crushed particles followed by complete hydriding which produces extensive cracking and fragmentation of the particles, thus generating fresh reaction surfaces. Activation is discussed in detail by *Schlapbach* in Chap. 2.

From a practical point of view, activation of most of the ambient temperature hydriding compounds (AB, AB_2, AB_5) can be readily achieved at room temperature and at H_2 pressures a few times the normal absorption plateau pressure. One exception to this is the AB compound TiFe which generally requires heating to about 400 °C for activation [5.5]. *Schlapbach* and *Riesterer* have reviewed the extensive literature on TiFe activation [5.55]. Again, from a user's point of view, it is fortunate that almost any ternary substitution to TiFe (e.g. mischmetal, Mn, Ni, Cr, Co, Al, Si, Zr, Nb, O) results in its ability to activate at room temperature without prior heating [5.16, 23, 40, 56–59].

5.2.7 Volume Change and Decrepitation

Because most of the hydrogen storage intermetallic compounds are inherently brittle, the volume changes during hydride/dehydride cycling result in cracking and, ultimately, particle size breakdown. This phenomenon, often called *decrepitation*, is a mixed blessing. The fine power, typically having surface areas of 0.2–2 m²/g, is desirable for high H_2 reactivity and impurity gettering ability. On the other hand, the fine powder can progressively pack in its container, resulting in high gas impedance, container stress bulging, and even container rupture [5.8, 60, 61]. Thus, from performance and safety points of view, engineering measures must be taken to avoid the packing resulting from decrepitation and cyclic volume changes.

One approach to avoid packing is to containerize the powder in H_2 permeable capsules [5.13, 21, 62]. Another is to include collapsable structures or inert fluffy material within the bed [5.63]. Alternatively, the container can be designed for elastic expansion [5.64]. The powder itself can be lubricated with inert additives such as fluorocarbon powders [5.65] or silicon oils [5.66]. A recent approach has been to incorporate the powder in a liquid slurry [5.67, 68]. Finally, there have been several techniques developed to pelletize the intermetallic powder using plastic, rubber or metal binders [5.69–77].

5.2.8 Heat of Reaction

All of the intermetallic hydrides discussed in this book abrorb H_2 exothermically and desorb it endothermically, i.e. the enthalpy change of hydriding, ΔH, is negative. This inevitable heat of reaction ranges from a nuisance for one designing and using a hydride storage container to the desired product of a hydride heat pump. In either extreme, it is one of the most important engineering properties that the device designer must deal with. Appropriate heat transfer means must be included in the reactor designs. ΔH varies markedly from intermetallic to intermetallic as is discussed by Griessen (Vol. I, Chap. 6). Values of ΔH, along with other useful engineering properties, are tabulated for various hydrogen–intermetallic systems in many of the reviews cited earlier [5.7, 8, 10–13, 15, 16, 18, 21–23, 78, 79].

5.2.9 Absorption/Desorption Kinetics

Intrinsic kinetics are reviewed in Chap. 4 by *Gerard* and *Ono*. For a given metal-hydrogen system, reaction rate constants are a function of pressure and temperature, often suggesting an optimum temperature/pressure regime for use [5.80, 81]. In general, intermetallic compounds show very rapid hydriding and dehydriding kinetics over rather wide temperature ranges, at least after the activation procedure is completed and as long as clean H_2 is used. Intrinsic kinetics are usually so fast that cyclic rates in practical devices are dictated mostly by how rapidly the heat of reaction can be added to or removed from the bed [5.23, 41, 60, 80–82]. A second factor that can limit engineering cycle rates is gas impedance in the bed [5.60]. Thus effective thermal conductivity and gas permeability represent the principal factors that the reactor designer must consider relative to rapid cycling requirements.

5.2.10 Heat Transfer (Effective Thermal Conductivity)

Metal powder beds have poor heat transfer properties. Intermetallic hydrides are no exception, with measured values of effective thermal conductivity seldom exceeding $1-2\,Wm^{-1}\,K^{-1}$ [5.82–87]. This led very early to the realization that a rapid charge/discharge container would require some means of enhancing heat transfer. The most obvious approach, widely used in the chemical industry, is to minimize heat transfer distance and maximize heat transfer surface area, for example using designs incorporating internal plates, fins, and tubing [5.88–92]. Another approach has been to use metal foam (e.g. Al) to nest the hydriding alloy, leading to about a factor of 7 improvement in charging rates [5.90].

The variables associated with the use of three-dimensional heat-transfer-enhancement networks (e.g. foams and rolled meshes) have been extensively studied by Suda et al. at Kogakuin University from both experimental and mathematical points of view [5.93]. Variables included intermetallic compound, network material, H_2 pressure, H concentration and packing density. Each variable was found to have some significance for the effective thermal conductivity. As shown in Fig. 5.3, network material (Cu, Al, Ni, stainless steel) and void fraction are especially important variables in determining effective thermal conductivity.

The most extensive experimental efforts in enhancing heat transfer have centered around the production of pressed composites that contain a high-heat-conductivity metal such as Al or Cu as a conductive skeleton. The aim is not only to increase effective thermal conductivity of the bed but also to provide mechanical stability. Much of the work has been done by *Ron* and co-workers, who dubbed these structures "porous metallic-matrix hydrides" (PMH) [5.75, 94–96]. To obtain the best mechanical stability, *Ron* fabricates his composites with the intermetallic powder in the *hydrided* (expanded) state, the hydrogen content being maintained by exposing the hydride to a poison such as SO_2.

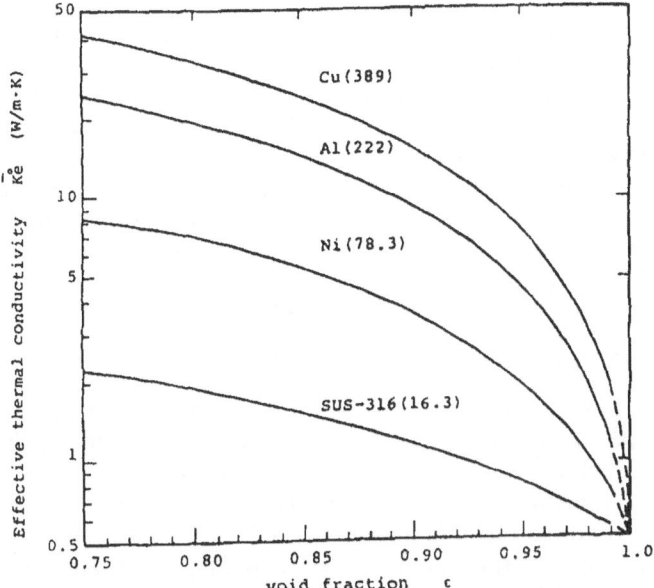

Fig. 5.3. Estimated effective thermal conductivity of three-dimensional metal hydride composite bed as a function of void fraction and network metal [5.93]. Thermal conductivities of network metals shown in parentheses (in units of $Wm^{-1}K^{-1}$). Effective thermal conductivity of hydride taken as $0.5\,Wm^{-1}K^{-1}$

The poisoned hydride can later be reactivated to provide good kinetics. Effective thermal conductivities as high as $32.5\,W\,m^{-1}\,K^{-1}$ have been projected with $LaNi_5$-20wt% Al pellets made in this manner [5.95].

Bernauer and co-workers have independently developed a cold pressing and sintering technique to make high conductivity composites of intermetallic powder and Al [5.71, 97, 98]. In this case, the intermetallic powders are handled in the unhydrided state, avoiding the need for surface poisoning. Effective thermal conductivities of 7–$9\,W\,m^{-1}\,K^{-1}$ are reported with TiFe-10wt % Al composites [5.71]. In addition to the above-mentioned groups, *Tuscher* et al. [5.99] and *Groll* et al. [5.100] have successfully employed the pressed composite approach for engineering size reservoirs.

A new approach called "microencapsulation" has been developed by *Ishikawa* et al. in Japan [5.76]. Fine powder is chemically coated with thin layers of porous Cu and then compacted into pellets.

Although the need for external heat transfer is usually an engineering burden, there are some applications where it can be largely avoided. For example, consider applications that require rapid cycling such as H_2 or D_2 separators, purifiers, or even short duty cycle storage tanks. Because the heat required for H_2 desorption is approximately the same as that generated during absorption, *Sandrock* and *Snape* proposed incorporating a heat storage medium within the bed so that the absorption heat can be used for desorption rather than be rejected [5.101]. The heat storage medium can be of either the latent (phase change) type or the sensible heat type. This "thermal ballast" allows the essentially adiabatic operation of a cyclic hydride/dehydride reactor. This concept was successfully used in a commercial scale H_2/hydride separation unit [5.73, 102] to be discussed later.

The mathematics of heat transfer is well established and used widely in designing industrial heat exchangers. A number of models incorporating mainly heat transfer, and to lesser extents mass transfer and reaction kinetics, have been applied to hydride beds [5.100, 103–108].

5.2.11 Effects of Gaseous Impurities

Some hydride devices use the same H_2 over and over, e.g. heat pumps, heat engines, actuators, temperature sensors, etc. Some use new H_2 for each cycle, e.g. storage containers, compressors, purifiers, etc. In the latter class, it is important to know the impurities present in the H_2 used and the effect of these impurities on cycle life. Although the fundamentals of impurity–hydride surface interactions are becoming more firmly established (Chap. 2), good engineering data are limited. Prior to about 1980 there were relatively few papers on impurity effects and even these gave only qualitative or semi-quantitative data on various intermetallics and impurity gasses [5.109–116]. Since 1980, extensive quantitative data of *Goodell* et al. on *AB* and *AB*$_5$ impurity effects have been published [5.117–122] along with several other useful contributions [5.123–130].

From an engineering point of view, the interactions between intermetallic hydrides and gaseous impurities are varied and complex. An attempt has been made to generalize impurity effects and to provide a quantitative base for cyclic life predictions [5.122]. There are four types of intermetallic–impurity interactions that can be observed: *poisoning, retardation, reaction*, and *innocuous*.

Table 5.1. Intermetallic–impurity interactions. After [5.122]

Type	Definition	$AB_5 = LaNi_5$ $AB = Ti(Fe, Mn)$ Examples
Poisoning	Rapid loss of H-capacity with cycling. Heterogeneous effect in bed with remaining unpoisoned material showing good kinetics.	H_2S & CH_3SH on AB_5 & AB CO on AB_5 & AB (low-temperature)
Retardation	Reduction in absorption/desorption kinetics without significant loss in ultimate capacity.	NH_3 on AB_5 & AB CO_2 on AB_5 CO on AB_5 & AB (high-temperature)
Reaction	Bulk corrosion loss leading to irreversible capacity loss. Unreacted portion of material often shows good kinetics.	O_2 on AB_5
Innocuous	No observable interactions.	N_2 & CH_4 on AB_5 & AB

These terms are defined in Table 5.1, along with some examples. The class and magnitude of the interaction can vary markedly with temperature and/or cycle time. For example, CO is a strong *poison* at room temperature for AB and AB_5 compounds but changes to a relatively mild *retardant/reactant* above 100°C [5.120]. For the case of O_2-$LaNi_5$ interactions, O_2 is a strong *poison/retardant* for the first hour or two. This is then followed by a dramatic *recovery*, after which O_2 becomes a reactant, slowly reducing hydrogen capacity [5.117–119].

It is possible to quantify hydrogen capacity as a function of the number of hydride/dehydride cycles with H_2 with a given impurity. Such data can then be generalized using an exponential damage model, the form of which is the same for any class of intermetallic impurity interaction [5.122].

$$\Delta(H/M) = \Delta(H/M)_0 \exp\left\{ - NA[(C - C_0)/100]^{-m} q\frac{r}{2} \Phi \right\}, \tag{5.5}$$

Table 5.2. Damage model parameters for various intermetallic-impurity interactions [5.122] 15 m absorption + 15 m disorption cycle

Intermetallic and Conditions	Impurity	Damage model Parameters	
		Φ	m
$TiFe_{0.85}Mn_{0.15}$ $T = 85$°C $P_a = 2758$ kPa	H_2S	728	1.0
	CO	692	1.0
	CO_2	692	1.0
	O_2	44.6	1.0
	NH_3	0.90	0.2
	N_2	0.56	0
	CH_4	0.86	0
	C_2H_4	0.92	0
$TiFe_{0.85}Mn_{0.15}$ $T = 25$°C $P_a = 6.55$ kPa	H_2S	1068	1.0
	CO	1320	1.0
	CO_2	1156	1.0
	O_2	169	1.0
	NH_3	12.6	0.5
$LaNi_5$ $T = 85$°C $P_a = 2068$ kPa	CH_3SH	1460	1.0
	H_2S	460	1.0
	CO	990	1.0
	O_2	3.90	0.5
	CO_2	1.03	0.2
	NH_3	1.22	0.2
	N_2	1.23	0
	CH_4	1.24	0
	C_2H_4	1.16	0
$LaNi_5$ $T = 25$°C $P_a = 345$ kPa	CH_3SH	1980	1.0
	H_2S	578	1.0
	CO	1733	1.0
	NH_3	8.6	0.5
	O_2	1.4	0.5
	CO_2	0.78	0.2
	C_2H_4	0.9	0

where

$\Delta(H/M)$ = hydrogen transfer capacity (hydrogen/metal atom ratio)
$\Delta(H/M)_0$ = original hydrogen capacity
N = number of cycless with impure H_2
C = impurity concentration
C_0 = impurity threshold level for damage
m = concentration power dependence
A = constant equal to $(0.001)^{(1-m)}$
$q = (H/M)_0$ = stoichiometry ratio for the appropriate AB_x intermetallic
r = number of operative impurity units per impurity molecule
Φ = damage function

This damage model has useful life prediction capabilities. This most important parameters in the model are the damage function Φ and the concentration exponent m, both of which have physical relationships to the actual damage mechanism [5.122]. Some care must be taken to experimentally determine Φ and m under conditions (e.g. temperature) roughly representative of the application. Values of Φ and m are given in Table 5.2 for a number of impurity interactions with $LaNi_5$ and $TiFe_{0.85}Mn_{0.15}$.

Careful hydride selection must be made for applications involving impure H_2. In most cases it will be impossible to completely avoid damage, but it is fortunate that many damaged alloys can be reactivated by appropriate techniques, usually involving heating and flushing with clean H_2. However, from an applications point of view, it is obvious that much more impurity information must be generated. The number of impurity combinations and levels possible with some of the applications to be discussed is much larger than the presently available data. This lack of data is especially troublesome for hydride applications involving H_2 separation from mixed gases and purification.

5.2.12 Cyclic Stability

In addition to the *extrinsic* impurity effects discussed above there is an *intrinsic* effect that can lead to loss of hydrogen capacity with extended cycling in pure H_2. This phenomenon involves disproportionation of the intermetallic to form a stable hydride. For example, with $LaNi_5$, the desired reversible reaction

$$LaNi_5 + 3H_2 \rightleftharpoons LaNi_5H_6 \tag{5.6}$$

is actually thermodynamically unstable relative to the less reversible disproportionation reaction

$$LaNi_5 + H_2 \longrightarrow LaH_2 + 5Ni. \tag{5.7}$$

There are many possible intermediate stages between reactions (5.6) and (5.7), including the formation of H-trapping sites [5.131] and other phases [5.132]. Because reaction (5.7) requires diffusion of metal atoms, a difficult process at low temperature, reaction (5.6) tends to dominate near room temperature. Thus the entire science and application of rechargeable intermetallic hydrides rests on the *metastable* equilibrium typical of reaction (5.6). One exception of this is Mg_2NiH_4 which is thermodynamically stable against disproportionation, at least at $300\,°C$ [5.16].

The queston facing the hydride device designer is how long he can expect the metastable reversible behavior, e.g. reaction (5.6), to hold, because disproportionation will result in H trapping and an effective loss of reversible capacity. The answer is not simple. There are numerous demonstrated examples of intermetallic disportionation or disproportionation related phenomena in the literature [5.18, 133–143]. Although the common denominator is increasing disproportionation with increasing temperature, there are marked differences from intermetallic to intermetallic as well as differences among reported investigations on the same compound. *Goodell* has reviewed the topic of cyclic stability of AB_5 compounds with the emphasis on $CaNi_5$, $LaNi_5$ and $La(Ni, Al)_5$ [5.143]. A brief review of the relatively disproportionation-prone $CaNi_5$ is also given by *Sandrock* et al. [5.144].

Although disproportionation is often detectable with extended cycling, even near room temperature, it appears that good engineering lives can be achieved. For example, $LaNi_{4.7}Al_{0.3}$ exhibited less than 5% H-capacity loss after cycling 1500 times at 85 °C [5.143]. It is also fortunate that the effects of disproportionation can be largely erased by annealing in H-free environments [5.133, 134, 137, 138, 141, 143]. Sometimes cyclic capacity loss is preceded by changes in plateau pressures. For example, the upper plateaus of AB compounds (e.g. TiFe) increase in pressure with cycling [5.41, 145, 146] and the normally single-plateau $LaNi_5$ develops a second plateau [5.143]. These cyclic plateau pressure effects may not be disproportionation phenomena but rather structural disorder or other strain effects produced by the large volume changes associated with cycling [5.146–148]. In any event, these plateau-pressure effects can also be eliminated by annealing [5.143, 147]. In summary, although long-term disproportionation and disproportionation-like effects must be considered in device design, the potential problems they may pose do not appear to be insurmountable.

5.2.13 Safety

Because of the low pressures and near-ambient temperature involved in the use of intermetallic hydrides, they are generally viewed to be a safer means of handling and storing hydrogen than cryogenic liquid and compressed gas. The principal safety factor of hydrides centers around potential pyrophoricity in a container rupture or routine maintenance situation. Quantitative data on intermetallic hydride safety are limited to the studies of *Lundin* and *Lynch* on $LaNi_5$ [5.149] and TiFe [5.150] along with some limited pyrophoricity tests on AB and AB_5 compounds by *Sandrock* [5.13] and some safety studies done for "Project Sunshine" by the National Chemical Laboratory for Industry (MITI) in Japan [5.151]. These tests show that the AB and AB_5 compounds are reasonably safe to handle, although the AB_5 compound can be mildly pyrophoric when suddenly exposed to air in the activated state. There are no published data on the safety of AB_2 compounds but our own experience, along with that of *Reilly* [5.152], suggests a wide variety of pyrophoric tendencies when in the activated state. Compounds high in Ti and Cr (e.g. $TiCr_{1.8}$) do not appear to be pyrophoric. Those high in Mn are very pyrophoric.

Most of the intermetallic compounds become fine powder upon hydriding. Care should be taken to avoid inhalation of the metallic dust when working with such materials. When designed properly, hydride devices should completely contain the intermetallic powder and potential problems should occur only when the device is scrapped or in the rare case of an accidental rupture.

A few comments on intermetallic hydride containment should be made. In addition to the usual stress considerations for pressure vessels, the use of hydrogen requires additional material considerations. For many materials, hydrogen environments can result in embrittlement, decreased fracture toughness, internal blistering at high temperature, etc. The literature on gaseous H_2 effects in materials is voluminous. Except for mentioning a few general surveys [5.153], no attempt will be made to summarize this complex topic. As will be seen in subsequent sections, most of the containment devices built to date have used austenitic stainless steel, copper or aluminum alloys which are largely immune to hydrogen effects at ambient temperature [5.153]. High temperature devices should be restricted to austenitic stainless steels. Low alloy steel has been used at ambient temperature, although care must be taken to use adequate safety factors and periodic hydrostatic tests. Many other materials are subject to embrittlement and other potential H_2 environmental problems and should *not* be used: high alloy or high strength steels (ferritic, martensitic and bainitic), titanium and its alloys, and some Ni-based alloys.

The question of whether atomic H emanating from intermetallic hydrides at contact points leads to special problems (as opposed to molecular H_2) has never been fully answered. However, limited tests with TiFe hydrides suggest that hydrogen diffusion rates into containment metals and associated embrittlement are the same as if gaseous H_2 alone was used [5.154]. It should be remembered, however, that the H_2 gas from a hydride is of extremely high purity which might tend to accelerate the embrittlement of susceptible materials. Thus, we feel it is wise at this point to avoid any containment material that is potentially susceptible to hydrogen environmental effects.

Remarks on the safety of the general use of hydrogen and some references are given in Sect. 1.3. Ways to eliminate hydrogen and oxygen from explosive mixtures by catalytic techniques are described in [5.155].

5.3 Specific Applications

The versatility of intermetallic hydrides has spawned an extraordinary variety of potential applications. These range from H_2 storage containers and H_2 compressors, which are already available commercially, to advanced thermodynamic and thermomechanical devices that can make use of low grade heat.

Applications cover the time span from the present-day industrial hydrogen sector to the proposed future "Hydrogen Economy" where H_2 would serve as a synthetic fuel and energy carrier. In this section, we will attempt to broadly outline the spectrum of specific intermetallic hydride application. The range of applications to be discussed can be seen at a glance from the table of contents for this chapter.

References to several earlier reviews will be given with the description of specific applications; additional reviews can be found in [5.156–157].

5.3.1 Hydrogen Storage

The reversible storage of hydrogen in the form of an intermetallic hydride has several advantages over conventional gaseous and liquid H_2 storage. As shown in Table 5.3 [5.28], hydrides offer pronounced volumetric advantages over compressed gas, along with much lower required pressure, a safety advantage. In fact, H densities achievable with hydride containers approach those of LH_2 dewars, a result of the fact that the H packing density within a given hydride crystal is usually significantly higher than that of LH_2. This, of course, can be achieved at ambient temperatures without the need of the cryogenic technology and the expensive and highly insulated containers required for LH_2 to minimize boil-off losses [5.13]. The major disadvantage of hydride storage over LH_2 storage is the relatively low hydrogen weight percentages achievable with today's technology. Mg-based alloys and intermetallics (Mg_2Ni) may be technically competitive with LH_2 if high temperature ($> 300\,°C$) heat is available to desorb the H_2. However, even the fairly heavy hydrides of AB, AB_5, and AB_2 intermetallics are competitive with compressed gas cylinders.

The major overall disadvantage of hydride storage over other options seems to be the relatively high cost of the intermetallic compounds and the need to invest in heat exchangers for the storage containers [5.158–161]. Although this may rule out the large-scale storage and distribution of hydrogen in the form of intermetallic hydrides, we believe there are some special situations where small scale hydride storage may be useful. These will be discussed under the general categories of stationary and vehicular storage.

Table 5.3. Comparison of H_2 storage parameters [5.28]

Mode	H Density[a]	Pressure [MPa at 25 °C]	Reversible[b] H_2 [wt%]	SCM H_2 per m^3 container[c]
LH_2	4.2	0.1 (20 K)	5.3	850
GH_2	1.0	20	1.3	176
$FeTiH_{1.7}$	5.5	0.48	1.4	550
$LaNi_5H_{6.7}$	7.6	0.17	1.1	755
MgH_2	7.6	<.00001	5.6	670

[a] H atoms per ml $\times 10^{-23}$
[b] Assumes container is 25% of hydride weight
[c] Assumes 50% void space in hydride container

a) Stationary Storage Units

Small portable hydride storage units are now available commercially and are successfully used, for example, in laboratory environments. In addition to the inherent safety (low pressure) advantage of hydride storage of H_2, hydrides also give excellent insurance that the hydrogen released is of very high purity. Reactive species, such as O_2, H_2O, CO, etc., tend to be gettered by the hydride and inert species, such as N_2, CH_4, etc., can be purged away during the first few percent of the discharge. Thus, hydride storage units are often used in specialized applications where "super-high-purity" H_2 is required (e.g., as gas chromatograph carrier gas or for high purity furnace atmospheres).

The first H_2/hydride storage unit was marketed as early as 1976 by the Billings Energy Corporation [5.162, 163]. This unit was an 11 cm diameter aluminum cylinder filled with TiFe or Ti(Fe, Mn) and had a H_2 storage capacity of about 2500 STL (liters, STP). The unit had no internal heat exchanger so that it was usually operated either from external ambient air heat exchange or with a heater jacket.

Ergenics, Inc. [5.51] currently manufacturers the three small hydride storage units shown in Fig. 5.4. These units range in capacity from 28 to 2550 STL. Depending on the particular model, air or water heat exchange is used. The intermetallic compounds used in these units are selected to match the particular temperature/pressure requirement of the individual customer. Usually AB_5 or AB compounds are used. Performance data for the model ST-90 loaded with $MmNi_{4.15}Fe_{0.85}$ (HY-STOR 209) are given in [5.60].

A finned-aluminum hydride container (40–158 STL H_2 capacity) is manufactured by Hydrogen Consultants, Inc. [5.164] and distributed by Baseline Industries, Inc. [5.165]. It is designed to be operated with simple ambient air heat exchange. An electrically heated $LaNi_5$ unit with 150 STL capacity was once manufactured by the Milton Roy Company [5.166], but is not now available. One of the earliest hydride storage units was a bakable all-metal unit developed by KFA Jülich for maximum H_2 output purity [5.167]. This unit, which uses a variant of TiFe, is still manufactured for sale [5.23]. Two lines of small storage systems have been introduced by HWT in Germany [5.168] and Suzuki Shokan in Japan [5.169].

A number of prototype large stationary hydride storage units have been built. Starting with the tabulations of *Baker* et al. [5.170], we have summarized in Table 5.4 most of the large (> 100 kg hydride) prototype storage units and their basic parameters (where available in the open literature [5.171–179]. As indicated by the blanks in Table 5.4, detailed data are not always available.

The first large unit was the now famous reservoir built in 1974 by Brookhaven National Laboratory for a New Jersey Public Service Electric and Gas electric peak shaving experiment [5.171, 180]. Since that time a number of large stationary storage units have been built using various designs and intermetallic storage compounds. The largest is the one constructed by Mannesmann, in cooperation with Daimler–Benz, under a European Community Contract [5.178]. This unit

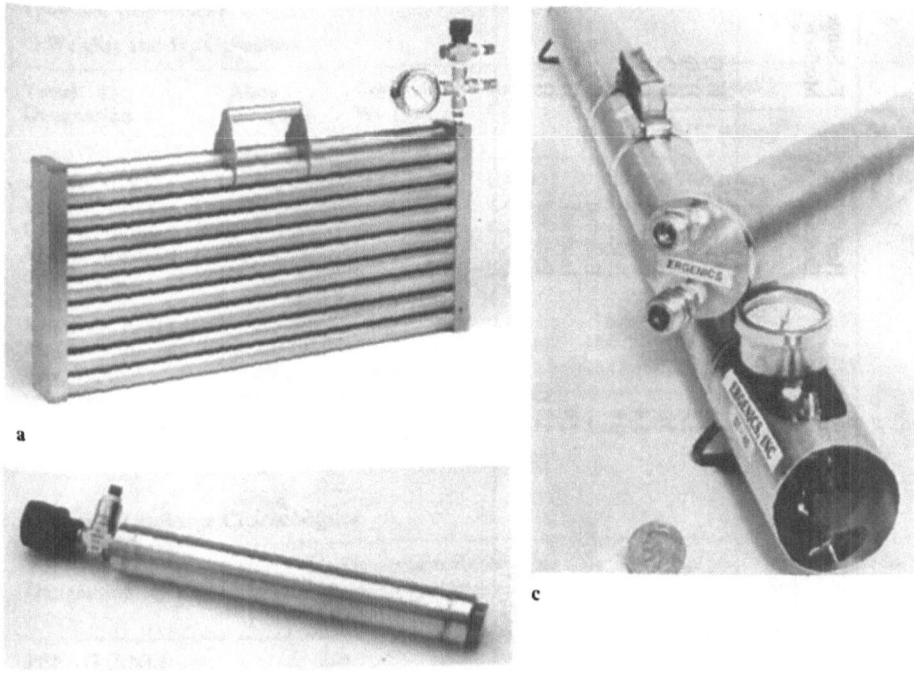

a

b

c

Fig. 5.4. Hydrogen hydride storage units commercially available from ERGENICS, Inc. [5.51], (a) air heat-exchanged ST-90 (2550 STL), (b) ST-1 (28 STL), (c) water heat-exchanged ST-45 (1275 STL)

contains about 10 tons of the multicomponent AB_2 compound $Ti_{0.98}Zr_{0.02} \cdot V_{0.43}Fe_{0.09}Cr_{0.05}Mn_{1.5}$. Aluminum flake is added to the storage compound to increase thermal conductivity [5.181]. Mannesmann makes large stationary and vehicular storage units [5.178]. Japan Chemical Engineering Ltd., designed a 1000 SCM (m^3 STP) hydrogen/hydride storage unit for Kogakuin University, Tokyo [5.182]. In addition, other organizations such as Ergenics [5.51] and HCI [5.164] make large storage units on a custom basis.

The Kogakuin University unit is really three storage containers consisting of three AB_5 compounds with different plateau pressures. Thus particular pressures can be chosen. Also, in effect, the unit can be used as a large thermal compressor (Sect. 5.3.5 will review hydride compressors).

In general, the large prototype units shown in Table 5.4 seem to have performed more or less satisfactorily. A better understanding of technical problems, such as heat transfer and expansion [5.183], has been achieved. The technical viability of large-scale storage of H_2 as a hydride has certainly been demonstrated, but the economics seem uncertain [5.158-161]. Large scale storage and transportation of H_2 as an intermetallic hydride may be viable only in those situations

Table 5.4a-c. Summary of large stationary hydride vessels. After [5.170]

a) General Design

Vessels Designation	Ref.	Storage Compound	Vessel Type Diameter [cm]	Material	Working Pressure [MPa]	Heat Exchange	Exchange Medium
PSE&G (BNL)	[5.171]	Ti(Fe, Mn)	30	SS Cyl	3.4	Internal	H_2O
VPTU-2 (BNL)	[5.172]	Ti(Fe, Mn)	66	Steel Cyl	3.4	Internal	H_2O
CEN Grenoble	[5.173]	TiFe		—	3.0	Internal	H_2O
BEC Homestead	[5.174]	Ti(Fe, Mn)	97	Steel Sphere	3.4	External	H_2O
GIRI Osaka	[5.175]	$MmNi_{4.5}Mn_{0.5}$	25	—	0.8	Internal	H_2O
Iwatani	[5.176]	$Mm(Ni, X)_5$	10	Finned Tubes	—	External	Air
Kawasaki K-3D	[5.177]	$LaNi_{4.8}Al_{0.2}$	—	SS Cyl	2.9	Internal	H_2O
Mannesmann EC	[5.178]	AB_2 Type	—	SS Tubes	5.0 (?)	External	H_2O
Kogakuin	[5.179]	$3\ AB_5s$	30–50	Steel Cyl	6.0 max.	Internal	H_2O

Table 5.4. (*Continued*)

b) Weights and H_2 Capacities

Vessel Designation	Alloy Wt. [kg]	Container Wt. [kg]	Ratio C/A	H_2 Storage Capacity		
				[kg]	[% Alloy]	[% System]
PSE&G (BNL)	400	164	0.41	6.4	1.60	1.13
VPTU-2 (BNL)	1739	2624	1.46	15.6	0.90	0.35
CEN Grenoble	900	209	0.23	—	—	—
BEC Homestead	1791	—	—	30.8	1.72	—
GIRI Osaka	106	—	—	1.44	1.36	—
Iwatani	combined 730		6.3	—	—	0.86
Kawasaki K-3D	993	395	0.4	16.6	1.67	1.20
Mannesmann EC	10000	—	—	168	1.68	—
Kagakuin	1200	400	0.3	18.0	1.50	1.13

c) Charge/Discharge Characteristics

Vessel Designation	H_2 Desorption Rates		Recharge Time [%-time, min]
	[l/min]	[l/min/kg alloy]	
PSE&G (BNL)	226	0.6	80%–120
VPTU-2 (BNL)	850	0.5	100%–220
CEN Grenoble	—	—	–
BEC Homestead	—	—	–
GIRI Osaka	50	0.5	100%–322
Iwatani	117	—	100%–600
Kawasaki K-3D	1800	1.8	100%–103
Mannesmann EC	—	—	—
Kogakuin	3000	2.6	100%–67

and societies (such as Japan, for example) where safety is of high importance and strictly regulated. A formally organized group at MITI Japan has been preparing special regulations regarding the uses of metal hydride containers for the storage and transportation of H_2.

A possible exception to the above economic concerns about large-scale storage and transportation of hydrogen as a hydride may be, represented by Mg-rich alloys and intermetallic compounds. It has been argued that Mg can be made cheaply from sea water and solar energy [5.184]. Reasonably good kinetics were occasionally achieved under practical conditions with Mg or Mg–Ni alloys [5.185–188]. Highly reactive hydrides of Mg and Mg alloys have recently been prepared by chemical precipitation or deposition from organo-metallic solutions [5.189] (cf. Vol. I, p. 40). Because of the very low room temperature plateau pressures exhibited by Mg-rich hydrides, schemes have been proposed for the transportation of hydrogen in low pressure bulk containers with subsequent high temperature liberation of the H_2 in separate high pressure vessels [5.187] or hot-zone, continuous-conveyor furnaces [5.188].

b) Fuel for Motor Vehicles

One of the most active and interesting uses for intermetallic hydrides has been as a carrier of hydrogen fuel for motor vehicles. The possibility of using metal hydrides as a vehicular fuel storage medium was first proposed by Brookhaven National Laboratory as early as 1967 [5.190, 191]. For historical interest, the first suggested system [5.191] is shown in Fig. 5.5. The main H_2 storage system consisted of a Mg-based alloy or intermetallic compound (Mg-5%Ni, Mg_2Ni, Mg_2Cu, etc.) that was heated by the exhaust gas. Because the Mg-based hydrides required substantial temperatures ($\sim 300\,°C$) for H_2 recovery, a separate higher pressure, start-up hydride was proposed. As shown in Fig. 5.5, this was the newly discovered (but not yet published [5.5]) TiFe hydride, the first of the practical AB intermetallic hydrides.

The scheme shown in Fig. 5.5 was not completely practical because there is not enough heat available from the exhaust gasses of an internal combustion

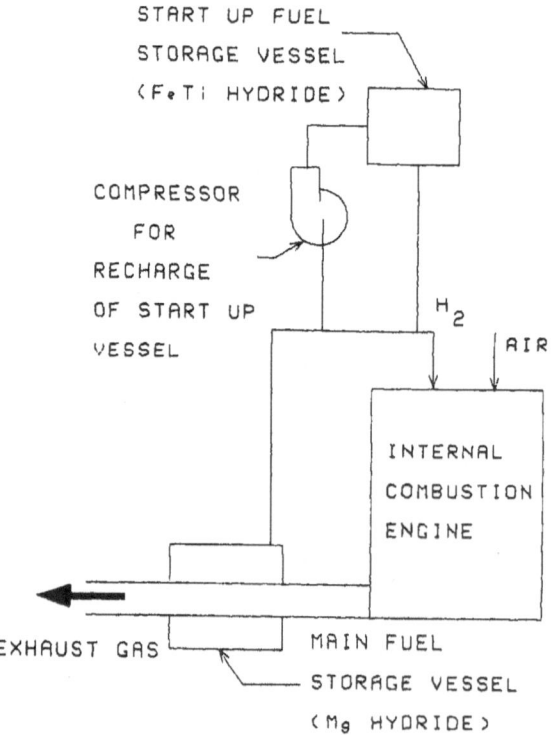

Fig. 5.5. Schematic flow diagram for hydrogen engine, hydride storage concept [5.191]

Table 5.5. Energy density automotive power sources [5.192]

Power Source	Energy Density [wh/kg]
Pb acid battery	22
Ni–Cd battery	44
Ag–Zn battery	110
Na–S battery	220
Li–S battery	220
FeTiH$_{1.95}$[a]	590
n-Octane[a]	13230

[a] Does not include container weight

engine to dissociate Mg-based hydrides at the rate necessary to fuel the engine (without supplemental burning of the H_2). Thus the first hydrogen/hydride fueled prototype vehicles were not built until 1974 when the full potential of the higher pressure TiFe hydrides was known [5.5, 192]. At that time there was considerable interest in electric vehicles. As shown in Table 5.5, *Reilly* et al. demonstrated that TiFeH$_{1.95}$ fuel storage for a relatively conventional internal combustion engine could be competitive on a weight basis with advanced batteries and electric motors [5.192]. The hydride concept was first demonstrated at Brookhaven with a small Wankel engine fuelled by TiFe hydride [5.192]. The years 1969–1974 also saw several reviews which discussed the pros and cons of vehicular hydrogen/hydride storage [5.88, 193–195].

Since 1974, a number of vehicles have been converted to H_2 operation using onboard intermetallic hydride storage of the hydrogen. Details of the hydride storage vessels used are summarized in Table 5.6 [5.170, 181, 196–205]. As can be seen, the vehicular hydride storage field has been dominated by three organizations: Billings Energy Corporation (BEC) of Provo, Utah and Independence, Missouri, USA; Daimler–Benz AG (DB) of Stuttgart, FRG; and Hydrogen Consultants Incorporated (HCI) of Littleton, Colorado, USA.

The first full-sized H_2 vehicle reported to be run from a hydride bed was a Chevrolet Monte Carlo automobile converted and tested by the Billings Energy Corp. in 1974 [5.206]. The hydride storage unit was based on TiFe and was used mainly as a backup to a main liquid hydrogen fuel tank. No details are available, so the TiFe tank for the Monte Carlo is not included in Table 5.6. The results were encouraging enough that Billings soon modified a number of vehicles that were successfully operated solely from hydride storage units: cars, busses, forklift trucks, small tractors, etc. (Table 5.6). All of the Billings vehicles used hydrides based on TiFe or Ti(Fe, Mn). At present, the Billings group seems to have relatively little activity in the H_2 vehicle area.

Daimler–Benz has maintained a strong interest in hydrogen/hydride automobiles and vans. By 1977, they had already converted and tested five vans

Table 5.6a–c. Summary of vehicular hydride systems. After [5.170]

a) General Design

Vessel	Ref.	Storage Compound	Vessel Type Diameter [cm]	Form	Working Pressure, MP	Heat Exchange	Exchange Medium
D-B low T	[5.196]	TiFe	30.5	Cyl.	5.0	Internal	Engine Coolant
BEC Omni	[5.201]	TiFe	66	Cyl.	3.4	Internal	Engine Coolant
BEC Forklift	[5.203]	TiFe	28	Cyl.	6.9	Internal	Engine Coolant
BEC Riverside Bus	[5.199]	Ti(Fe, Mn)	20.3	Al Cyl.	5.2	External	Engine Coolant
BEC Postal Jeep	[5.200]	Ti(Fe, Mn)	18.4	Al Cyl.	4.5	External	Engine Coolant
BEC Tractor	[5.202]	Ti(Fe, Mn)	18.4	Al Cyl.	4.1	External	Engine Coolant
HCl AC50 Forklift	[5.170]	$LaNi_5$	2.9	Cu Tubes	1.0	External	Engine Coolant
HCl Dodge D-50	[5.170]	Ti(Fe, Mn)	2.9	Al Tubes	3.4	External	Engine Coolant
D-B Combo	[5.170]	TiFe	3.6	SS Tubes	—	External	Engine Coolant
D-B High T	[5.196]	Mg_2Ni	3.6	SS Tubes	—	External	Exhaust Gas
BEC Pontiac	[5.197]	TiFe	4.4	SS Tubes	3.4	External	Exhaust Gas
BEC Provo Bus	[5.198]	TiFe	7.6	SS Tubes	3.4	External	Exhaust Gas
D-B Combo	[5.196]	$Mg_2Ni + FeTi$	3.6	SS Tubes	—	External	Exhaust Gas
D-B Berlin Van	[5.181]	AB_2	5.0	SS Tubes	—	External	H_2O-Exhaust Heated
D-B Berlin Car	[5.181]	AB_2	5.0	SS Tubes	—	External	H_2O-Exhaust Heated
Electrolyzer Tractor	[5.204]	$MnNi_{4.5}Al_{0.5}$	6.3	SS Tubes	0.7	External	Exhaust Gas
HCl-Eimco Mine (Main)	[5.205]	$MnNi_{4.5}Al_{0.5}$	2.9	SS Tubes	2.8	External	Engine Coolant
HCl-Eimco Mine (Cold Start)	[5.205]	$MnNi_{4.17}Fe_{0.83}$	2.9	SS Tubes	2.8	External	Engine Coolant

Table 5.6. (*Continued*)

b) Weight & H_2 Capacities (E = Estimated)

Vessel	Alloy Wt. [kg]	Container Wt. [kg]	Ratio C/A	H_2 Storage Capacity [kg]	[% Alloy]	[% System]
D-B Low T	200	—	—	4.0	2.0 (?)	—
BEC Omni	156E	26	.17	2.5	1.6E	1.4E
BEC Forklift	91	100	1.09	1.5E	· 1.6E	0.78E
BEC Riversible Bus	90	23E	.25	1.57	1.74	1.39
BEC Postal Jeep	65	25	.39	0.68	1.05	0.76
BEC Tractor	48	18	.37	0.77	1.60	1.17
HCI AC50 Forklift	270	180	.67	3.4	1.26	0.76
HCI Dodge D-50	318	115	.36	5.0	1.57	1.15
D-B Combo	246	154	.63	5.5	2.24	1.38
D-B High T	55	51E	.93	1.7E	3.0E	1.7E
BEC Pontiac	198	135	.68	1.8	0.91	0.54
BEC Provo Bus	508	200	.39	6.3	1.24	0.89
D-B Combo	246	154	.63	5.5	2.24	1.38
D-B Berlin Van	81	61	.75	1.5	1.85	1.06
D-B Berlin Car	81	61	.75	1.5	1.85	1.06
Electrolyzer Tractor	80	74	.92	1.0	1.25	0.65
HCI-Eimco Mine (Main)	407	274	.67	5.2	1.28	0.76
HCI-Eimco Mine (Cold Start)	68	46	.67	0.81	1.19	0.71

c) Charge/Discharge Characteristics (E = Estimated)

System	H_2 Desorption [l/min]	[l/min/kg]	Recharge Time [% in time, min]	Comments
D-B Low T	900	4.5	75%–10, 100%–45	From room temperature
BEC Omni	—	··	80%–20	—
BEC Forklift	—	—	—	5 vessels onboard
BEC Riverside Bus	120E	1.3	75%–120E	10 vessels onboard
BEC Postal Jeep	—	—	75%–120E	6 vessels onboard
BEC Tractor	—	—	—	—
HCI-AC50 Forklift	—	—	90%–12	—
HCI Dodge D-50	1100	3.5	90%–30	—
D-B Combo	—	—	—	3 vessels onboard
D-B High T	—	—	—	1 vessel onboard
BEC Pontiac	640(?)	3.2	80%–10; 100%–60	Tap H_2O to charge
BEC Provo Bus	1300	2.6	80%–15; 100%–60	Tap H_2O, 2 vessels
D-B Combo	—	—	—	3 vessels onboard
D-B Berlin Van	800	9.9	100%–10	4 vessels on board (max of 2 used at any one time)
D-B Berlin Car	800	9.9	100%–10	2 vessels onboard
Electrolyzer Tractor	—	—	90%–30	—
HCI-Eimco Mine (Main)	—	—	90%–15	12 modules onboard
HCI-Eimco Mine (Cold Start)	—	—	90%–15	2 modules onboard

[5.26]. They have continued up to the present time, as summarized in the large number of Daimler–Benz review papers on their activities [5.17, 26, 98, 181, 196, 207–209]. From this body of work, a number of hydride innovations can be cited:

1. The use of combined TiFe and Mg_2Ni storage beds to make use of waste engine heat from both the cooling water and exhaust gas (also suggested by BNL [5.191] and Billings [5.206]).
2. The use of air/hydride heat exchangers within the passenger compartment to provide internal cooling.
3. The development of new AB_2 hydrides and their use in vehicles.

At present the principal activity of the Daimler–Benz group centers around a demonstration program for Berlin involving five dual gasoline/H_2-fueled passenger vehicles and five H_2-fueled delivery vans [5.181, 209]. The AB_2 storage compound $Ti_{0.98}Zr_{0.02}V_{0.43}Fe_{0.09}Cr_{0.05}Mn_{1.5}$ (blended with some Al for improved thermal conductivity) is being used.

While the Billings and Daimler–Benz vehicle work has concentrated mainly on passenger cars and busses, the work at Hydrogen Consultants, Inc. has concentrated mainly on industrial vehicles such as forklifts and mine vehicles. Indoor industrial vehicles are probably a more logical and nearer-term option than road vehicles. The low-pollution advantage of H_2 engines (especially relative to diesel fuel) is obvious for indoor or mine operation. The weight disadvantage of hydride storage is not very important for forklifts and mine vehicles because they are usually heavily ballasted with counterweights anyway.

Although industrial H_2-powered vehicles seem to have growing viability, the future of passenger-carrying H_2 road vehicles seems to be rather uncertain. At the time of writing, petroleum is in plentiful supply and oil prices are relatively low. Even if oil becomes in short supply again, the present fuel distribution system will probably provide strong incentives for an alternative liquid carrier such as methanol which can either be burned directly or cracked onboard to form H_2 [5.210].

H_2-powered passenger vehicle research continues. It should be noted that there is a growing tendency toward LH_2 vehicles, as evidenced by the number of LH_2 vehicles at the 5th World Hydrogen Energy Conference in Canada. *Carpetis* [5.211] argues that the lower weight associated with LH_2 storage (in comparison to hydride storage) offsets the higher cost of LH_2 production (in comparison to GH_2), so that for vehicle ranges greater than 200 km LH_2 is preferred over hydrides. Below 200 km, especially when safety is considered, hydrides seem to have the economic edge. Finally, new lightweight designs of composite high-pressure vessels may offer competition to both hydrides and LH_2 for passenger vehicles [5.212]. Developments of hydriding alloys with high H_2 content under moderate temperature conditions are awaited.

5.3.2 Purification and Separation

As mentioned in Sect. 5.3.1, intermetallic hydrides offer a convenient source of very high purity hydrogen. The use of elemental hydride/dehydride reactions for the purification or separation of hydrogen has been known for many decades. For example, a favorite source of super-high-purity H_2 has been uranium hydride beds [5.1]. As early as the end of the 19th century, *Morely* used a Pd-hydriding/dehydriding cycle to separate H_2 from H_2-N_2 mixtures [5.213]. The advent of intermetallic hydrides has greatly increased the number of practical hydride formers that can be used for H_2 purification and separation.

We should distinguish between *purification* and *separation*. By *purification*, we mean the removal of less than a few thousand ppm impurities from H_2 by a combination of chemisorbing active impurities and purging away inert impurities. This can be done in a conventional "dead-end" reactor. By *separation*, we mean the removal of H_2 from a mixture containing from 1–90% non-hydrogen impurities. Because of localized inert gas blanketing, separation usually requires a "flow-through" reactor.

The first quantitative H_2 *purification* effects associated with an intermetallic compound (TiFe) were demonstrated by *Wenzl* and *Klatt* [5.167]. They designed an all-metal, "ultra-high-vacuum-technology" TiFe storage container that purified 99.9% H_2 to 99.9999%. Purification was accomplished by discarding the first 10% of the discharge. Levels of O_2, N_2, CO, CO_2 and CH_4 were reduced to <0.1 ppm. H_2O was reduced to 0.2 ppm but only if the input gas was first dried by cold trapping. Similar results were obtained by *Gamo* et al. [5.141] using a $TiMn_{1.5}$ purifier. They were able to purify 99.99% H_2 to 99.9999% (excepting H_2O). Purging curves are reproduced in Fig. 5.6. Note that with fresh $TiMn_{1.5}$ a throw-away purge of only 3% is needed to reduce impurity levels to below detectability (Fig. 5.6a). With extended cycling *Gamo* et al. noted a partial loss of H_2 capacity, which is accompanied by an increased purge requirement (Fig. 5.6b). The application of "pre-purification" using molecular

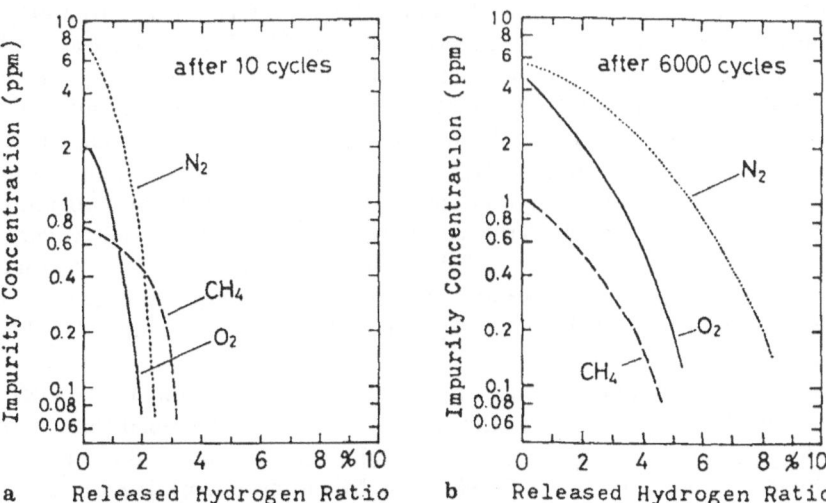

Fig. 5.6. Comparison of the impurity concentrations in hydrogen released from $TiMn_{1.5}$ hydride after (**a**) 10 and (**b**) 6000 absorption/desorption cycles [5.141]

sieves has resulted in improved hydride purifier designs from China [5.214, 215]. Laboratory scale and industrial scale purifiers are commercially available from HWT [5.168].

A number of laboratories (including our own) use hydride storage to assure that the H_2 ultimately used is of "hydride/dehydride" purity. There is no question that these storage devices do result in "super-high-purity" H_2. The output streams must be carefully filtered to minimize the number of fine intermetallic hydride particles carried out with the H_2. These storage devices are cycled only once a day or even less frequently.

The first partially successful H_2 *separation* experiments to use intermetallic compounds were performed by *Reilly* and *Wiswall* in the early 1970s [5.83, 109, 110, 216]. Using La(Ni, Cu)$_5$ compounds, they were able to separate H_2 from mixtures containing up to 24% CO_2 and 2% CO, although poisoning problems associated with CO required operation above 100 °C. More recently *Gidaspow* et al. successfully separated H_2 from H_2–CH_4–N_2 mixtures using TiFe and Ti(Fe, Ni) compounds [5.217, 218]. Difficulties were noted when CO_2 was included. Similar successful experiments were run by Blytas using polymer-bonded LaNi$_5$ pellets [5.69] and by *Block* and *Bahs* with Ti–Mn intermetallics [5.126]. *Meyerhoff* demonstrated the separation of H_2 from dissociated NH$_3$ (75%H_2–25%N_2–trace NH$_3$) using LaNi$_{4.7}$Al$_{0.3}$ pellets [5.219].

The above separation experiments were done on a relatively small laboratory scale. A much larger industrial demonstration resulted from a joint Ergenics, Inc.–Air Products and Chemicals, Inc. project on H_2 recovery from waste gas streams [5.102]. An automated hydride unit was successfully operated for 6 months, separating H_2 from an ammonia purge gas stream within the Air Products New Orleans NH$_3$ plant. Ammonia purge gas is a low value by-product of the NH$_3$ synthesis process and is normally burned for its fuel value. Its composition is approximately (in vol.%) 60H_2–20N_2–12CH_4–5Ar–1 to 3NH$_3$, with none of the non-H_2 species being poisons for LaNi$_5$. The LaNi$_5$ was used in the form of mechanically stable, metal (Ni) ballasted pellets [5.73], which also allowed for an approximately adiabatic pressure swing design with little or no external heat transfer required. Three alternating flow-through type reactors were used for continuous H_2 separation and recovery. As shown in Fig. 5.7, the breakthrough front was used to determine when the hydriding reaction was complete, after which the reactor was depressurized for H_2 recovery. The 60% H_2 stream was upgraded to 99% with recoveries in the 90–93% range [5.102]. Recently, *Wang* et al. have also shown that it is possible to recover H_2 from industrial ammonia "tail-gas" by separating the H_2 with mischmetal-nickel hydrides [5.220].

Because ammonia purge gas is a relatively "clean" stream as far as intermetallic hydrides are concerned, the commercial scale demonstration was technically successful. There are many other industrial streams that contain known hydride poisons such as CO, H_2S, etc. Much more work is required to successfully apply hydride separation techniques to such streams. Even with the "clean" streams there can often be economic competition from other techniques

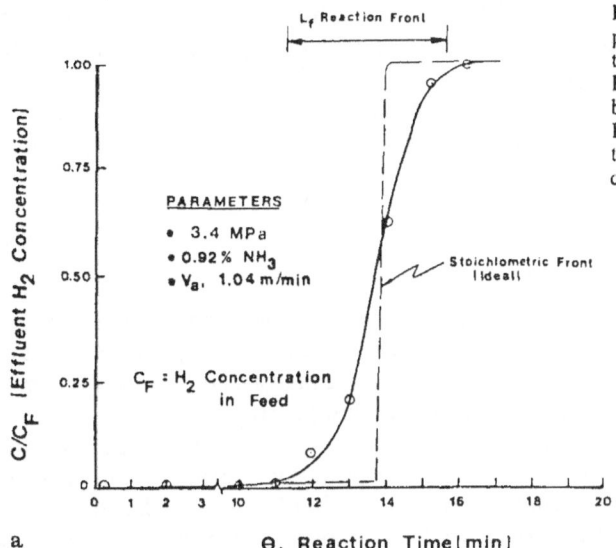

Fig. 5.7. Breakthrough phenomena associated with the separation of H_2 from an H_2–NH_3 mixture using thermally ballasted $LaNi_5$ [5.102] (a) Efficient H_2 concentration vs time; (b) Typical H_2 breakthrough curve and reaction front

a Θ, Reaction Time [min]

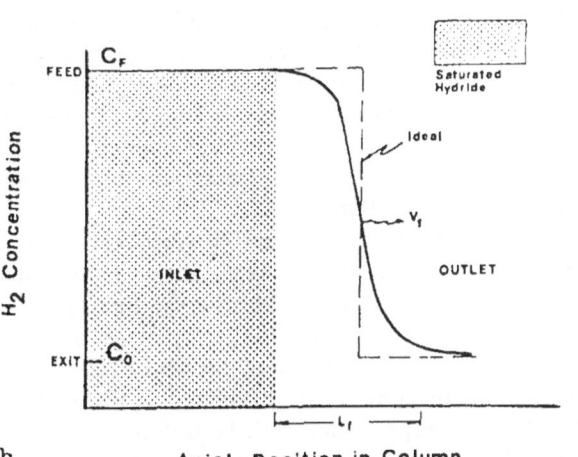

b Axial Position in Column

such as polymer membrane diffusion [5.221], and adsorption on carbon or zeolites [5.222, 223]. Comparatively speaking, hydrides have an advantage at relatively low H_2 partial pressures [5.224].

5.3.3 Isotope Separation

Intermetallic hydride formers can be used to affect separations among the various isotopes of hydrogen: protium (H), deuterium (D), and tritium (T). For example, applications might include the separation of D from natural hydrogen (0.016% D, balance H) or the removal of radioactive T from non-radioactive D and/or H. Hydrides (elemental and intermetallic) have been used, at least on a laboratory scale, for isotope separation. We will briefly review some of the principles and experiments performed.

The effectiveness of a given isotope separation process is usually quantified by the *separation factor* α. For the case of hydride processes the separation factor may be defined as

$$\alpha = \frac{R_S/H_S}{R_G/H_G}, \tag{5.8}$$

where R_S and R_G are the concentrations of the heavy isotope in the solid and gas phases, respectively, and H_S and H_G are the corresponding protium concentrations. In practice, α is a function of at least three factors: equilibrium thermodynamics, bulk diffusion rates, and surface exchange processes.

Equilibrium separation factors can be derived from PTC data, e.g., the plateau pressure of the deuteride vs the corresponding hydride. From entropy considerations, it is expected that the deuteride should show a slightly higher dissociation pressure than the hydride [5.2]. This, often called the *normal isotope effect*, represents the situation where the solid phase has a slightly higher affinity for H than D (or similarly T). Usually the effect is fairly small, for example in the case of uranium [5.2, 225]. In some cases, however, the effect can be large, for example the pronounced normal isotope effect observed with Pd [5.226, 227]. The *inverse isotope effect* represents the situation where the heavier hydrogen isotope results in a lower plateau pressure, i.e., the deuteride and tritide are more stable than the corresponding hydride. For pure metals, the best-known example of this is the upper plateau of vanadium [5.228].

Intermetallic compounds exhibit both normal and inverse equilibrium isotopic effects. Sometimes little or no isotopic effect is seen. A striking example of the variety of isotopic effects that can be seen is shown by the three-plateau $CaNi_5$-H_2 (D_2) system shown in Fig. 5.8 [5.144]. The first plateau shows a normal isotopic effect, the second almost no effect, and the third plateau an inverse effect. It should be noted that the isotope effect (normal vs inverse) can change with temperature. The effective slope of the van't Hoff diagram (ΔH) is often different for H_2 and D_2 (or T_2) so that crossovers can occur with temperature. This is well documented for AB_5 compounds [5.229–231] and is probably a general phenomena for most intermetallics [5.232]. Ti_2Ni was found to show significantly different absorption equilibrium conditions for hydrogen and deuterium [5.231].

In addition to the equilibrium isotopic effect discussed above, the "dynamic" processes of relative diffusion and surface exchange are important factors determining the effective separation coefficient. From classical diffusion theory, the relative diffusion coefficients D are given by relative isotopic masses M according to the ratio $D_H/D_R = \sqrt{M_R/M_H}$. Thus H (mass 1) would be expected to diffuse $\sqrt{2}$ times faster than D (mass 2). In fact, H does diffuse more rapidly than D in body-centered-cubic metals such as V but the $\sqrt{2}$ ratio is seldom obeyed

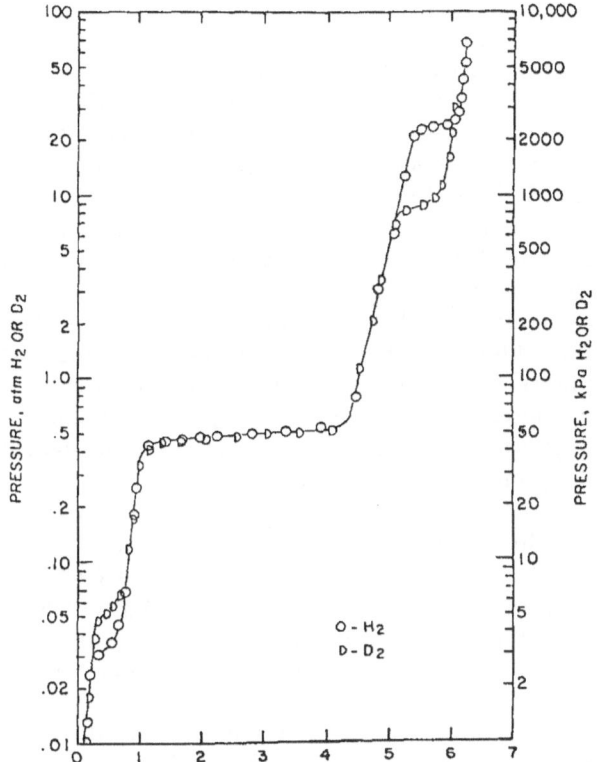

Fig. 5.8. Comparison of the H_2 and D_2 static desorption isotherms of $CaNi_5$ at 25 °C [5.144]

very well [5.233]. However, for face-centered-cubic metals, such as Pd, it has been found that D actually diffuses faster than H, at least at temperatures of practical interest [5.233]. For intermetallic compounds, which can have a variety of complex structures, relative isotopic diffusion data are scarce and generalizations are difficult at present.

The least understood dynamic factor affecting isotopic separation is surface exchange, i.e., catalytic dissociation and recombination phenomena of species such as H_2, HD, D_2, HT, etc. (cf. Chap. 2). It is clear this is an important factor influenced heavily by surface structure and conditions [5.81, 234]. In summary, the actual isotopic separation achieved is a complex function of at least three equilibrium and dynamic processes. In the case of tritium, radioactive decay results in ^{3}He formation in the lattice and induces further complications [5.235, 236].

Although there is no hydride/dehydride commercial-scale process for the separation of hydrogen isotopes, there have been a number of promising laboratory-scale separations. The chromatographic separation of H and D in a Pd column was successfully demonstrated by *Gluekauf* and *Kitt* in 1957, long before the advent of intermetallic hydrides [5.237]. More recently, *Aldridge* has demonstrated chromatographic separation of D from dilute H–D mixtures using AB_5 compounds of the type $La(Ni, Co)_5$ and $Ca(Ni, Cu)_5$ [5.238]. Perhaps reflecting the uncertainties associated with the fundamental surface and bulk separation properties discussed above, *Aldridge* found strong intermetallic composition effects. Single pass separation factors of 1.3–3.6 were achieved [5.238].

Extensive work toward the establishment of hydride/dehydride techniques for the separation of T from H–T mixtures has been done at Brookhaven National Laboratories. *Tanaka* et al. examined a number of Ti intermetallics [5.232]. Separation factors α ranged from 0.74 (preferential retention of T in gas phase) to 2.05 (preferential absorption of T in solid phase) and again were composition dependent. The extensive experimental and modelling work of *Hill* and co-workers should also be acknowledged [5.239–241]. Differing somewhat from the early chromatographic separation techniques, they used pressure swing and thermal swing processes, often in cascades for multiple partial enrichment. Although the experimental work was done on pure V, the results and theoretical analysis should be applicable to intermetallic compounds.

It is clear that hydrides offer the potential for commercial isotope separation. Further economic and process studies must be made. For example, if we think about heavy water production (separation of D_2 from normal hydrogen) we must make economic comparisons with the many other D_2O production techniques available and proposed [5.242], we well as assess the future demand in comparison to existing production facilities, e.g., the Girdler–Sulfide plants already in existence.

Finally, there remain some unusual technical uncertainties concerning the use of intermetallic compounds for isotope separation. Perhaps the most famous is that associated with the intermetallic TiNi. In 1977 *Buchner* reported in a patent that TiNi would readily absorb H_2 but almost no D_2 [5.243]. If correct, this would mean that TiNi was the perfect D-separator and possessed unique properties never before seen with a hydride former. However, attempts to reproduce this result by *Reilly* [5.244] and *Sandrock* [5.245] failed. TiNi was found to absorb H_2 or D_2 with little overall difference. Finally, a study by *Rummel* [5.246] shed further light on the TiNi–D_2 discrepancy. *Rummel* found that TiNi was indeed very selective to H_2 absorption (over D_2) on the *first cycle* but that the selectivity was largely lost after cycling 20 times. After 20 cycles *Rummel* estimated an α value of 0.77 for D–H separation, close to the Brookhaven measurement for TiNi of $\alpha = 0.74$ for T–H separation [5.232]. The reasons for the strange variation of H–D selectivity with cycling are unknown but the anomalous behavior serves to illustrate the uncertainties still associated with the use of intermetallic hydride formers for isotope separation.

5.3.4 Hydrogen Getters

The widely used term *getter* signifies a material used for the scavenging of unwanted gases by surface or bulk reaction. Zr-based getters have been used commercially in high vacuum applications for many years, e.g. television and other electron tubes, glow-discharge lamps, high-vacuum equipment, etc. [5.247]. Such applications usually involve the removal not only of H_2 but also of other gaseous species such as CO, N_2, H_2O, and the getters were originally optimized for such mixtures. The growth of fusion technology has resulted in a increasing need for getters specifically useful for hydrogen and its isotopes, particularly tritium [5.23].

Traditionally, the specific gettering of hydrogen isotopes has been done with uranium getter beds [5.7, 13, 225, 248]. Intermetallic compounds also offer interesting possibilities: Commercially available Zr-based multigas getters are metallurgically oriented toward intermetallic compounds [5.249]. These typically consist of Zr–Al compounds (Zr_3Al_2 and Zr_5Al_3) Zr_2Ni, Zr_2Fe, and the Laves phase $Zr(V, Fe)_2$. Often α–Zr is microstructurally present in addition to the intermetallic. These materials have a good ability to getter hydrogen isotopes. They usually require heating to the order of $500\,^\circ C$ [2.249] although some show reasonably good H_2 absorption characteristics as low as $200\,^\circ C$. With high temperature preactivation, some gettering ability is obtained at $25\,^\circ C$. There is particular interest in the $Zr(V, Fe)_2$ Laves phase intermetallic getters [5.249–253]. The mechanism of activation and gettering is described in Chap. 2. Some hydrogen absorbing getter alloys are used for the storage and handling of tritium [5.23, 254].

5.3.5 Hydrogen Compression

The use of intermetallic hydrides and heat to compress H_2 is today a commercial reality with two models currently being marketed by ERGENICS, Inc. [5.51]. The concept of hydrogen/hydride compression is simple: H_2 (or its isotopes) is absorbed at low temperature (low pressure) and desorbed at high temperature (high pressure). The pressure increase on the hydride bed is given by the van't Hoff relation, (5.1) or (5.2).

The first hydride compressor reported was constructed about 1971 at Brookhaven National Laboratory and was based on VH_2 [5.255]. By using a Hg column, fluids other than H_2 could be pumped, in particular T_2. This compressor is still used. The first intermetallic hydride compressor was built at Philips around 1973 and based on $LaNi_5$ [5.256]. That unit, along with an improvement [5.89], was used in a hydrogen liquifier system to be discussed later in Sect. 5.3.10. Another AB_5 hydrogen-compressor based on $MmNi_5$ (Mm = mischmetal) was reported by Brookhaven in 1976 [5.257]. The unit was charged by applying 6.9–13.8 MPa H_2 at $0\,^\circ C$ and then heated to $100\,^\circ C$ where 19.3 MPa H_2 pressure was produced. With a separate "cryogenic pressure generator", the H_2 pressure could be increased to > 70 MPa. An attempt to make a relatively large commercial "proof of concept" unit was completed by Hydrogen Consultants, Denver Research Institute, and Ergenics Inc. [5.258]. This unit contained 30 kg of $LaNi_{4.5}Al_{0.5}$ and resulted in disappointing performance and efficiency. About the same time a small compressor, using compacted $CaNi_5$, was also demonstrated in Austria [5.259]. A $LaNi_5$ compressor

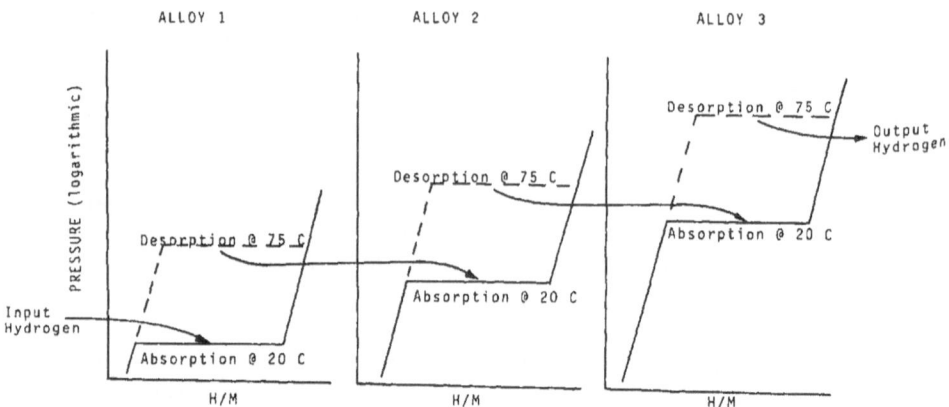

Fig. 5.9. Principle of a three-stage hydride hydrogen compressor operating between 20 °C and 70 °C [5.266]

was demonstrated and characterized in Japan for use in a water desalination process (to be discussed Sect. 5.3.8) [5.260] and a small prototype unit has been made in China [5.261].

The first commercial hydrogen/intermetallic-hydride compressors were developed by *Golben* through improved heat/mass transfer and automation designs [5.266]. *Golben* has pioneered the practical design of multi-stage compressors. As shown schematically in Fig. 5.9, several hydrides can be used to "step" the pressure to quite high levels using only limited temperature swings [2.266]. At present Ergenics, Inc. markets two commercial hydrogen/hydride compressors [5.51]: (1) a single stage, electrically heated unit which compresses H_2 from 0.5 to 3.5 MPa at a rate of 6 STL/min, and (2) a four-stage hot-water-heated (75 °C) unit that compresses H_2 from 0.4 to 4.1 MPa at a rate of about 40 STL/min. The exact performance is a function of the available heating and cooling water temperatures.

Hydride compressors have a number of advantages over mechanical compressors. They are quiet and vibration-free and can be operated on low grade heat such as industrial waste heat and solar heat. They have good potential for low production and maintainance costs. Output H_2 is highly purified, as discussed earlier, but of course care must be taken to avoid the introduction of poisons and reactants that might damage the intermetallic surfaces. Hydride compressors have the potential for good thermal efficiencies, especially at moderate input temperatures [5.267], and form the basis for a number of thermodynamic systems such as heat pumps, refrigerators, and heat engines. These applications will be discussed in subsequent sections.

5.3.6 Heat Storage

Hydriding/dehydriding reactions are associated with substantial heats of reaction, on the order of 20–60 kJ/mol H_2 for typical intermetallic hydrides.

This led *Libowitz* to suggest that dehydriding/hydriding might be a good thermal storage technique [5.268, 269]. For example, solar energy might be used to endo-thermically dissociate an intermetallic hydride and the resultant H_2 stored in the gas phase. Then, at night, the intermetallic could be exothermically rehydrided to provide needed heat. Instead of storing the H_2 as a gas, a separate higher pressure hydride former can be used [5.268, 269]. As shown in Table 5.7 [5.23], the volume required is substantially less than in most conventional specific heat and latent heat storage techniques. A disadvantage, however, is the relatively high cost of hydride heat storage, mainly a result of the cost of the alloy or inter-metallic compounds used [5.269]. *Alefeld* et al. have argued that the relatively low cost of zeolites favors a zeolite-H_2O heat storage system, especially when coupled with heat pumping or refrigeration [5.270]. The recently developed precipitation of very reactive MgH_2 from an organic solution using an organo-transition metal catalyst [5.189] has opened a way to cheap metal hydride heat storage [5.271, 272].

There have been relatively few demonstrations of pure hydride heat storage. Two have been made in Japan where energy costs are high compared to much of the world, thus providing a greater economic incentive. *Ono* [5.273] has built and studied a small prototype heat storage system using 1.5 kg of TiFe. *Yonezu* et al. [5.274] built a heat storage system that utilized $LaNi_5$ and $CaNi_5$ (7 kg total). The latter group used liquid $\rightleftarrows$ gas heat pipes for efficient heat transfer.

Table 5.7. Volume of heat-storage systems for 100 kWh (approximate heat consumption per winter day of a German one-family house) [5.23]

Principle	Material[a]	Volume [l]	Temperature [°C]	Effective Density [kgl⁻¹]
Heterogeneous Evaporation	Metals + H_2			
	$FeTiH_{1.9-0.1}$	370	−20 to 60	3.7
	$CaNi_5H_{6-0}$	190	0–150	5.3
	$MgH_{2-0.1}$	150	250–350	0.9
	LiH	80	500–1000	0.5
	Zeolite + H_2O	150	0–100	1.3
Latent Heat of Melting		1075	0	1
	H_2O	2400	40–80	...
	Paraffin	1020	32	1.5
	Glauber's Salt $(Na_2SO_4 \cdot 10H_2O)$	200	800	2
	LiF			
Specific Heat	H_2O	1670	40–90	1
	MgO	167	150–860	3.6
	He(g, 100 bar)	60000	$\Delta T = 100$	...
Electrochemical	Pb Battery	2500	25	2
Burning	Oil	10	...	0.8

[a] For solid hydrides, the bed density has been assumed to be about 1.5 times smaller than the X-ray density

A demonstration storage unit of 10 kW thermal power based on MgH_2 is planned [5.275].

Whereas the above work was directed toward solar energy storage, a hydride heat storage system for vehicles has been developed by Daimler–Benz in Germany [5.276]. The system is designed to provide immediate heat for the engine and passenger compartment when the vehicle is cold. With the vehicle fully cold, a valve is opened to release hydrogen from a low temperature (high pressure) hydride into a previously dehydrided high-temperature (low-pressure) bed thus providing "instant" heat. Once the vehicle is in operation, exhaust heat is used to drive the hydrogen back to the low temperature bed and the valve closed. This effectively stores heat for the next cold start.

5.3.7 Heat Pumps and Refrigerators

Heat pumps and refrigerators have been an active applications-development area for intermetallic hydrides since the mid 1970s. The activities have covered a broad international spectrum, including, among others, The Netherlands, USA, Japan, Sweden, West Germany, Israel, Austria, and France.

In brief, a hydride heat pump (including refrigerators) consists of two coupled beds of hydride formers (usually intermetallic compounds). By transferring H_2 back and forth between the two beds in a closed thermodynamic system, the cyclic exothermic hydriding and endothermic dehydriding enthalpy changes can be used to produce heating or cooling. The cycle can be driven by mechanical energy or thermal energy. There are basically one mechanically driven system (heating or cooling) and three thermally driven systems (temperature upgrading, heat upgrading, and refrigeration). There are various subdivisions of these four types as confirmed by the large number of patents for hydride heat pumps [5.277–284]. We will briefly and schematically discuss each type using ideal (hysteresis-free and plateau-slope-free) ln P vs 1/T plots (van't Hoff diagrams) to describe the generalities of the thermodynamic cycles. Then we will summarize the numerous prototypes that have been built and some of the non-ideal hydride operating characteristics that must be considered when designing a heat pump system. The thermodynamics of heat pumps has been summarized in detail by *Dantzer* and co-workers [5.285].

Heat and mass transfer are often limiting factors in metal hydride beds. Studies of the thermal conductivity and modelling of heat and mass transfer of metal hydride beds started relatively late, but led to technologically interesting improvements [5.29, 30, 100, 285, 287, 288].

a) Mechanically Driven

The mechanically driven hydride heat pump is conceptually the most simple and was the first hydride type conceived (by researchers at a U.S. Navy Lab) and published [5.277, 286]. The operation of a typical system is shown in Fig. 5.10. Two identical hydride formers are connected by a reversible mechanical pump

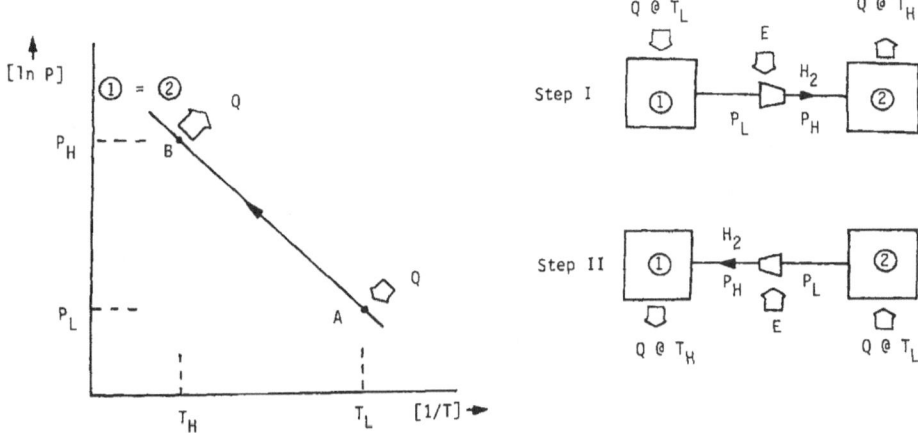

Fig. 5.10. Principles of a mechanically driven hydride heat pump (or refrigerator) using two beds of the same intermetallic hydride former coupled by a mechanical compressor

or compressor. By pumping H_2 from bed 1 (at pressure P_L and temperature T_L) to bed 2 (increased pressure P_H) a high temperature T_H is produced. When bed 2 is saturated, the roles of the beds are then reversed and H_2 is pumped from bed 2 (with T_L heat input) to produce T_H at bed 1. In effect, mechanical energy is used to *pump* heat from T_L to T_H. Note that this, like all other hydride heat pumps represents a batch-like cyclic process. Heat transfer means must be used to transfer the required heat at the appropriate time to and from each reactor.

If T_L is say room temperature or a relatively low solar heating temperature then T_H will be higher and we have a heat pump. However, if we choose a hydride and pressure cycle such that T_H is near room temperature (or another convenient sink temperature), then the resultant T_L will be below room temperature and we have a refrigerator.

Although historically and conceptually instructive, the mechanically driven hydride heat pump has never been pursued. This is probably because it requires a mechanical or electrical input. As shown below, the thermally driven systems virtually obviate the need for the input of any energy other than heat.

b) Thermally Driven, Temperature-Upgrading

The concept of a thermally driven, temperature-upgrading hydride heat pump was first developed by *Terry* in about 1976 [5.279], although an electrically assisted system of similar concept was independently developed by *van Mal* and *Ferguson* at about the same time [5.280]. At least two coupled hydride forming reactors, containing different intermetallics, are used. The idea of such a heat pump is to produce heat at a temperature higher than the input temperature without electrical or mechanical input. Such a system could have value for

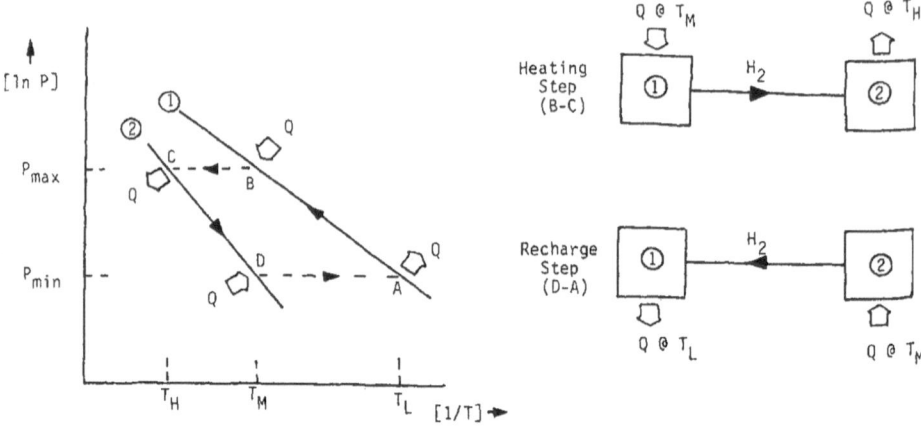

Fig. 5.11. Principles of a thermally driven, temperature-upgrading heat pump using two different intermetallic hydride formers

the upgrading of solar heat or largely wasted, low-grade heat rejected by industry and power plants. For short reviews see [5.287].

We shall illustrate the classic cycle for a typical temperature-upgrading heat pump. To do that we will use the schematic van't Hoff diagrams for two hypothetical ideal hydriding alloys shown in Fig. 5.11. Remember that on a standard van't Hoff diagram temperature increases towards the left. The two intermetallics (let us call them 1 and 2) have different van't Hoff lines and are contained in separate reactors connected by a hydrogen gas line. The idea will be to apply medium temperature (T_M) heat, reject some of that heat to a low temperature (T_L) sink (e.g., ambient), and successfully raise some of that heat to a much more useful high temperature (T_H). The two hydriding intermetallics as shown in Fig. 5.11 are (1) a relatively low temperature or higher pressure hydride former, and (2) a higher temperature or lower pressure hydride former.

Let us trace the complete heat pump cycle ABCDA on Fig. 5.11. We start the cycle at point A (T_L, P_{min}) with intermetallic 1 in the completely hydrided state and intermetallic 2 in the completely dehydrided state. Using our waste heat source at T_M, we heat intermetallic 1 to T_M ($A \rightarrow B$) whereupon its H_2 pressure increases to P_{max}. If we continue to put heat to intermetallic 1 at T_M (say one heat unit Q) then the hydrogen can be driven from hydride 1 to intermetallic 2 thus hydriding intermetallic 2 ($B \rightarrow C$). The process of hydriding intermetallic 2 is exothermic and produces a high temperature T_H, which is what we are after. Thus about $1Q$ of high quality T_H heat is produced during the $B \rightarrow C$ heating step. After this stage is complete, intermetallic 2 is completely hydrided and intermetallic 1 is completely dehydrided. We must now perform a recharge step to complete the cycle. Intermetallic hydride 2 is allowed to cool from T_H to T_M ($C \rightarrow D$) and intermetallic 1 is cooled from T_M to T_L. One Q is put into bed 2 to drive the H_2 back to intermetallic 1 ($D \rightarrow A$). One Q of heat is rejected to the T_L low temperature heat sink. The cycle is now complete. Note that about $2Q$ of intermediate temperature (T_M) heat have been put in, about one Q of low temperature (T_L) heat rejected, and about one Q of valuable high temperature (T_H) heat created (overall maximum efficiency or coefficient of performance of about 50%).

Because this cycle is a "batch" process we must have two sets of reactors operating 180° out of phase to provide continuous heat pumping. We must also provide some plumbing and valving to pipe the hydrogen and heat around for maximum benefit. If we wish, we can add more stages to the heat pump [5.8, 29].

c) Thermally Driven, Heat-Upgrading

This hydride heat pump cycle was first developed at Brookhaven by *Cottingham* around 1975 [5.278, 289] and is shown in Fig. 5.12. The cycle is similar to that described above in that two distinct intermetallics are used. However, it is designed to use high temperature heat to pump low temperature heat to an intermediate temperature. Note that hydride 1 is dissociated at P_{min} using heat at T_L and the H_2 used to hydride intermetallic 2. This process $(C \rightarrow D)$ generates heat Q at T_M. Hydride 2 is now heated to T_H, its pressure rises to P_{max}, and the H_2 allowed to flow back to intermetallic 1 $(A \rightarrow B)$ resulting in about another heat unit Q being generated at T_M. Thus approximately $1Q$ at T_H is combined with $1Q$ at T_L to produce $2Q$ at T_M, i.e., heat has been upgraded from about $1Q$ to $2Q$ but at a sacrifice in temperature. There are useful situations for the more efficient and reversible management of heat in power plants and space heating by this heat-upgrading cycle [5.27, 278, 290].

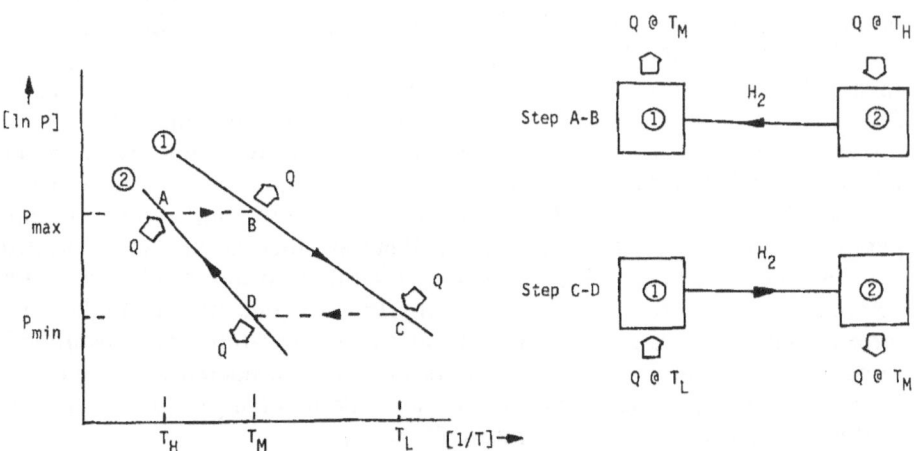

Fig. 5.12. Principles of thermally driven, heat-upgrading heat pump and refrigerators using two different intermetallic hydride formers

d) Thermally Driven, Refrigeration

The concept of thermally driven hydride refrigerators was independently developed at about the same time by several groups: *Gruen* and *Sheft* [5.291], *Terry* [5.279], *van Mal* and *Ferguson* [5.280] and possibly others. The refrigeration cycle is identical to the heat-upgrading cycle shown in Fig. 5.12, except that intermetallics 1 and 2 are chosen such that H_2 transfer $C \rightarrow D$ results in T_L being substantially below ambient temperature. Note that it is exactly the opposite of the temperature upgrading cycle (Fig. 5.11). About $1Q$ input at T_H, coupled with about $2Q$ rejected T_M (usually an ambient sink), results in $1Q$ of heat being removed (cooling) at some subambient temperature T_L. Such a cycle is particularly useful for solar-driven air-conditioning or the conversion of industrial waste heat to chilled water.

e) Experience with Prototypes

Hydride heat pumps and refrigerators are not yet commercially available. However, numerous prototype units have been constructed and tested. The first was the now-famous HYCSOS system, an extensive experimental system built at Argonne National Laboratory to study solar energy heating, cooling, storage, and power generation with intermetallic hydrides [5.90, 142, 291–294]. This was followed by several other prototypes built by groups from many countries [5.14, 29, 30, 49, 271, 285, 295–307], including the present authors' laboratories. A broad summary of these systems, including the intermetallics used and temperatures achieved, is given in Table 5.8. It is impossible to fully describe in this review the details of the many systems listed in Table 5.8. Rather we will provide a short summary of the design and operating considerations that came out of this work.

The design and operation of a hydride heat pump or refrigerator requires some careful considerations that may not be obvious from our simplified description of the thermodynamic cycles shown in Figs. 5.10–12. First of all, we have shown the van't Hoff diagrams as lines representing each intermetallic hydride. Invariably, hysteresis and plateau slope actually broaden these lines into bands. Secondly, we have shown the H_2 transfers as horizontal lines at constant pressure. There are always pressure-driving-force considerations that must be taken into account during transfer flows. Thirdly, there are temperature gradients associated with heat transfer that limit the actual input temperatures or temperatures achieved (T_L, T_M, T_H). Further, there are thermal masses associated with the intermetallic and containment system that lead to sensible heat losses and reduction in operating efficiency. Finally, there are uncertainties in the long-term stability of intermetallics, especially at high temperature where disproportionation can occur. All of these factors must be considered in designing a hydride heat pump or refrigerator to produce a desired temperature from given heat source and heat sink temperature [5.308].

In spite of the above cautions, many of the systems listed in Table 5.8 have been made to work well. Thermodynamic [5.285, 293, 308–310] and economic

Table 5.8. Summary of hydride heat pump and refrigerator demonstrations. After [5.14]

Organization	Ref.	Type[a]	Intermetallic compounds used	Temperatures, °C T_H	T_M	T_L
Solar Turbines International (USA)	[5.296]	R	$LaNi_{4.5}Al_{0.5}–LaNi_5$	93	29	4.4–10
		HP	$MmNi_{4.15}Fe_{0.85}–LaNi_5$	140–180	50–96	20–36
Studsvik Energeteknik AB (Sweden)	[5.49] [5.295]	HP	$LaNi_5–$	70	42	15
		HP	$LaNi_{4.7}Al_{0.3}–TiFe_{0.8}Ni_{0.2}$	—	—	—
			$LaNi_{4.9}Al_{0.1}–TiFe_{0.8}Ni_{0.2}$	—	—	—
Argonne N.L. (USA)	[5.90, 293]	R	$LaNi_5–CaNi_5$	—	—	5–15
		R	$MmNi_{4.15}Fe_{0.85}–LaNi_{4.7}Al_{0.3}$	110	40–50	
Sekisui (Japan)	[5.298]	R	$LaNi_x–La(Ni, Al)_x$	92–111	30	15
Daimler–Benz AG (FRG)	[5.297]	R[b]	$Ti_{0.8}Zr_{0.2}CrMn–LaNi_5$	125	55	–2
		HP	$MmNi_{4.5}Al_{0.5}–LaNi_{4.7}Al_{0.3}$	120	55	15
		HP	$MmNi_{4.15}Fe_{0.85}–MmNi_{4.5}Al_{0.5}$	64	36	12
		R	$Ti_{0.9}Zr_{0.1}CrMn–LaNi_5$	150	50	–25
Chuo Denki Kogyo (Japan)	[5.299]	R	$La(Ni, Al)_5–Mm(Ni, Fe)_5$	140–160	32	0–15
Japan Metals & Chemicals-Kogakuin U. (Japan)	[5.300]	R	$Mm(Ni, Al)_5–Mm(Ni, Fe)_5$	120–160	35	0–20
Kogakuin U. (Japan)	[5.302]	HP[c]	$La(Ni, Al)_x–Mm(Ni, Al)_5–Mm(Ni, Fe)_5$	120–150	80–100	0–40
Kubota, Ltd. (Japan)	[5.301]	HP	$LaNi_5–LaNi_{4.7}Al_{0.3}$	87–90	70	15
			$LaNi_{4.7}Al_{0.3}–MmNi_{4.5}Al_{0.5}$	85	65	25
Ergenics, Inc. (USA)	[5.303, 304]	HP	$LaNi_5–LaNi_{4.7}Al_{0.3}$	120	30	–4
		R	$LaNi_{4.75}Al_{0.25}–MmNi_4Fe$			
Technion (Israel)	[5.305]	R	$LaNi_{4.7}Al_{0.3}–MmNi_{4.15}Fe_{0.85}$	150	35	7
Vienna Univ. (Ausstria)	[5.306]	HP	$LaNi_{4.7}Al_{0.3}–LaNi_5$	79	65	13
		HP	$LaNi_{4.7}Al_{0.3}–MmNi_{4.5}Al_{0.5}$	92	80	13

[a] HP = Heat Pump; R = Refrigerator
[b] Estimated performance. Prototypes not tested.
[c] Two-Stage Heat Pump

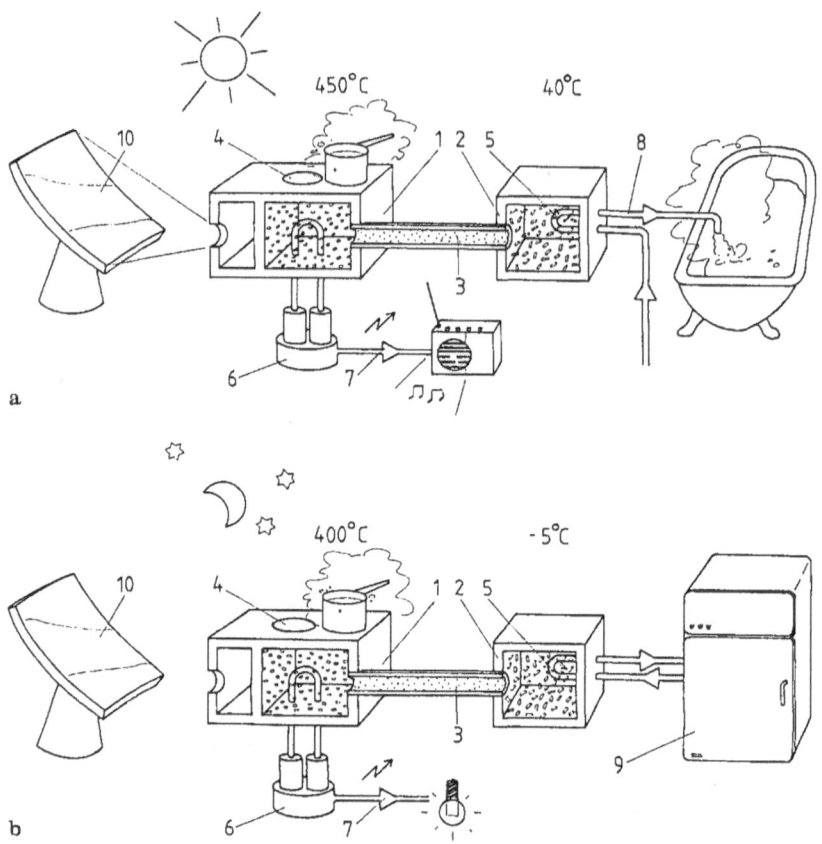

Fig. 5.13 a, b. Solar-hydride system incorporating heat storage, refrigeration and power generation. (a) Power-station daytime operation [5.271]. (b) Power-station night operation [5.271]. 1: high temperature storage unit, 2: low temperature storage unit, 3: hydrogen, 4: stove, 5: heat exchanger, 6: Stirling generator, 7: electricity, 8: hot water, 9: refrigerator, 10: solar collector

[5.311, 312] analyses of hydride heat pumps and refrigerators suggest that they have some potential advantages over competing heat-driven systems and thus chances for commercialization. However, it has been pointed out by one of us (*Suda*) that for commercial reality to be achieved, existing intermetallic compounds need to be improved, especially in the cost and H-capacity areas, and more rapid heat transfer reactors are needed [5.29, 100, 285, 287, 288]. Targets for *stable and reversible* alloy capacity and cost are 1.5 wt%H and $8.5/wt%H$_2$ available/kg alloy, respectively. The target for effective thermal conductivity is $10 \, Wm^{-1} K^{-1}$. There is hope that these targets can be achieved and hydride heat pumps and refrigerators will be added to the growing list of commercial devices that make use of intermetallic hydrides.

Another economically interesting thermochemical system for the storage of solar heat was developed by *Kleinwächter* et al. [5.271, 272]; it is based on highly reactive MgH_2 powder [5.189]. The system captures solar heat by a fixed-focus solar mirror and produces high temperature heat, cooling, warm water and, with the help of a Stirling engine, also electricity (Fig. 5.13).

5.3.8 Heat Engines

A hydrogen/hydride compressor can be coupled in a closed circuit with an expansion engine to create a heat engine, in essence a device to convert heat energy directly into mechanical energy in a Brayton-like cycle. The first hydride device to accomplish this was patented by *Winsche* in 1970 [5.313]; however, this was not a pure hydride heat engine because the device was really a hydride heater that intermittantly heated a water intermediate to steam to provide the mechanical power. One of the first pure hydride/hydrogen heat engines (with only H_2 as the working fluid) was conceptualized by *Terry* and *Schoeppel* early in 1975 resulting in subsequent patents [5.314, 315]. Independently, about the same time, *Powell* et al. (Brookhaven National Lab) developed similar hydride/ dehydride thermodynamic cycles and recognized the value of such heat engines in improving the efficiency of power plants by utilizing the waste heat normally produced [5.316–318].

The above studies and patents were apparently not realized in practice either by *Terry* and *Schoeppel* or by the Brookhaven group (except for the previously cited laboratory pump of *Reilly* et al. [5.255]. However, extensive prototype work has been done at Sandia National Laboratory and Ergenics in the US and the National Chemical Laboratory for Industry in Japan. Work at Sandia concentrated on the utilization of "flat-plate" solar energy, especially for solar-powered water pumping [5.319–321]. AB_5-type intermetallics were used with various diaphragm and bladder pump designs. Using a $23\leftrightarrow68\,°C$ cycle an efficiency of 2.4% was measured, compared to a maximum theoretical efficiency of 6.8% [5.321]. A piston engine, driven by a $LaNi_5$ compressor, was built and tested by *Ono* et al. in Japan [5.273, 322]. With a $20\leftrightarrow80\,°C$ cycle this unit showed the significantly higher efficiency of 7.7% compared to 15.5% as a theoretical maximum [5.322]. This unit is now being used to pump salt water for a reverse-osmosis desalinization appearatus [5.260]. Recently, *Golben* of Ergenics, Inc., developed a hydride cycle water pump which was successfully operated using solar heat in a Texas field demonstration [5.323]. A demonstration in Romania was also reported [5.324].

Hydride heat engines are not yet commercial but appear to have the potential for efficient conversion of low-grade heat to mechanical energy. Most of the design constraints associated with compressors and heat pumps (low hysteresis, good thermal conductivity, disproportionation resistance, etc.) also apply to heat engines.

5.3.9 Temperature Sensors and Actuators

Because the dissociation pressure increases markedly with temperature (5.1, 2), hydrides can be used to sense temperature and actuate devices such as valves or switches. Although not based on an intermetallic hydride, the most successful of this family of devices is an aircraft fire detector that has been commercially produced by Systron–Donner since 196 [5.325]. The sensor is typically a Ti hydride sealed in a long stainless steel tube in series with a pressure switch. The tube is flexible enough to be wound around fire- or overheat-sensitive regions of the aircraft, such as the engine. When even a short length of the sensor is heated, the H_2 evolved increases the pressure enough to close the pressure switch indicating to the pilot that a fire or overheat situation is present. The units automatically reset when cooled by the reabsorption of the H_2. More than 75,000 units have been manufactured to date and are used on many of the world's military and commercial aircraft [5.325].

Because of the high pressures exhibited by heated intermetallic hydrides, valves can be opened or fire extinguisher seals punctured using piston or bellows drivers. This has been the basis of several actuators developed and patented by *Golben* for the automatic release of N_2 or powder fire supressents [5.326–328]. Again, these devices are automatically resettable because the H_2 gas is absorbed into the intermetallic when cooled. Studies by *Welter* and *Witt* have shown that hydride actuators are superior to organic fluid actuators because premature recondensation in cold sections is avoided [5.329, 330]. *Welter* suggests applications such as automobile thermostats. He has also developed a hydride temperature sensor that makes use of the resistance change in a $NbH_{0.49}$ wire [5.330]. In Japan, Sekisui Chemicals, Ltd. has developed hydride actuators for opening water valves and greenhouse windows [5.331].

It is obvious that intermetallic hydrides can be used as thermometers through the P–T van't Hoff relation (5.1, 2). However, the hysteresis must be carefully taken into account and only intermetallic hydrides with a very flat plateau should be used, e.g., $LaNi_5$. The pressure generated will generally be representative of the hottest part of the hydride bed.

5.3.10 Liquid H_2 Applications

Intermetallic hydrides have applications in the area of cryogenic (liquid) hydrogen. One of the early (1973) Philips demonstrations for a $LaNi_5$ thermal compressor was to provide high H_2 pressure for expansion through a Joule–Thompson (J–T) valve, thus resulting a thermally driven hydrogen refrigerator (liquefier) [5.256]. Interest in this concept has been re-established by *Jones* et al. of the Caltech Jet Propulsion Laboratory for space vehicle applications [5.332–334]. Using an improved J–T expansion valve and $LaNi_5$ compressors, extensive simulated life tests have been successfully run [5.333]. Designed for deep space interplanetary vehicles, the compression heat sources would be radioisotope thermal generators with the J–T cooling used to prevent LH_2 boil-off from propellent tanks and provide 20 K cooling for infrared detectors.

Another early Philips device was called a "cold-accumulator" (cold battery) [5.8, 22]. This consisted simply of a $LaNi_5$ resorvoir (at ambient temperature) in closed series with a LH_2 dewar. As H_2 boils off it is absorbed in the $LaNi_5$

bed, thus maintaining constant pressure and temperature of the LH_2 until it is fully vaporized. Using heat on the $LaNi_5H_x$ bed and a cryogenerator, the H_2 can be periodically reliquefied. The vibration-free nature of the cold accumulator is especially desirable for *IR* detectors [5.8].

The general problem of boil-off losses during storage and transfer of LH_2 is substantial [5.335]. For example, about 30% of the LH_2 delivered to Kennedy Space Center is lost during storage and transfer [5.336]. During the space shuttle operations in the 1990s this represents a LH_2 loss of about 12×10^6 liters/year. Intermetallic absorbers offer an opportunity to capture this boil-off for reliquefaction. In fact, it has been successfully demonstrated by Ergenics, Inc. that NASA KSC low pressure, high-volume boil-off surges occurring during LH_2 delivery can be captured in an intermetallic absorber bed [5.336]. The AB_5 compound $LaNi_{4.6}Al_{0.4}$ was used for that demonstration. In a general sense, the growing aerospace and industrial market for LH_2 should spawn more use for intermetallic hydrides in the areas described above.

5.3.11 Catalysts

The rapidity with which many intermetallic compounds will absorb hydrogen at room temperature indicates that they are good catalysts for the $H_2 \rightarrow 2H$ dissociation reaction. This catalytic activity, coupled with the implication that they should have high concentrations of atomic H on their surface, suggest that they should be useful as catalysts for a number of commercial reactions, e.g. methanation, ammonia synthesis, hydrogenation, etc. In fact, there is abundant historic evidence that hydriding intermetallics do have good catalytic activity for such reactions. The first demonstrations of this were the hydrogenation of nitrobenzene, cyclohexene, and actophenone carried out at the Philips Lab before 1971 using a $LaNi_5$ catalyst [5.337].

The most extensive work on the H_2 catalytic activity of intermetallic compounds has been done by *Wallace* and coworkers, concentrating on intermetallics between rare earth elements (or actinides) and transition metals (usually Ni, Co, Mn, or Fe) [5.337–342]. Examples include the following: NH_3 synthesis using AB_2, AB_3, AB_5, A_2B_7, and A_2B_{17} rare earth intermetallics some of which showed higher specific surface activity than conventional NH_3 catalysts [5.339]; methanation with $LaNi_5$ and other *RE* intermetallics [5.340]; methanation with ANi_5 catalysts ($A = $ Th, U, Zr) showing composition effects on H_2S poisoning resistance [5.341]; and methanation using various AB_2 catalysts showing wide composition effects on activity [5.342]. More thorough reviews of the extensive work by *Wallace* et al. are given elsewhere [5.337, 338]. It should be noted that other groups have also seen promising catalytic effects with intermetallics. *Atkinson* and *Nicks* showed that catalysts based on mischmetal-nickel alloys and intermetallics had methanation activity superior to a commercial catalyst with the optimum activity occurring near the intermetallic $MmNi_5$ [5.343, 344]. More recently *Barrault* et al. have confirmed the good CO hydrogenation activity of $RENi_5$ catalysts [5.345, 346]. *Ozyagcilar* reported NH_3 synthesis using Ti hydride "supported" by intermetallic compounds [5.347]. *Hirata* reported the synthesis of various hydrocarbons from CO and H_2 using $FeTi_{1.14}O_{0.03}$ [5.348].

All of the above experiments showed that substantial temperatures (usually $> 300\,°C$) were required for good catalytic activity. For the H_2–CO (methanation) and H_2–N_2 (ammonia synthesis) experiments it was generally found that the A_xB_y intermetallics decomposed into A-oxides or nitrides leaving high surface area elemental B to which the high catalytic activity was attributed. Thus, the resultant structures were really not the original intermetallic compounds or hydrides but rather more like conventional "oxide-supported" transition metal catalysts (e.g. Al_2O_3–Ni). However, it is argued that the production of such a structure by the high temperature decomposition of the intermetallic results in a superior catalyst [5.344, 345]. Furthermore, *Schlapbach* et al. have shown that impurity-induced surface segregation of intermetallics can occur even at near-ambient temperatures [5.55, 114, 129, 349, 350] and that this may be used to lengthen the lifetimes of hydrogenation catalysts [5.351].

In contrast to the relatively high temperature catalytic studies cited above, there is also evidence of catalytic behavior at near ambient temperatures, more closely representing the catalytic properties of the intermetallic hydride itself. For example, ethylene has been hydrogenated to ethane ($C_2H_4 + H_2 \rightarrow C_2H_6$) at low temperatures using $LaNi_5$ and $LaNi_5H_x$ [5.352] and $ZrNiH_{2.8}$ [5.353]. *Soga* et al. found C_2H_4 hydrogenation activity as low as $-84\,°C$, with the $LaNi_5H_x$ being far more active than the unhydrided $LaNi_5$ [5.352]. This implies direct surface reaction from the atomic H leaving the hydride lattice. *Breda* and *Jonville* of Battelle Geneva have developed a process for the hydrogenation of organics by the "gas-free" electrolytic transfer of H from an aqueous electrolyte to the organic via an intermetallic hydride (e.g., $TiNiH_x$) electrode [5.354]. Finally, in the area of NH_3 synthesis, *Lewis* has successfully made NH_3 at temperatures as low as $100\,°C$ from gaseous H_2 and N_2 using as catalysts the hydrides of $CaNi_5$ and Mg_2Ni [5.355].

In summary, the hydridable intermetallics clearly have interesting catalytic properties for a number of reaction types. To our knowledge, none have actually been commercially used to date.

5.3.12 Batteries and Electrochemical Catalysts

Intermetallic hydrogen storage compounds have been studied for a number of battery and other electrochemical applications. The battery applications center around the so-called nickel–hydrogen battery, a high reliability, rechargeable system first developed for satellite applications [5.356]. The electrode reactions are as follows:

$$\text{Negative: } H_2 + 2OH^- \leftrightarrows 2H_2O + 2e^- \tag{5.9}$$

$$\text{Positive: } 2NiOOH + 2H_2O + 2e^- \leftrightarrows 2Ni(OH)_2 + 2OH^- \tag{5.10}$$

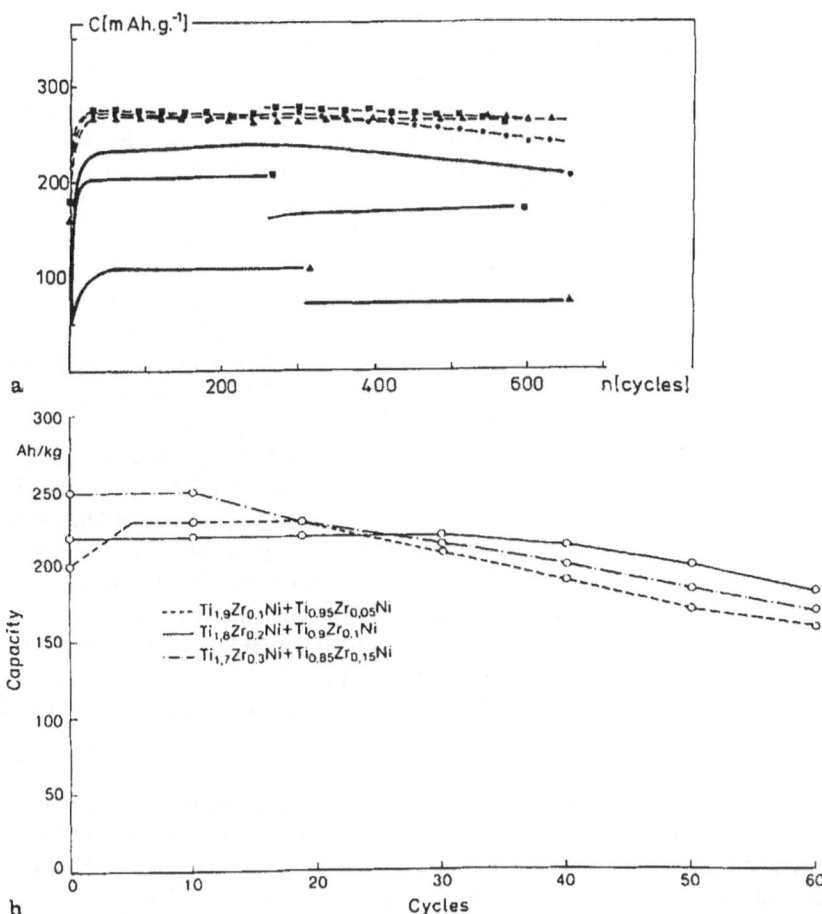

Fig. 5.14. (a) The total storage capacity (dashed curves) and the high-rate capacity (solid lines) of several multicomponent electrodes as a function of the cycle number and for various rates. Charge rate: 1.25 C if $n \leqq \approx 100$ and 2.5 C if $n > \approx 100$; discharge rate: 1.25 C if $n \leqq \approx 300$ and 2.5 C if $n > \approx 300$; (●) $La_{0.7}Nd_{0.3}Ni_{2.5}Co_{2.4}Al_{0.1}$; (■) $LaNi_2Co_3Al_{0.1}$; (▲) $La_{0.9}Ti_{0.1}Ni_{2.5}Co_{2.0}Al_{0.5}$ [5.362]. **(b)** Decrease of capacity of Ti–Ni electrodes with Zr substitution as a function of number of charge/discharge cycles [5.17]

with the net reaction being

$$2NiOOH + H_2 \leftrightarrows 2Ni(OH)_2. \tag{5.11}$$

The right ($\rightarrow$) reaction is discharging and the left ($\leftarrow$) reaction is charging. Thus with charging in a sealed cell a substantial gaseous H_2 pressure results, of as much as 3.5 MPa [5.356].

The charging pressure can be reduced by using intermetallic H_2 storage compounds thus reducing pressure vessel requirements and overall cell weight. One method is to simply absorb the H_2 gas product in say a small $LaNi_5$ pellet, a concept first promoted by *Earl* et al. at COMSAT [5.357]. The second method is to electrochemically transfer the H produced during charging directly from the KOH electrolyte to an intermetallic storage electrode. The concept of a Pd–H storage electrode was developed 25 years ago by *Lewis* et al. [5.226]. This approach was first demonstrated with intermetallic compounds by *Justi* et al. around 1970 using TiNi and Ti_2Ni electrodes [5.358] and in 1973 using $LaNi_5$ electrodes [5.359]. A capacity of 100 mAh/g $LaNi_5$ was reached which was far below the theoretical capacity of 372 mAh/g corresponding to $LaNi_5H_6$ reached easily with gaseous hydrogen. *Percheron* and coworkers [5.360] were able to raise the capacity to 335 mAh/g in substituted $LaNi_5$ alloys. A rapid decrease of the capacity with cycling and at elevated temperature was noticed. *Willems* et al. [5.362, 363] finally developed AB_5 multicomponent electrodes which could be recharged up to 600 times (Fig. 5.14a). Battelle Geneva [5.17] developed electrodes based on Ti–Ni alloy. Capacities of up to 250 mAh/g were reached with two-phase Ti–Ni–Zr alloys; however, cyclic life was not really satisfactory (Fig. 5.14b). The use of metal hydrides for storing hydrogen in a Ag–hydrogen battery was studied by *Schultze* and *Antoine* [5.364].

Bittner and *Badcock* [5.365, 366] and *Willems* [5.362] reviewed both the hydride/gas and hydride/electrode battery storage experience up to 1984.

Meanwhile the pollution related to the rapidly growing market for small size rechargeable Ni–Cd batteries with the toxic heavy metal Cd electrode was noticed. Metal hydride electrodes could ideally replace the Cd electrode. Major activities in the direction of developing a commercial Ni–metal-hydride rechargeable battery occurred in Japan and the US and the annual number of metal hydride electrode patents increased drastically after 1984. Electrodes made from AB_5 compounds, $C14/C15$ AB_2 Laves phase type alloys or $Ti(Zr)Ni_x$ alloys reached electrode capacities between 200 and 350 mAh/g and energy densities of a fuel cell which are 30 to 60% higher than those of conventional Ni–Cd cells. The Ni-electrode technology of the Ni–Cd cells is used and a KOH solution serves as electrolyte. Corrosion of the metal in KOH solution was considerably reduced in multicomponent alloys with reduced lattice expansion upon hydride formation and chemical etching [5.367].

Current research and development activities concentrate on the following areas:

—Optimization of the AB_5-type alloy composition (for recent reviews see [5.367–370]). Partial substitution of Ni by other 3d elements or Al affects the cycle life. Smaller volume expansion, slower decrepitation rate and lower Vickers hardness lower the capacity and rise the cycle life in agreement with Willems [5.362]. The alloy $LaNi_{2.5}Co_{2.5}$ shows good durability. The addition of small amounts of Si, Al, Zr and Ti greatly influence self-discharge rate and low temperature dischargeability. These additives improve cycle life [5.371].

—Microstructure and oxygen content of the cast alloy are important. High alloy cooling rate results in small grain size and long cycle life, suggesting the importance of grain boundaries. 10 at.% oxygen in Ti–Ni alloys decreases the capacity by about 30% [5.17].

—Studies of passivation and corrosion of AB_5-type electrodes in fuel cells and of the effect of overcharging and partial or full discharging. In each cycle the oxidized layer at the surface of AB_5 particles in the electrode is crushed as a result of the volume expansion of the active powder followed by an oxidation of the newly formed clean AB_5 surface by a reaction with the electrolyte [5.114, 372].

The corrosion with the electrolyte varies strongly with the potential [5.373]; however, surface analysis of $LaNi_{5-x}Al_x$ electrodes revealed no passivating Al-oxide film. Oxygen evolved at the Ni electrode during overcharging causes oxidation of the metal hydride electrode [5.374] (cf. Chap. 2).

—AB_5 electrode and fuel cell manufacturing: Microencapsulation of the alloy powder with a thin Cu layer [5.76] was reported to improve charging rate, self-discharge characteristics and decrepitation and thus cycle life [5.375]. Overpressure due to overcharging of sealed cells can be limited by Cu coating of the metal hydride electrode [5.374] and by overdimensions of the Ni electrode. For compacting the alloy powder PTEE binder (3 wt.%) gave very satisfactory results. Flexible electrodes with 15% self-discharge after 30 days storage at 20 °C and 5% capacity decay after 1000 cycles were reported [5.376].

—Solid state batteries using ion conducting polymers as electrolyte are being developed.

—$C14$-type and $C15$-type AB_2 Laves phase compounds [5.377, 378], V–Zr–Ti–Ni–Cr multiphase alloys [5.379, 380] and Zr–Ni–Ti alloys [5.381] have been tested as metal hydride electrodes. Capacities of up to 360 mAh were reported. The most severe problems concern the surface and activation/corrosion and, for the multiphase alloys, the reproducibility of the alloy microstructure.

—Chemical etching or special oxidation treatments are used to form surface layers on the electrode powder, which are passivated to some extent to prevent rapid corrosion, but contain active species like Ni [5.367, 379]. Specific selective oxidation phenomena seem to be as important for surface activation as they are for the activation of the alloy for gaseous hydrogen sorption [5.114]; see also Chap. 2.

—The deposition of catalytically active second phase materials at the surface to improve electron transfer, exchange current, kinetics [5.381a].

—Rechargeable cells based on amorphous alloy electrodes have been tested [5.367, 382]. The continuous melt spinning of amorphous ribbons would simplify electrode fabrication considerably.

Some Japanese companies and one US company started to commercialize these batteries for small computer and electronic applications in the early 1990s. The performance of an AA size battery [5.383] and of a C size battery [5.379, 384]

Table 5.9. Specification of the Ni-metal hydride rechargeable battery of AA size, AB_5 type alloy [5.383], and C size, Ti–Ni–Zr–V–Cr alloy [5.379, 384]

Celll	AA	C
Voltage [V]	1.2	1.2
Capacity [Ah]	1.070	3.5
Energy density [Wh/l]	175	174
[Wh/kg]	54	53
Max. continuous discharge current [A]	3	15
Discharge time [h]	1.5–4.5	1–5
Charge/discharge cycle	500 times (0–45˙C)	500 times
Weight [g]	24	80

are given in Table 5.9; for comparison with energy density of other battery types see Table 5.5. Further references can be found in [5.385–387].

The hydrogen storage electrode battery described above is really a reversible combination of a fuel cell and an electrolyzer, so perhaps it is not surprising that the intermetallic compounds successfully used in such batteries are often found to be good electrochemical catalysts for individual fuel cells and electrolyzers. For example, the fact that rare-earth/nickel AB_5 intermetallics made good fuel cell catalysts was published in a 1968 patent [5.388] which was originally filed in 1965, well before the discovery of the extraordinary AB_5 hydriding properties [5.4]. The use of AB_5 intermetallics as storage electrodes for solid electrolyte fuel cells has been proposed [5.389]. Finally, it has also been found that AB_5 intermetallics are excellent electrocatalysts for hydrogen evolution cathodes in electrolyzers [5.390, 391].

5.3.13 Permanent Magnet Production

Hydrogen embrittlement of construction alloys has been known and studied for many decades and is usually considered as a highly undesirable and risky phenomenon. The most extreme case, hydrogen decrepitation, results in the complete pulverization of brittle hydrogen storage intermetallic compounds. It turned out the first large scale application of intermetallic hydride technology was in the powder grinding of permanent magnet alloys.

Over the past twenty years three new families of exciting rare earth permanent magnets have been developed: $SmCo_5$, $Sm_2(Co, Fe, Cu, Zr)_{17}$ and $Nd_2Fe_{14}B$. An essential property of these magnetic materials is their uniaxial magneto-crystalline anisotropy with a preferred c-axis of magnetic alignment. The alloys cast from the melt are powdered and the individual powder grains become aligned in a magnetic field before being compacted. Perfect alignment can be achieved if each grain is unconstrained and a single crystal.

All three families of permanent magnet alloys form hydrides [5.392]. The large volume change upon hydride formation results in the decrepitation of the bulk material by intergranular or transgranular fracture. Parameters like hydrogen pressure, grain size of the cast alloy and pressure or temperature cycling allow the control of the hydrogen decrepitation process so that subsequent milling is unnecessary or easier. For magnets of the $SmCo_5$ family hydrogen desorbs after the decrepitation stage whereas for the Sm_2Co_{17} family and for $Nd_2Fe_{14}B$-type alloys the powder stays in the hydride form until vacuum sintering starts [5.392–395].

Hydrogen decrepitation and the reducing hydrogen atmosphere during pulverization and sintering result in much cleaner grain surfaces as compared to ball-milled powder. Detrimental effects on the magnetic properties due to surface and bulk oxygen are strongly reduced [5.392, 395, 396].

5.3.14 Other Applications

In this concluding section we summarize a few other intermetallic hydride applications that are at best only indirectly related to the above categories.

In the area of the direct conversion of solar energy to electric power, *Salomon* has proposed coupling a solar heated $LaNi_5$ or $CaNi_5$ hydride bed (i.e., a solar-hydride compressor) to a protonic conductor membrane [5.397]. A hydrogen pressure gradient across a protonic conductor results in a usable electric potential. In addition, protonic conductors can be used for the direct electrochemical H-charging of intermetallics [5.398].

In the area of solar photovoltaic electric storage, Texas Instruments Co. has developed a residentially oriented system that involves the photoelectrolytic decomposition of HBr into H_2 and Br_2 which is stored for later recombination in an H–Br fuel cell [5.399]. A fine, atomized, doped Si is used for photoelectrolysis and the H_2 is stored in $CaNi_5$ hydride beds. Relatively good cyclic life of the hydride was achieved provided the temperature was kept below 60 °C [5.400]. Using a more direct photoelectric storage technique *Clark* et al. successfully coupled a CdSe photoelectrode with a $LaNi_3Co_2$ hydride storage electrode [5.401].

For remote power generation in arctic conditions, a high pressure $MmNi_5$ hydrogen reservoir was coupled to a fuel cell during a joint Brookhaven National Laboratory–US Army program [5.7]. More recently, a hydride-fueled, small (200 W) portable fuel cell has been developed at Ergenics [5.402]. This battery unit is of the solid electrolyte type and is considered competetive with rechargeable batteries for many applications.

In one of the first large H_2-storage demonstrations, electric energy storage (peak shaving) was demonstrated in the form of an electrolyzer–TiFe storage tank–Fuel cell combination during a joint Brookhaven–New Jersey Public Service Electric and Gas Co. program [5.180].

A system to store wind energy has been devised by Kawasaki Heavy Industries in Japan [5.403]. Windmills compress air to generate heat which is in turn used to endothermically desorb H_2 from an O-modified TiFe hydride bed. The H_2 is stored in the gas phase and rereacted with the TiFe to generate heat on windless days, which will then be used in a greenhouse. A demonstration system was built at Ogata Village, Akita Prefecture, in 1985.

Completing our survey of potential intermetallic hydride applications that range from distant space applications to everyday home devices, we cite a catalytic ignition system developed for home ranges and emergency power systems which uses a TiFe hydride reservoir [5.24]. Although the unit is completely independent of electric power, it has apparently never displaced electric igniters.

5.4 Concluding Remarks

We have tried to provide a reasonably thorough review of the many applications that have been considered for intermetallic hydrides along with property considerations that must be made to make the application technically viable. Surely we have not covered all possibilities. It is hoped and expected that many of the readers of this review will have other applications and inventions in mind. Because the intermetallic hydride field is young, with only about 20 years of applications activity, we expect that the spectrum of potential applications has only just begun to develop.

Acknowledgements. We wish to especially thank our colleagues at Ergenics, Inc., Kogakuin University and Fribourg University for their help and encouragement in preparing this review. In a broader sense, however, we owe most of our gratitude to the community of "World Hydriders" whose work serves as the basis of this chapter. The "Intermetallic Hydriders" are a close-knit group and we apologize in advance to our friends whose applications or favorite papers were inadvertantly omitted. L.S. acknowledges financial support by NEFF and BEW.

References

5.1   W.M. Mueller, J.P. Blackledge, G.G. Libowitz: *Metal Hydrides* (Academic, New York 1968)
5.2   G.G. Libowitz: *The Solid State Chemistry of Binary Metal Hydrides* (Benjamin, New York 1965)
5.3   G.G. Libowitz, H.F. Hayes, T.R.P. Gibb: J. Phys. Chem. **62**, 76 (1958)
5.4   J.H.N. van Vucht, F.A. Kuijpers, H.C.A.M. Bruning: Philips Res. Reports **25**, 133 (1970)
5.5   J.J. Reilly, R.H. Wiswall: Inorganic Chem. **13**, 218 (1974)
5.6   D.P. Gregory: Scientific American **228**, 13 (Jan. 1973)
5.7   R. Wiswall: "Hydrogen Storage", in *Hydrogen in Metals II Application Oriented Properties*, ed. by G. Alefeld, J. Volkl, Topics Appl. Phys. **29**, 201 (1978)
5.8   H.H. van Mal: *Stability of Ternary Hydrides and Some Applications*, PhD Thesis, Technishe Hogeschool Delft (1976)
5.9   J.J. Reilly, G.D. Sandrock: Scientific American **242**, 118 (Feb. 1980)
5.10 S.C. Garg, A.W. McClaine: "Metal Hydrides for Energy Storage Applications", Tech. Note N-1393 (Navy Civil Engineering Laboratory, Port Hueneme, CA 1975)
5.11 H.W. Newkirk: "Hydrogen Storage by Binary and Ternary Intermetallics for Energy Applications—A Review", Rpt. URCL-52110 (Lawrence Livermore Laboratory, Livermore, CA 1976)
5.12 J.J. Reilly: "Metal Hydrides as Hydrogen Storage Media and Their Applications", in *Hydrogen: Its Technology and Implications, Vol. II*, Ed. by K.E. Cox, K.D. Williamson, Jr. (CRC, Cleveland 1977) pp. 13–48
5.13 G.D. Sandrock: "Hydrogen Storage" in *Hydrogen Energy*, Rpt. 199 (Royal Swedish Academy of Engineering Sciences, Stockholm 1981), pp. 67–151
5.14 S. Suda: "Recent Advancements of Metal Hydride Heat Pump Developments" Chemical Engineering (Japan) **18**, 1–29 (1983)

5.15 K.C. Hoffman, J.J. Reilly, F.J. Salzano, C.H. Waide, R.H. Wiswall, W.E. Winsche: Int. J. Hydrogen Energy 1, 133 (1976)

5.16 J.J. Reilly: Z. Phys. Chem. NF 117, 155 (1979)

5.17 H. Buchner: *Energiespeicherung in Metallhydriden* (Springer, Vienna 1982)

5.18 R.L. Cohen, J.H. Wernick: Science 214, 1081 (1981)

5.19 E. Snape, E.L. Huston, G.D. Sandrock: "Development of Solar-Hydrogen Systems Using Metal Hydrides", in Proc. 2nd Miami Int. Conf. on Alternative Energy Sources (Hemisphere, Washington 1981) pp. 3569–3585

5.20 E. Snape, F.E. Lynch: Chemtech 10, 768 (1980)

5.21 G.D. Sandrock, E.L. Huston: Chemtech 11, 754 (1981)

5.22 K.H.J. Buschow: "Hydrogen Absorption in Intermetallic Compounds", in *Handbook of the Physics and Chemistry of Rare Earths*, ed. by K.A. Geschneidner, L. Eyring, Vol. 6 (Elsevier, Lausanne 1984) Chap. 47, pp. 1–104

5.23 H. Wenzl: Int. Metals Rev. 27, 140 (1982)

5.24 J.H. Swisher, E.D. Johnson: J. Less-Common Met. 74, 301 (1980)

5.25 S. Suda (Ed.): *Hydrogen Storage Alloys—Developments and Applications to Energy Systems* (Ohyou Gijuto Shuppan, Tokyo, 1984) in Japanese

5.26 H. Buchner: "The Hydrogen/Hydride Energy Concept", in Ref. [5.31], pp. 569–599

5.27 G. Alefeld: "Introduction", in *Hydrogen in Metals II*, ed. by G. Alefeld, J. Völkl, Topics Appl. Phys. Vol. 29 (Springer, Berlin, Heidelberg 1977) pp. 2–10

5.28 E.L. Huston: "Liquid and Solid Storage of Hydrogen", in Proc. 5th World Hydrogen Energy Conf. (Pergamon, Oxford 1984) pp. 1171–1186

5.29 S. Suda: "Recent Development of Hydride Energy Systems in Japan", in Proc. 5th World Hydrogen Energy Conf. (Pergamon, Oxford 1984) pp. 1201–1211; H. Bjurström, S. Suda: Int. J. Hydrogen Energy 14, 19 (1989)

5.30 S. Suda, X.-L. Wang: "Metal Hydrides", Int. J. Hydrogen Energy 15, 569 (1990)

5.31 A.F. Andresen, A.J. Maeland (Eds.): *Hydrides for Energy Storage*, Proc. of an International Symposium held in Geilo, Norway, Aug. 1977 (Pergamon, Oxford 1978)

5.32 G.G. Libowitz, G.D. Sandrock (Eds.): *Metal Hydrides—1980*, Proc. of the International Symposium on the Properties and Applications of Metal Hydrides, Colorado Springs USA, April 1980 (Elsevier, Lausanne 1980). Also published as Vols. 73 and 74 of J. Less-Common Metals

5.33 W.E. Wallace, T. Schober, S. Suda (Eds.): *Metal Hydrides—1982*, Proc. of the International Symposium on the Properties and Applications of Metal Hydrides, Toba, Japan, June 1982 (Elsevier, Lausanne 1983). Also published as Vols. 88 and 89 of J. Less Common Metals

5.34 M. Ron, Y. Josephy, L. Jacob, D. Shaltiel (Eds.): *Metal Hydrides—1984*, Proc. of the International Symposium on the Properties and Applications of Metal Hydrides, Eilat, Israel, April 1984 (Elsevier, Lausanne 1985). Also published as J. Less-Common Met. 103, 104 (1985)

5.35 A. Percheron-Guegan, M. Gupta (Eds.): *Metal Hydrides—1986*, Proc. of the International Symposium on the Properties and Applications of Metals Hydrides, Maubuisson, France, May 1986 (Elsevier, Lausanne 1986). Also published as J. Less-Common Metals 129–131 (1986)

5.36 G.D. Sandrock: "Development of Low Cost Nickel-Rare Earth Hydrides for Hydrogen Storage", Proc. 2nd World Hydrogen Energy Conf. (Pergamon, Oxford 1978) pp. 1625–1656

5.37 Y. Osumi, H. Suzuki, A. Kato, K. Oguro, M. Nakane: "Hydrogen Absorption-Desorption Characteristics of Mischmetal-Nickel Alloys", Proc. 3rd World Hydrogen Energy Conf. (Pergamon, Oxford 1981), pp. 865–879

5.38 C.E. Lundin, F.E. Lynch: "A Detailed Analysis of the Hydriding Characteristics of LaNi$_5$", in Record of the 10th Intersociety Energy Conversion Engineering Conference, (Inst. of Electrical and Electronics Engineers, New York 1975), pp. 1380–1384

5.39 J.J. Reilly, R.H. Wiswall: Inorg. Chem. 7, 2254 (1968)

5.40 G.D. Sandrock: "The Metallurgy and Production of Rechargeable Hydrides", in Ref. [5.31], pp. 353–393

5.41 P.D. Goodell, G.D. Sandrock, E.L. Huston: J. Less-Common Met. 73, 135 (1980)

5.42 P.D. Goodell, G.D. Sandrock, E.L. Huston: "Microstructure and Hydriding Studies of AB$_5$ Hydrogen Storage Compounds", Rpt. SAND 79-7095, (Sandia National Laboratory, Albuquerque, NM 1980)

5.43 M. Yamaguchi, I. Yamamoto, T. Ohta: Trans. Japan Inst. of Metals (Suppl.) 21, 317 (1980)

5.44 C.N. Park, T.B. Flanagan: J. Less-Common Met. 94, L1 (1983)

5.45 E. Tuscher, P. Weinzierl: J. Less-Common Met. 104, 329 (1984)

5.46 Y. Josephy, M. Ron: Z. Phys. Chem. NF **147**, 233 (1986)
5.47 M. Nagel, Y. Komazaki, S. Suda: J. Less-Common Met. **120**, 45 (1986)
5.48 M. Nagel, Y. Komazaki, Y. Matsubara, S. Suda: J. Less-Common Met. **123**, 47 (1986)
5.49 G. Anevi, L. Jansson, D. Lewis: J. Less-Common Met. **104**, 341 (1984)
5.50 S. Suda, N. Kobayashi: "Evaluating Metal Hydride Energy Conversion Systems", in *Metal-Hydrogen Systems*, ed. by T.N. Veziroglu (Pergamon, Oxford 1981) pp. 649–656
5.51 ERGENICS, 247 Margaret King Avenue, Ringwood N.J. 07456, USA
5.52 Aldrich Chemical Co., P.O. Box 355, Milwaukee, WI 53201 USA
5.53 a) Chuo Denki Kogyo Co., Ltd., No. 1-1-12 Toranomon, Minato-ku, Tokyo, Japan
      b) Japan Metals and Chemicals Co., Ltd., 8-4 Koami-cho, Nihombashi, Chuo-ku, Tokyo, Japan
      c) Santoku Metal Industry Co., Ltd., Awaya Bldg. 4-23, Ginza 8-chome, Chuo-ku, Tokyo, Japan
5.54 a) MPD Technology Ltd., Wiggin St., Birmingham B16 OAJ, UK
      b) GFE Gesellschaft für Electrometallurgie mbH, Höfener Strasse 45, D-8500 Nürnberg 80, Germany
5.55 L. Schlapbach, T. Riesterer: Appl. Phys. A **32**, 169 (1983)
5.56 M.H. Mintz, S. Vaknin, S. Biderman, Z. Hadari: J. Appl. Phys. **52**, 463 (1981)
5.57 T. Sasai, K. Oku, H. Konno, K. Onouwe, S. Kashu: J. Less-Common Met. **89**, 281 (1983)
5.58 M. Amano, Y. Sasaki, R. Watanabe, M. Shibata: J. Less-Common Met. **89**, 513 (1983)
5.59 B. Teh-you, J. Pin-sun, S. Lan-yin, G. Biao: "Addition of Various Elements for Improving the Property and Characteristics of TiFe Hydrogen Storage Material", Proc. 4th World Hydrogen Energy Conf. (Pergamon, Oxford 1982) pp. 1239–1243
5.60 F.E. Lynch: J. Less-Common Met. **74**, 411 (1980)
5.61 M. Kawamura, S. Ono, Y. Mizuno: "Stress Induced in Metal Hydrogen Powder Bed by Hydriding Reaction", *Metal-Hydrogen Systems* (Pergamon, Oxford 1982) pp. 489–500
5.62 P.P. Turillon, G.D. Sandrock: "Hydrogen Container", U.S. Pat. 4 133 426, Jan. 9, 1979
5.63 P.P. Turillon, G.D. Sandrock: "Hydride Storage Containment", U.S. Pats. 4 134 490 and 4 134 491, Jan. 16, 1979
5.64 F.R. Block, A. Dey, H. Kappes, K. Reith: J. Less-Common Met. **131**, 329 (1987)
5.65 N.R. Baker, F.E. Lynch: "Hydrogen Sorbent Flow Aid Composition and Containment Thereof", U.S. Pat. 4 600 525, July 15, 1986
5.66 Wang Qi-dong, Wu Jing, Chen Chang-pin, Li Zhou-peng: J. Less-Common Met. **131**, 399 (1987)
5.67 J.J. Reilly, J.R. Johnson: J. Less-Common Met. **104**, 175 (1984)
5.68 R.D. Holstvoogd, K.J. Ptasinski, W.P.M. van Swaaij: Chem. Eng. Sci. **41**, 867 (1986)
5.69 G.C. Blytas: "Hydrogen Sorbent Composition and its Use", U.S. Pat. 4 036 944, July 19, 1977
5.70 H. Buhl, S. Will: "Form Retaining Hydrogen Storage Material", U.S. Pat. 4 110 425, Aug. 25, 1978
5.71 J. Toepler, O. Bernauer, H. Buchner: J. Less-Common Met. **74**, 385 (1980)
5.72 E.E. Eaton, C.E. Olsen, H. Sheinberg, W.A. Steyert: Int. J. Hydrogen Energy **6**, 609 (1981)
5.73 P.S. Rudman, G.D. Sandrock, P.D. Goodell: J. Less-Common Met. **89**, 437 (1983)
5.74 N.J. Bridger, B.J. Miles: "A System for Absorbing and Desorbing Hydrogen, and Hydridable Materials Therefore", U.K. Patent Application GB2086362A, 12 May 1982
5.75 M. Ron, D.M. Gruen, M.H. Mendelsohn, J. Sheft: "Method of Preparing Porous Metal Hydride Compacts", U.S. Pat. 4 292 265, Sept. 29, 1981
5.76 H. Ishikawa, K. Oguro, A. Kato, H. Suzuki, E. Ishii: J. Less-Common Met. **120**, 123 (1986)
5.77 H. Uchida, T. Ebisawa, K. Terao, N. Hosoda, Y.C. Huang: J. Less-Common Met. **131**, 365 (1987)
5.78 E.L. Huston, G.D. Sandrock: J. Less-Common Met. **74**, 435 (1980)
5.79 G.C. Carter, F.L. Carter: "Metal Hydrides for Hydrogen Storage: A Review of Theoretical and Experimental Research, and Critically Compiled Data", in Proc. Miami International Symposium on Metal-Hydrogen Systems (Pergamon, Oxford 1981) pp. 503–529
5.80 S. Suda, Y. Komazaki, N. Kobayashi: "Mixing Effects of Metal Hydrides on Equilibrium Behavior and Reaction Kinetics", Proc. 3rd World Hydrogen Energy Conf. (Pergamon, Oxford 1981) pp. 2169–2183
5.81 P.S. Rudman: J. Less-Common Met. **89**, 93 (1983)
5.82 P.D. Goodell: J. Less-Common Met. **74**, 175 (1980)
5.83 J.J. Reilly, R.H. Wiswall, Jr.: "Hydrogen Storage and Purification Systems II", Rept BNL 19436 (Brookhaven National Laboratory, Upton, NY 1974)

5.84   M. Onischak, D. Dharia, D. Gidaspow: "Heat Transfer Analysis of Metal Hydrides in Metal-Hydrogen Secondary Batteries", Proc. 1st World Hydrogen Energy Conf. (U. of Miami, Coral Gables, FL 1976) pp. 7B-37-50
5.85   S. Suda, N. Kobayashi, K. Yoshida: Int. J. Hydrogen Energy 6, 521 (1981)
5.86   Y. Ishido, M. Kawamura, S. Ono: Int. J. Hydrogen Energy 7, 173 (1982)
5.87   E. Suissa, I. Jacob, Z. Hadari: J. Less-Common Met. 104, 287 (1984)
5.88   D.L. Cummings, G.L. Powers: Ind. Eng. Chem., Process Des. Develop. 13, 182 (1974)
5.89   O. Boser, D. Lehrfeld: "The Rate Limiting Processes for the Sorption of Hydrogen in LaNi₅", Record 10th Intersociety Energy Conversion Engineering Conf. (IEEE, New York 1975) pp. 1363–1369
5.90   I. Sheft, D.M. Gruen, G. Lamich: "HYCSOS: A Chemical Heat Pump and Energy Conversion System Based on Metal Hydrides", Rept. ANL-79-8, (Argonne Nat. Laboratory, Argonne, IL 1979)
5.91   R.L. Woolley: "Method and Apparatus for Providing Increased Thermal Conductivity and Heat Capacity to a Pressure Vessel Containing a Hydride-Forming Metal Material", U.S. Pat. 4 187 092, Feb. 5, 1980
5.92   M. Nagel, Y. Komazaki, S. Suda: J. Less-Common Met. 120, 35 (1986)
5.93   S. Suda, Y. Komazaki, N. Kobayashi: J. Less-Common Met. 89, 317 (1983)
5.94   M. Ron: "Metal Hydrides of Improved Heat Transfer Characteristics", Proc. 11th Intersociety Energy Conversion Engineering Conf. (AIChE, New York 1976) pp. 954–960
5.95   M. Ron, D. Gruen, M. Mendelsohn, I. Sheft: J. Less-Common Met. 74, 445 (1980)
5.96   Y. Josephy, Y. Eisenberg, S. Perez, A. Ben-David, M. Ron: J. Less-Common Met. 104, 297 (1984)
5.97   J. Topler, O. Bernauer, H. Buchner, H. Saufferer: J. Less-Common Met. 89, 519 (1983)
5.98   H. Buchner: Int. J. Hydrogen Energy 9, 501 (1984)
5.99   E. Tuscher, P. Weinzierl, O.J. Eder: Int. J. Hydrogen Energy 8, 199 (1983)
5.100 M. Groll, W. Supper, U. Mayer, O. Brost: Int. J. Hydrogen Energy 12, 89 (1987)
5.101 G.D. Sandrock, E. Snape: U.S. Pat. 4 566 281, Jan. 28, 1986
5.102 J.J. Sheridan III, F.G. Eisenberg, E.J. Greskovich, G.D. Sandrock, E.L. Huston: J. Less-Common Met. 89, 447 (1983)
5.103 W.S. Yu, E. Suuberg, C. Waide: "Modelling Studies of Fixed-Bed Metal-Hydride Storage Systems", The Hydrogen Economy Miami Energy (THEME) Conference (U. of Miami, Coral Gables, FL 1974) pp. S4-22-35
5.104 P.W. Fisher, J.S. Watson: "Modeling Solid Hydrogen Storage Beds", Rept. ORNL/TH-7561 (Oak Ridge Nat. Lab., Oak Ridge, TN 1981)
5.105 L.A. Mahoney: "Theory and Analysis of the Behavior of a Packed Bed of Metal Hydrides", MSc. Thesis (U. of California, Davis, CA 1984)
5.106 I.A. El Osery: Int. J. Hydrogen Energy 8, 191 (1983)
5.107 G.G. Lucas, W.L. Richards: Int. J. Hydrogen Energy 9, 225 (1984)
5.108 S. Suda, N. Kobayashi, E. Morishita, N. Takemoto: J. Less-Common Met. 89, 325 (1983)
5.109 J.J. Reilly, R.H. Wiswall, Jr.: "Alloys for Isolation of Hydrogen", U.S. Pat. 3 825 415 July 23, 1974
5.110 J.J. Reilly, R.H. Wiswall, Jr.: "Hydrogen Storage and Purification Systems", Rept. BNL 17136, (Brookhaven Nat. Lab, Upton, NY 1972)
5.111 R.A. Guidotti, G.B. Atkinson, M.M. Wong: J. Less-Common Met. 52, 13 (1977)
5.112 D.M. Gualtieri, K.S.V.L. Narasimhon, T. Takeshita: J. Appl. Phys. 47, 3423 (1976)
5.113 E.J. McHenry: "A New Design for a Low-Pressure Ni–H₂ Cell", Paper presented at 1977 Fall Mtg. of the Electrochemical Soc., Extended Abstract 77-2 (Electrochem. Soc., Princeton, NJ 1977) p. 98
5.114 L. Schlapbach, A. Seiler, F. Stucki and H.C. Siegmann: J. Less-Common Met. 73, 145 (1980); H.C. Siegman, L. Schlapbach, C.R. Brundle: Phys. Rev. Lett. 40, 972 (1978)
5.115 W.E. Wallace, R.F. Karlicek, Jr., H. Immamura: J. Phys. Chem. 83, 1708 (1979)
5.116 L. Schlapabach: J. Phys. F 10, 2477 (1980)
5.117 P.D. Goodell, G.D. Sandrock: "Metallurgical Studies of Hydrogen Storage Alloys: Surface Poisoning of LaNi₅, FeTi, and Fe₀.₈₅Mn₀.₁₅Ti by O₂,CO₂ and H₂O", Rept. DOE/CS/BNL-51174 (U.S. Dept. of Energy, Washington DC 1980)
5.118 G.D. Sandrock, P.D. Goodell: J. Less-Common Met. 73, 161 (1980)
5.119 P.D. Goodell: J. Less-Common Met. 89, 45 (1983)
5.120 F.G. Eisenberg, P.D. Goodell: J. Less-Common Met. 89, 55 (1983)
5.121 E.L. Huston, Ed.: "Development of a Commerical Metal Hydride Process for Hydrogen Recovery", Rept. BNL 35440 (two vols.), (Brookhaven Nat. Lab., Upton, NY 1984)

5.122 G.D. Sandrock, P.D. Goodell: J. Less-Common Met. **104**, 159 (1984)
5.123 A. Yoshikawa, K. Yagisawa: Trans. Japan Inst. Metals (Suppl.) **21**, 393 (1980)
5.124 A.B. Goncharuk, S.N. Endrzheevskaya: Poroshkovaya Metallurgiya **8**, 77 (1980)
5.125 F.R. Block, H.J. Bahs: J. Less-Common Met. **89**, 77 (1983)
5.126 F.R. Block, H.J. Bahs: J. Less-Common Met. **104**, 223 (1984)
5.127 S. Ono, Y. Ishido, J. Kitagawa: Trans. Japan Inst. of Metals (Suppl.) **21**, 357 (1980)
5.128 B. Vigeholm, J. Kjoller, B. Larsen, A.S. Pedersen: J. Less-Common Met. **104**, 141 (1984)
5.129 L. Schlapbach, J.C. Achard, A. Percheron-Guegan: "Surface Segregation in $LaNi_{5-x}Al_x$ and its Implication on the Cyclic Life Time for Hydrogen Storage", Proc. 3rd International Congress on Hydrogen and Materials (Paris, 1982)
5.130 G.L. Holleck, J.R. Driscoll, B.E. Paul: J. Less-Common Met. **74**, 379 (1980)
5.131 M.J. Benham, D.K. Ross: Z. Phys. Chem. NF **147**, 219 (1986)
5.132 T. Matsumoto, A. Matsushita: J. Less-Common Met. **123**, 135 (1986); V.Z. Mordkovich, N.N. Korostyshevsky, Yu.K. Baychtok, E.I. Mazus, N.V. Dudakova, V.P. Mordovin: Int. J. Hydrogen Energy **15**, 723 (1990)
5.133 K. Yamanaka, H. Saito, M. Someno: Nippon Kagaku Kaishi **8**, 1267 (1975)
5.134 R.L. Cohen, K.W. West, K.H.J. Buschow: Solid State Commun. **25**, 293 (1978)
5.135 K.H.J. Buschow: J. Less-Common Met. **51**, 173 (1977)
5.136 R.L. Cohen, K.W. West, J.H. Wernick: J. Less-Common Met. **70**, 229 (1980)
5.137 R.L. Cohen, K.W. West: J. Less-Common Met. **95**, 17 (1983)
5.138 B.E. Sirovich, I. Ginsburgh: "Method of Regenerating Disproportionated Hydrides", U.S. Pat. 4 302 436, Nov. 24, 1981
5.139 M.H. Mintz, Z. Hadari, M.P. Dariel: J. Less-Common Met. **74**, 287 (1980)
5.140 H. Oesterreicher, K. Ensslen, A. Kerlin, E. Bucher: Mater. Res. Bull. **15**, 275 (1980)
5.141 T. Gamo, Y. Moriwaki, N. Yanagihara, T. Iwaki: J. Less-Common Met. **89**, 495 (1983)
5.142 I. Sheft, D.M. Gruen, G.J. Lamich: J. Less-Common Met. **74**, 401 (1980)
5.143 P.D. Goodell: J. Less-Common Met. **99**, 1 (1984)
5.144 G.D. Sandrock, J.J. Murray, M.L. Post, J.B. Taylor: Mater. Res. Bull. **17**, 887 (1982)
5.145 M. Amano, Y. Sasaki: Trans. Japan Inst. of Metals (Suppl.) **21**, 329 (1980)
5.146 S.J.C. Irvine, I.R. Harris: "Effect of Induced Disorder on the Hydrogenation Behavior of the Phase ZrCo", in Ref. [5.31] pp. 431–466
5.147 J.J. Reilly, J.R. Johnson, J.F. Lynch, F. Reidinger: J. Less-Common Met. **89**, 505 (1983)
5.148 J. Schefer, L. Schlapbach, J.J. Reilly: Progress Report Neutron Diffraction, AF-SSP-118, 1982
5.149 C.E. Lundin, F.E. Lynch: "Solid-State Hydrogen Storage Materials for Applications to Energy Needs", First Annual Report for Contract AFOSR F44620-74-C-0020, (University of Denver, Denver, CO 1975)
5.150 C.E. Lundin, F.E. Lynch: "The Safety Characteristics of FeTi Hydride", Proc. 10th Intersociety Energy Conversion Engineering Conference (IEEE, New York 1975) pp. 1386–1390
5.151 "Safety Studies on Hydrogen Explosions and Accidents", Final Report of Sunshine Project: 1974–1983 (Hydrogen Energy Division), National Chemical Laboratory for Industry, Agency of Industrial Science and Technology (MITI, Tokyo 1984)
5.152 J.J. Reilly, Brookhaven National Laboratory, Upton, NY, 1984: Private Communication
5.153 J.G. Hust: "Survey of Materials for Hydrogen Service: in *Selected Topics on Hydrogen Fuel*, NBS Special Publication 419 (National Bureau of Standards, U.S. Government Printing Office, Washington 1975) pp. 4.1–37; J.E. Campbell: "Effects of Hydrogen Gas on Metals at Ambient Temperature", DMIC Report S-31 (Defense Metals Information Center, Columbus, OH 1970); I.M. Bernstein, A.W. Thompson: *Hydrogen in Metals*, (American Society for Metals, Metals Park, OH 1974)
5.154 H.G. Nelson: "Hydrogen-Metal Interactions", Proc. of the 1977 DOE Chemical Energy Storage and Hydrogen Energy Systems Contracts Review, Rept. JPL 78-1 (Jet Propulsion Laboratory, Pasadena, CA 1978) pp. 141–145; S.L. Robinson: "Hydrogen Compatibility of Structural Materials for Energy Storage and Transmission Applications", Rept. SAND77-8296 (Sandia Nat. Labs., Livermore, CA 1978)
5.155 K. Ledjeff: Int. J. Hydrogen Energy **12**, 361 (1987); M. Haruta, H. Sano: Int. J. Hydrogen Energy **7**, 801 (1982)
5.156 R.G. Barnes (ed.): "Hydrogen Storage materials", Materials Science Forum, vol. **31** (Trans. Tech., Switzerland 1988)
5.157 D. Behrens (ed.): "Wasserstofftechnologie", Dechema (1986)

5.158 S.L. Robinson, J.J. Iannucci: "Technologies and Economics of Small-Scale Hydrogen Storage", Rpt. SAND79-8646 (Sandia Nat. Labs., Livermore, CA 1979)
5.159 C. Carpetis: Int. J. Hydrogen Energy 5, 423 (1980)
5.160 C. Carpetis: Int. J. Hydrogen Energy 7, 191 (1982)
5.161 G. Eklund, O. von Krusenstierna: Int. J. Hydrogen Energy 8, 463 (1983)
5.162 R.M. Hartley (Ed.): Hydrogen Progress, 2nd Quarter 1977 (Billings Energy Corp., Provo, UT)
5.163 J.H. Kelly, R. Hagler, Jr.: Int. J. Hydrogen Energy 5, 35 (1980)
5.164 Hydrogen Consultants, Inc., 12420 N. Dumont Way, Littleton CO 80125 USA
5.165 Baseline Industries, Inc. P.O. Box 649, Lyons, CO 80540 USA
5.166 J.C. McCue: J. Less-Common Met. 74, 333 (1980)
5.167 H. Wenzl, K.H. Klatt: "The Use of FeTi-Hydride for Production and Storage of Suprapure Hydrogen", in Ref. [5.31] pp. 323–327
5.168 HWT Gesellschaft für Hydrid- und Wasserstofftechnik mbH, Postfach 10 08 27, 4330 Mülheim a.d. Ruhr, FRG
5.169 Suzuki Shokan Co., 3-1 Kojimachi, Chiyoda-ku, Tokyo 102, Japan
5.170 N. Baker, L. Huston, F. Lynch, L. Olavson, G. Sandrock: "A Clean Internal Combustion Engine for Underground Mining Machinery", Final Phase I Report for U.S. Bureau of Mines Contract H0202034 (Eimco Mining Machinery International, Salt Lake City, UT 1981) pp. 127–129
5.171 G. Strickland, J.J. Reilly, R.H. Wiswall, Jr: "An Engineering-Scale Storage Reservoir of Iron Titanium Hydride", Proc. of the Hydrogen Economy Miami Energy (THEME) Conference (Univ. of Miami, Coral Gables, FL 1974) pp. S4-9–S4-21
5.172 M.J. Rosso: "Hydrogen-Technology Advanced-Component Test System (HYTACTS)," Chemical/Hydrogen Energy Storage Systems Program Semi-annual Contracts Review (Brookhaven Nat. Lab., Upton, NY 1980)
5.173 P. Guinet, P. Perroud, J. Rebiere: Int. J. Hydrogen Energy 5, 609 (1980)
5.174 R.E. Billings: "Modification and Operation of the Hydrogen Homestead Hydride Vessel Energy Storage System", Final Report for Contract BNL 481417-S (Billings Energy Corp., Independence, MO, USA)
5.175 H. Suzuki, Y. Osumi, A. Kato, K. Oguro, M. Nakane: J. Less-Common Met. 89, 545 (1983)
5.176 Iwatani Information 2 (No. 3) (Iwatani and Co., Ltd. Tokyo, Japan May/June 1982) pp. 1, 2
5.177 S. Ono, H. Kanazawa, H. Toma: "Development of a Large-Sized Hydrogen Storage Vessel Using Metal Hydrides", Proc. 5th World Hydrogen Energy Conf. (Pergamon, Oxford 1984) pp. 1349–1357
5.178 "Hydrogen—The Energy Source of the Future", Rept. CO2 (Mannesmann Röhrenwerke, Dusseldorf, BRD 1984)
5.179 S. Suda: Unpublished data (Kogakuin Univ., Tokyo, Japan 1984)
5.180 J.M. Burger, P.A. Lewis, R.J. Isler, F.J. Salzano, J.M. King, Jr.: "Energy Storge for Utilities via Hydrogen Systems", Proc. 9th Intersociety Energy Conversion Engineering Conf. (ASME, New York 1974) pp. 428–434
5.181 R. Povel, J. Töpler, G. Withalm, C. Halene: "Hydrogen Drive in Field Testing", Proc. 5th World Hydrogen Energy Conf. (Pergamon, Oxford 1984) pp. 1563–1577
5.182 Japan Chemical Engineering, Ltd., 3–12–23 Kitahorie, Nishi-ku, Osaka-shi, Osaka, Japan
5.183 G. Strickland, J. Milau, M.J. Rosso: "Some Observations on the Effects of the Volumetric Expansion of Iron-Titanium Hydride on Vessels Built at BNL", Rept. BNL 23130 (Brookhaven Nat. Lab. Upton, NY 11973, USA 1977)
5.184 H.K. Abdel-Aal: "The Prospects of Low-Cost Magnesium Hydrides—An Economic Study", Proc. 3rd World Hydrogen Energy Conf. (Pergamon, Oxford 1981) pp. 849–864
5.185 J.J. Reilly, R.H. Wiswall Jr.: Inorganic Chem 7, 2254 (1968)
5.186 A.S. Pedersen, J. Kjoller, B. Larsen, B. Vigeholm: "On the Hydrogenation Mechanism in Magnesium I", Proc. 5th World Hydrogen Energy Conf. (Pergamon, Oxford 1984) pp. 1269–1277; B. Vigeholm, J. Kjoller, B. Larsen, A.S. Pedersen: J. Less-Common Met. 104, 141 (1984)
5.187 N. Nishimiya, A. Suzuki, S. Ono: "A Novel Batchwise Hydrogen Transmitting System Using Metal Hydrides", Proc. 3rd World Hydrogen Energy Conf. (Pergamon, Oxford 1981) pp. 917–929
5.188 N.J. Bridger: "Improvements in or Relating to Hydrogen from a Hydride Material: U.K. Patent Specification 1 568 374, 29 May 1980
5.189 B. Bogdanovic: Int. J. Hydrogen Energy 12, 863 (1987); B. Bogdanovic, K.H. Claus, S. Gürtzgen, B. Spliethoff, U. Wilczok: J. Less-Common Met. 131, 163 (1987); H. Imamura, M.

Takashima: Int. J. Hydrogen Energy **15**, 911 (1990) and references therein; A.T. Raissi, D.K. Slattery, M.J. Axelrod, R. Zidan: Proc. 8th World Hydrogen Energy Conference (1990), Hawaii

5.190 W. Winsche, R.H. Wiswall, J. Reilly, T. Sheehan, C. Waide, K. Hoffman: "Metal Hydrides as a Source of Fuel for Vehicular Propulsion", Rpt. BNL-11754 (Brookhaven Nat. Lab., Upton, NY, USA 1967)

5.191 K.C. Hoffman, W.E. Winsche, R.H. Wiswall, J.J. Reilly, T.V. Sheehan, C.H. Waide: "Metal Hydrides as a Source of Fuel for Vehicular Propulsion", Paper 690232 (Society of Automative Engineers, New York 1969)

5.192 J.J. Reilly, K.C. Hoffmann, G. Strickland, R.H. Wiswall: "Iron-Titanium Hydride as a Source of Hydrogen Fuel for Stationary and Automotive Applications", 26th Power Sources Symposium, Paper BNL-18651 (Brookhaven Nat. Lab., Upton, NY, USA April 1974)

5.193 J.J. Reilly, R.H. Wiswall, Jr., K.C. Hoffman: "Metal Hydrides as a Source of Hydrogen Fuel", American Chemical Soc. Meeting, Paper BNL-14804 (Brookhaven Nat. Lab., Upton, NY, USA Sept. 1970)

5.194 A.L. Austin: "A Survey of Hydrogen's Potential as a Vehicular Fuel", Rpt. UCRL-51228 (Lawrence Livermore Lab., Livermore, CA, USA 1972)

5.195 L.O. Williams: Cryogenics **693** (Dec. 1973)

5.196 H. Buchner: "Results of Hydride Research and Consequences for the Development of Hydride Vehicles", Proc. of the 4th International Symposium on Automotive Propulsion Systems, Rpt. CONF-770430/1 (U.S. Dept. of Energy, Washington, 1978) pp. 746–758

5.197 D.L. Henriksen, D.B. MacKay, V.R. Anderson: "Prototype Hydrogen Automobile Using a Metal Hydride", Proc. 1st World Hydrogen Energy Conf. (Univ. of Miami, Coral Gables, FL, USA 1976) pp. 7C-1–7C-12

5.198 R.E. Billings: Int. J. Hydrogen Energy **3**, 49 (1978)

5.199 R.L. Woolley: "Design Considerations for the Riverside Hydrogen Bus", Proc. 2nd World Hydrogen Energy Conf. (Pergamon, Oxford 1978) pp. 1829–1849

5.200 V.R. Anderson: "Hydrogen Energy in United States Post Office Delivery Systems", Proc. 2nd World Hydrogen Energy Conf. (Pergamon, Oxford 1978) pp. 1879–1901

5.201 "A Car with a Future" Hydrogen Progress (Billings Energy Corp., Independence, MO, USA Spring 1979) pp. 17–21

5.202 R.E. Billings, R.L. Wooley, B.C. Campbell, J.H. Ruckman, V.R. Anderson: "The Hydrogen Homestead", Hydrogen Progress (Billings Energy Corp., Provo, UT, USA Fall 1977) pp. 28–36

5.203 R.E. Billings: "Conversion of Hybrid MERADCOM Forklift from Gaseous Hydrogen Fuel Storage to a Hydride Hydrogen Storage System", Project Summary for Contract BNL-491266-S (Billings Energy Corp., Independence, MO, USA May 1980)

5.204 D. Davidson, M. Fairlie, A.E. Stuart: "Development of a Hydrogen-Fuelled Farm Tractor", Proc. 5th World Hydrogen Energy Conf. (Pergamon, Oxford 1984) pp. 1623–1630

5.205 L.G. Olavson, N.R. Baker, F.E. Lynch, L.C. Mejia: "Hydrogen Fuel for Underground Mining Machinery", SAE Paper 840133 (SAE, Warrendale, PA, USA 1984)

5.206 R. Billings: "Hydrogen Storage in Automobiles Using Cryogenics and Metal Hydrides", Proc. of the Hydrogen Economy Miami Energy (THEME) Conf. (U. of Miami, Coral Gables, FL 1974) pp. S8–51 to 61

5.207 H. Buchner: Int. J. Hydrogen Energy **8**, 373 (1983); **7**, 259 (1982)

5.208 H. Buchner: Prog. Energy Comb. Sci. **6**, 331 (1980)

5.209 O. Bernauer: Int. J. Hydrogen Energy **14**, 727 (1989)

5.210 J.G. Finegold, J.T. McKinnon, M.E. Karpuk: "Dissociated Methanol as a Consumable Hydride for Automobiles and Gas Turbines", Proc. 4th World Hydrogen Energy Conf. (Pergamon, Oxford 1982) pp. 1359–1369

5.211 C. Carpetis: Int. J. Hydrogen Energy **7**, 61 (1982)

5.212 R. Gordon: "Composite Pressure Vessels for Gaseous Hydrogen Powered Vehicles", Proc. 5th World Hydrogen Energy Conf. (Pergamon, Oxford 1984) pp. 1225–1236

5.213 E.W. Morley: Smithsonian Contributions to Knowledge **29**, 56 (1892–1903)

5.214 Wang Qi-dong, Wu Jing, Chen Chang-pin, Lou Wei-iang, Fang Tian-Shui: "Improvement on Metal Hydride Suprapure Hydrogen Purifier with Oxygen-Removing Molecular Sieve and Double-Valve Blow-Off Technique", Proc. 6th World Hydrogen Energy Conf. (Pergamon, new York 1986) pp. 881–892

5.215 Wu Yan-min: "A New Ultrapure Hydrogen Purifier", Proc. 6th World Hydrogen Energy Conf. (Pergamon, New York 1986) pp. 893–897

5.216 J.J. Reilly, R.H. Wiswall Jr.: "Separation of Hydrogen from Other Gases", U.S. Patent 3 793 435 (February 19, 1974)

5.217 D. Gidaspow, Y. Liu: "Hydrogen Separation and Compression Through Hydride Formation and Dissociation by Low-Level Heat", Proc. 11th Intersociety Energy Conversion Engineering Conf. (AIChE, New York 1976) pp. 920–925

5.218 V. Cholera, D. Gidaspow: "Hydrogen Separation and Production from Coal-Derived Gases Using Fe_xTiNi_{1-x}", Proc. 12th Intersociety Energy Conversion Engineering Conf. (Am. Nuclear Soc., LaGrange Park, IL, USA 1977) pp. 981–986

5.219 R.W. Meyerhoff: "Hydrogen from Ammonia", U.S. Pat. 4 544 527, Oct. 1, 1985

5.220 Wang Qi-dong, Wu Jing, Chen Chang-pin, Ye Zhou: J. Less-Common Met. 131, 321 (1987)

5.221 J.M.S. Henis, M.K. Tripodi: Science 220, 11 (1983)

5.222 S. Sircar, J.W. Zondlo: "Hydrogen Purification by Selective Adsorption", U.S. Patent 4 077 779, March 7, 1978

5.223 J. Zwiebel, D.B. Broughton, D.T. Camp (Eds.): Gas Purification by Absorption (AIChE, New York 1973)

5.224 E.L. Huston, J.J. Sheridan III: "Applications for Rechargeable Metal Hydrides", in Industrial Applications of Rare Earth Elements, ACS Symposium Series 164 (Am. Chem. Soc., Washington 1981) pp. 223–250

5.225 S. Imoto, T. Tanabe, K. Utsunomiya: Int. J. Hydrogen Energy 7, 597 (1982)

5.226 F.A. Lewis: The Palladium Hydrogen System. (Academic, London 1967)

5.227 R. Lässer, K.H. Klatt: Phys. Rev. B 28, 748 (1983)

5.228 R.H. Wiswall Jr., J.J. Reilly: Inorg. Chem. 11, 1691 (1972)

5.229 A. Biris, R.V. Bucur, P. Ghete, E. Indrea, D. Luper: J. Less-Common Met. 49, 477 (1976)

5.230 D. Dayan, M.P. Dariel: Mater. Res. Bull. 16, 137 (1981)

5.231 A. Yawny, G. Friedlmeier, J.C. Bolcich: Int. J. Hydrogen Energy 14, 587 (1989)

5.232 J. Tanaka, R.H. Wiswall, J.J Reilly: Inorg. Chem. 17, 498 (1978)

5.233 J. Völkl, G. Alefeld: "Diffusion of Hydrogen in Metals", in Hydrogen in Metals I, Basic Properties, Topics Appl. Phys. 28, 321 (1978)

5.234 G. Sicking, P. Albers, E. Magomedbekov: J. Less-Common Met. 89, 373 (1983)

5.235 R. Lässer: J. Less-Common Met. 131, 263 (1987); T. Schober, C. Dieker, R. Lässer, H. Trinkaus: J. Less-Common Met. 131, 293 (1987)

5.236 J.L. Flament, G. Lozes: Proc. 3rd Int. Conf. "Hydrogen and Materials", Paris 1982, vol. 1, Inst. Sup. Mater. Constr. Mech., St. Ouen, France

5.237 E. Gluekauf, G.P. Kitt: "Chromatographic Separations of Hydrogen Isotopes", in Vapor Phase Chromatography (Butterworths, London 1957) p. 422

5.238 F.T. Aldridge: "Chromatographic Hydrogen Isotope Separation", U.S. Pat. 4 276 060, June 30, 1981; J. Less-Common Met. 108, 131 (1985)

5.239 Y.W. Wong, F.B. Hill: A.I.Ch.E.J. 25, 592 (1979)

5.240 Y.W. Wong, F.B. Hill, Y.N.I. Chan: Separation Sci. Tech. 15, 423 (1980)

5.241 F.B. Hill, V. Grzetic: J. Less-Common Met. 89, 399 (1983)

5.242 H.K. Rae (Ed.): Separation of Hydrogen Isotopes, (Am. Chem. Soc., Washington 1978)

5.243 H. Buchner: "Method for Preparation of Deuterium by Isotope Separation", U.P. Pat. 3 940 912, Mar. 2, 1976

5.244 J.J. Reilly: Brookhaven Nat. Lab., Upton, NY USA, Unpublished data (1977)

5.245 G.D. Sandrock: ERGENICS, Inc., Wyckoff, NJ, USA, Unpublished data (1977)

5.246 W. Rummel: Siemens Forsch.—u. Entwick.—Ber. 10, 371 (1981)

5.247 P. della Porta: J. Vac. Sci. Technol. 9, 532 (1972)

5.248 H.S. Cullingford, M.G. Wheeler, J.W. McMullen: "A Hydrogen Storage Bed Design for Tritium Test Assembly", in Metal-Hydrogen Systems, (Pergamon, Oxford 1982) pp. 601–617

5.249 C. Boffito, F. Doni, L. Rosai: J. Less-Common Met. 104, 149 (1984); C. Boffito, B. Ferrario, P. della Porat, L. Rosai: J. Vac. Sci. Technol. 18, 1117 (1981)

5.250 St-101 and St 707 Non-Evaporable Getters, (S.A.E.S. Getters, S.p.A, Milan, Italy 1977, 1987)

5.251 F. Doni, C. Boffito, B. Ferrario: J. Vac. Sci. Technol. A4, 2474 (1986)

5.252 M.H. Mendelsohn, D.M. Gruen: J. Less-Common Met. 74, 449 (1980); D. Shaltiel, L. Jacob, D. Davidov: J. Less-Common Met. 53, 117 (1977)

5.253 F. Meli, Z. Sheng, I. Vedel, L. Schlapbach: Vacuum 41, (1990)

5.254 S. Bredendiek-Kämp, H. Klewe-Nebenius, G. Pfennig, M. Bruns, M. Devillers, H.J. Ache: Fesenius Z. Anal. Chem. 335, 669 (1989); R.D. Penzenhorn, M. Devillers, M. Sirch: J. Nucl. Mater. 170, 217 (1990)

5.255 J.J. Reilly, A. Holtz, R.H. Wiswall Jr.: Rev. Sci. Inst. **42**, 1485 (1971)

5.256 H.H. van Mal: Chemie-Ing.-Techn. **45**, 80 (1973)

5.257 J.J. Reilly, R.H. Wiswall: "Hydrogen Storage and Purification III, Rpt. BNL 21322 (Brookhaven Nat. Lab, Upton, NY, USA 1976)

5.258 F.E. Lynch, R.A. Nye, P.P. Turillon: "Hydride Chemical Compressor", Phase I Final Report to Contract BNL 484822-S (Ergenics, Inc., Wyckoff, NJ, USA 1981)

5.259 E. Tuscher, O.J. Eder, P. Weinzierl: "A Chemical Compressor Based on Compacted Metal Hydrides", in *Metal-Hydrogen Systems*, (Pergamon, Oxford 1982)

5.260 K. Nomura, E. Akiba, S. Ono: J. Less-Common Met. **89**, 551 (1983)

5.261 Wang Qi-dong, Wu Jing, Chen Chang-pin, Au Ming, Fang Tian-shui: "Research and Development on High Pressure Suprapure Hydrogen Hydride Compressors", Proc. 6th World Hydrogen Energy Conf. (Pergamon, New York 1986) pp. 872–880

5.266 P.M. Golben: "Multi-Stage Hydride–Hydrogen Compressor", Proc. 18th Intersociety Energy Conversion Engineering Conf. (A.I.Ch.E., New York 1983) pp. 1746–1753

5.267 R.W. Meyerhoff: "Efficiency of Hydrogen Compression by Means of Hydrides", Proc. 2nd Miami Int. Sym. on Alternative Energy Sources, (Hemisphere, Washington 1981) pp. 3555–3567

5.268 G.G. Libowitz: "Metal Hydrides for Thermal Energy Storage", Proc. 9th Intersociety Energy Conversion Engineering Conf. (Am. Soc. Mech. Eng., New York 1974) pp. 322–325

5.269 G.G. Libowitz, Z. Blank: "An Evaluation of the Use of Metal Hydrides for Solar Thermal Energy Storage", Proc. 11th Intersociety Energy Conversion Engineering Conf. (A.I.Ch.E., New York 1976) pp. 673–680

5.270 G. Alefeld, P. Maier-Laxhuber, M. Rothmeyer: "Thermochemical Heat Storage and Heat Transformation with Zeolites as Absorbents", Technologies and Their Commercialization, Vol. 1 (Springer, Berlin, Heidelberg 1981) pp. 796–820

5.271 B. Bogdanovic, B. Spliethoff, A. Ritter: Z. Phys. Chem. NF **164**, 1497 (1989) and VDI-Berichte **725**, 37 (1989); J. Kleinwächter, B. Bogdanovic, H. Spliethoff, A. Ritter: Proc. 4th Int. Symp. on Research Development and Applications of Solar Thermal Technology, Santa Fe, 1988; Bomin Solar, D-7850 Lörach, BRD

5.272 M. Wierse, R. Werner, M. Groll: J. Less-Common Metals **172–174**, 1111 (1991)

5.273 S. Ono: "Solar Energy Storage by Metal Hydride", in *Solar-Hydrogen Energy Systems*, (Pergamon, Oxford 1979) pp. 193–224

5.274 I. Yonezu, K. Nasako, N. Honda, T. Sakai: J. Less-Common Met. **89**, 351 (1983); K. Nasako, T. Yonesaki, I. Yonezu, S. Fujitani, T. Saito, M. Moroto, M. Osumi, N. Furukawa: Proc. ISES Solar World Congress, Kobe, Japan (1989)

5.275 B. Bogdanovic: Private communication

5.276 "Hydride System Able to Warm Cold Engines", Design News, Aug. 8, 1983, p. 17; H. Buchner, H. Säufferer: "Parking Heater and Method Using Hydrides in Motor Vehicles Powered by Hydrogen", U.S. Pat. 4 214 699, July 29, 1980

5.277 A.W. McClaine: "Method and Apparatus for Heat Transfer, Using Metal Hydrides", U.S. Pat. 4 039 023, August 2, 1977

5.278 J.G. Cottingham: "Hydride Heat Pump", U.S. Pat. 4 044 819, August 30, 1977

5.279 L.E. Terry: "Hydrogen-Hydride Absorption Systems and Methods for Refrigeration and Heat Pump Cycles", U.S. Pat. 4 055 962, Nov. 1, 1977 (Reissued as Re. 30 840, Jan. 5, 1982)

5.280 H.H. van Mal, E.T. Ferguson: "Cyclic Desorption Refrigerator and Heat Pump, Respectively", U.S. Pat. 4 111 002, Sept. 5, 1978

5.281 W.H. Bowman, B.E. Sirovich: "Moving Bed Hydride/Dehydride Systems", U.S. Pat. 4 178 987, Dec. 18, 1979

5.282 L. Terry: "Hydrogen-Hydride Absorption Systems and Methods for Regrigeration and Heat Pump Cycles", U.S. Pat. 4 188 795, Feb. 19, 1980

5.283 B.E. Sirovich: "Hydride Heat Pump", U.S. Pat. 4 200 144, Apr. 29, 1980

5.284 D.M. Gruen, P.R. Fields: "System for Thermal Energy Storage, Space Heating and Cooling and Power Conversion", U.S. Pat. 4 262 739, Apr. 21, 1981

5.285 P. Dantzer, E. Orgaz: J. Chem. Phys. **85**, 2961 (1986); P. Dantzer, E. Orgaz: Int. J. Hydrogen Energy **11**, 797 (1986); E. Orgaz, P. Dantzer: J. Less-Common Met. **131**, 385 (1987)

5.286 S. Wolf: "Hydrogen Sponge Heat Pump", Proc. 10th Intersociety Energy Conversion Engineering Conf. (IEEE, New York 1975) pp. 1348–1351

5.287 M. Groll, R. Werner: IEA Task VII, "Hydrogen Storage, Conversion and Safety", Workshop Osaka, Japan (1989) unpublished; M. Groll, W. Supper, R. Werner: Z. Phys. Chem. NF **164**, 1485 (1989); R. Werner, M. Groll: "Design aspects of metal hydride heat transformers" in

Récent progrès en Génie des Procédés: Pompes à chaleurs, Vol. **2–88**, no. 5 p. 320; R. Werner, M. Groll: J. Less-Common Met. **172–174**, 1122 (1991)

5.288 D.-W. Sun, S.-J. Deng: Int. J. Hydrogen Energy **15**, 331 (1990)

5.289 J.G. Cottingham: "A Hydride Heat Pump to Enhance Solar Energy Collection and Storage and for Waste Heat Scavenging", Rpt. BNL-19914 (Brookhaven Nat. Lab., Upton, NY, USA Mar. 1975)

5.290 G. Alefeld: "A Metal Hydrogen Heat Pump as Topping Process for Power Generation", Proc. 2nd World Hydrogen Energy Conf. (Pergamon, Oxford 1978) pp. 1947–1957

5.291 D.M. Gruen, I. Sheft: "Metal Hydride Systems for Solar Energy Storage and Conversion", Proc. of the Workshop on Solar Energy Storage Subsystems for the Heating and Cooling of Buildings, Rpt. NSF-RA-N075-041 (National Science Foundation, Washington, 1975) pp. 96–99

5.292 I. Sheft, D.M. Gruen, G.J. Lamich, L.W. Carlson, A.E. Knox, J.M. Nixon, M.H. Mendelsohn: "HYCSOS: A System for Evaluation of Hydrides as Chemical Heat Pumps", in Ref. [5.31] pp. 551–567

5.293 D.M. Gruen, F. Schreiner, I. Sheft: Int. J. Hydrogen Energy **3**, 303 (1978)

5.294 D.M. Gruen, M. Mendelsohn, I. Sheft, G. Lamich: "Materials and Performance Characteristics of the HYCSOS Chemical Heat Pump and Energy Conversion Systems", Proc. 2nd World Hydrogen Energy Conf. (Pergamon, Oxford 1978) pp. 1931–1946

5.295 G. Anevi, D. Lewis: "HEPTA 5. A Summary of Exploratory Work on Potential Uses of Metalic Hydrides", Rpt. ET-81/118 (Studsvik Energiteknik AB, Nykoping, Sweden 1981)

5.296 D.A. Rohy, T.A. Argabright, G.W. Wade: "Metal Hydride Heat Pump", Proc. 17th Intersociety Energy Conversion Engineering Conf. (IEEE, New York 1982) pp. 1160–1165

5.297 H. Buchner: BMFT Status Seminar (Bundes ministerium für Forschung, BRD 1979) p. 338

5.298 N. Nishizaki, M. Miyamoto, K. Miyamato, K. Yoshida, Y. Nataka: "Experimental Studies on Metal Hydride Heat Pump for Residential Heating, Air Conditioning, and Hot Water Supply" Proc. 17th Falll Meeting (Soc. Chem. Eng. Japan, Sendai, Japan 1983) p. 18

5.299 K. Yamamura, R. Ishikawa, S. Houno, T. Kanno, T. Kumagaya: "Development of Metal Hydride Heat Pump for Industrial Waste Heat Recovery", Proc. 17th Fall Meeting (Soc. Chem. Eng. Japan, Sendai, Japan 1983) p. 21

5.300 Y. Matsubara, Y. Kamazaki, M. Nagel, T. Amaiwa, S. Suda: "Development of Metal Hydride Air-Conditioner for Industrial Uses", Proc. 17th Fall Meeting (Soc. Chem. Eng. Japan, Sendai, Japan 1983) p. 19

5.301 Y. Yanai, M. Shichiri, S. Tuboi, T. Imamura, Y. Mizuno: "Testing Results of a Metal Hydride Heat Pump", Proc. 17th Fall Meeting (Soc. Chem. Eng. Japan, Sendai, Japan 1983) p. 20

5.302 S. Suda: J. Less-Common Met. **104**, 209 (1984)

5.303 P.P. Turillon: "Hydride Heat Pump for Industrial Waste Heat Recovery", Final Rept. for New York State ERDA Contract No. ER-215-78/79 (ERGENICS, Inc., Wyckoff, NJ, USA 1984)

5.304 L.E. Terry, G.D. Sandrock, E.L. Huston, R.W. Meyerhoff: "Solar Powered Hydride Refrigerator", Unpublished Data (ERGENICS, Inc., Wyckoff, NJ, USA 1981)

5.305 M. Ron: J. Less-Common Met. **104**, 259 (1984)

5.306 E. Tuscher, P. Weinzierl, O.J. Elder: Int. J. Hydrogen Energy **9**, 783 (1984)

5.307 K. Nasako, I. Yonezu, S. Fujitani, A. Furukawa, T. Yonesaki, T. Sakai: MRS Int. Meeting on Adv. Mat. **2**, 113 (1989); I. Yonezu, S. Fujitani, K. Nasako, A. Furukawa, T. Yonesaki, N. Furukawa: MRS Int. Meeting on Adv. Mat. **2**, 149 (1989)

5.308 M. Ron, Y. Josephy: Z. Phys. Chem. NF **147**, 241 (1986); **164**, 1475 (1989)

5.309 H. Abelson, J.S. Horowitz: "A Thermodynamic Analysis of a Metal Hydride Heat Pump", Proc. 15th Intersociety Energy Conversion Engineering Conf. (Am. Inst. of Aeronautics and Astronautics, New York 1980) pp. 936–945

5.310 T. Nishizaki, K. Miyamoto, K. Yoshida: J. Less-Common Met. **89**, 559 (1983)

5.311 R. Gorman, P.S. Moritz: "Design Study and Cost-Effectiveness of the Metal Hydride Solar Heat Pump and Power System (HYCSOS)", Proc. 14th Intersociety Energy Conversion Engineering Conf. (Am. Chem. Soc., Washington 1979) pp. 109–113

5.312 R. Gorman, W.L. Akridge: "Hydride Heat Pump System for Building Air Conditioning Using High Temperature Solar Input", Rpt. for Argonne N.L.P.O. No. 898403 and SN No. 97164 (TRW, Inc., McLean, VA, USA)

5.313 W.E. Winsche: "Intermittent Power Source", U.S. Pat. 3 504 494, Apr. 7, 1970

5.314 L.E. Terry, R.J. Schoeppel: "Hydride-Dehydride Power System and Methods", U.S. Pat. 3 943 719, Mar. 16, 1976

5.315 L.E. Terry, R.J. Schoeppel: "Power Cycles Based Upon Cyclical Hydriding and Dehydridingg of a Material", U.S. Patents 4 199 827, Apr. 22, 1980; 4 311 014, Jan. 19, 1982; 4 397 153, Aug. 9, 1983

5.316 J.R. Powell, F.J. Salzano, W-S. Yu, J.S. Milau: "High Efficiency Power Conversion Cycles for Central Station and Peaking Power Plants", Paper 75-WA/Ener-2 (Am. Soc. Mech. Eng., New York 1975)

5.317 J.R. Powell, F.J. Salzano, W-S. Yu, J.S. Milau: "High Efficiency Power Conversion Cycles Using Hydrogen Compressed by Absorption on Metal Hydrides", Proc. 10th Intersociety Energy Conversion Engineering Conf. (IEEE, New York 1975) pp. 1339–1347

5.318 J.R. Powell, F.J. Salzano: "Hydride Compressor", U.S. Pat. 4 085 590, Apr. 25, 1978

5.319 T.E. Hinkebein, C.J. Northrup, A.A. Heckes: "Closed-Cycle Hydride Engines", Rept. SAND78-2228 (Sandia Nat. Labs. Albuquerque, NM, USA 1978)

5.320 A.A. Heckes, T.E. Hinkebein, C.J.M. Northrup: "Hydride Engines", Proc. 14th Intersociety Energy Conversion Enginnering Conf. (Am. Chem. Soc., Washington 1979) pp. 743–746

5.321 C.J.M. Northrup Jr., A.A. Heckes: J. Less-Common Met. **74**, 419 (1980)

5.322 K. Nomura, Y. Ishido, S. Ono: Energy Conversion **19**, 45 (1979)

5.332 "Hydride Pump Uses Low-Grade Heat", World Water, June 1986, p. 47

5.324 P. Ghete, R. Sarbu, R. Lupu, R. Lupu, A. Biris, C. Bratu: "Water Pump with Metallic Hydrides", Proc. 6th World Hydrogen Energy Conf. (Pergamon, New York 1986) pp. 920–932

5.325 D.E. Warren, K.A. Faughman, R.A. Fellow, J.W. Godden, B.M. Seck: J Less-Common Met. **104**, 375 (1984)

5.326 P.M. Golben: "Metal Hydride Actuation Device", U.S. Pat. 4 282 931, Aug. 11, 1981

5.327 P.M. Golben: "Thermally Activated Metal Hydride Sensor/Actuator, U.S. Pat. 4 377 209, Mar. 22, 1983

5.328 P.M. Golben: "Fast-Acting Self-Resetting Hydride Actuator", U.S. Pat. 4 385 494, May. 31, 1983

5.329 J-M. Welter, J-D. Witt: Regelungstechnische Praxis **25**, 51 (1983)

5.330 J.M. Welter: J. Less-Common met. **104**, 251 (1984)

5.331 K. Shinoda, K. Oguma: "Metal Hydride Thermal Sensors", Proc. 15th Fall Meeting (Soc. Chem. Eng. Japan, Kanazawa, Japan 1981) p. 427

5.332 G.A. Klein, J.A. Jones: "Molecular Absorption Cryogenic Cooler For Liquid Hydrogen Propulsion Systems", AIAA Paper 82-0830 (Am. Inst. of Aeronautics and Astronautics, New York 1982)

5.333 J.A. Jones, P.M. Golben: "Life Test Results of Hydride Compressors for Cryogenic Refrigerators", AIAA Paper 84-0058 (Am. Inst. of Aeronautics and Astronautics, New York 1984)

5.334 J.A. Jones: "Cryogenic Hydride Refrigeration", Videotape AVC-038-84CID (Jet Propulsion Laboratory, Pasadena, CA, USA 1984)

5.335 D.A. Mathis: *Hydrogen Technology for Energy* (Noyes Data Corp, Park Ridge, NJ, USA 1976) pp. 55–72

5.336 M.J. Rosso, M. Golben: J. Less-Common Met. **131**, 283 (1987)

5.337 Netherlands Patent 6 912 908, cited by W.E. Wallace: "Rare Earth and Actinide Intermetallics as Hydrogenation Catalysts", in Ref. [5.31] pp. 501–514

5.338 W.E. Wallace, A. Elattar, H. Imamura, R.S. Craig, A.G. Moldovian: "Intermetallic Compounds: Surface Chemistry, Hydrogen Absorption and Heterogeneous Catalysis" in *Science and Technology of Rare Earth Materials* (Academic, New York 1980) pp. 329–351

5.339 T. Takeshita, W.E. Wallace, R.S. Craig: J. Catal. **44**, 236 (1976)

5.340 V.T. Coon, T. Takeshita, W.E. Wallace, R.S. Craig: J. Phys. Chem. **80**, 1878 (1976)

5.341 A. Elattar, T. Takeshita, W.E. Wallace, R.S. Craig: Science **196**, 1093 (1977)

5.342 A. Elattar, W.E. Wallace, R.S. Craig: "Intermetallic Compounds as Catalysts for Hydrogenation of Carbon Oxides", in *The Rare Earths in Modern Science and Technology* (Plenum, New York 1978) pp. 87–92

5.343 G.B. Atkinson, L.J. Nicks: J. Catal. **46**, 417 (1977)

5.344 G.B. Atkinson, L.J. Nicks: "Preparation and Use of High Surface Area Transition Metal Catalysts", U.S. Pat. 4 071 473, Jan. 31, 1978

5.345 J. Barrault, D. Duprez, A. Percheron-Guegan, J.C. Achard: J. Less-Common Met. **89**, 537 (1983)

5.346 J. Barrault, A. Guilleminot, A. Percheron-Guegan, V. Paul-Boncour, J.C. Achard: J. Less-Common Met. **131**, 425 (1987)

5.347 M.N. Ozyagcilar: "Method of Making Ammonia", European Pat. Appl. 0 034 403 (European Patent Office, The Hague, Aug. 26, 1981)
5.348 T. Hirata: J. Less-Common Met. **124**, 11 (1986)
5.349 L. Schlapbach, A. Seiler, F. Stucki: Mater. Res. Bull. **13**, 697 (1978)
5.350 L. Schlapbach, A. Seiler, F. Stucki: Mater. Res. Bull. **13**, 1031 (1978)
5.351 L. Schlapbach, A. Seiler, F. Stucki: Mater Res. Bull. **14**, 785 (1979)
5.352 K. Soga, H. Imamura, S. Ikeda: J. Phys. Chem. **81**, 1762 (1977)
5.353 P.A. Chernavskii, V.V. Lunin: Kinetics and Catalysis **24**, 769 (1983)
5.354 F. Breda, P. Jonville: "Hydrogen Transfer by Metal Hydride Between Aqueous Medium and Organic Compound", U.S. Pat. 4 120 763, Oct. 17, 1978
5.355 D. Lewis: "Method of Producing Ammonia", U.S. Pat. 4 325 931, Apr. 20, 1982
5.356 M. Klein, B.S. Baker: "Nickel–Hydrogen Battery System", Proc. 9th Intersociety Energy Conversion Engineering Conf. (Am. Soc. Mech. Eng., New York 1974) pp. 118–122
5.357 M.W. Earl, J.D. Dunlop: COMSAT Tech. Rev. **3**, 437 (1973)
5.358 E.W. Justi, H.H. Ewe, A.W. Kalberlak, N.M. Saridakis, M.H. Schaeffer: Energy Conversion **10**, 183 (1970); H. Ewe, Ph.D. thesis, Technical University Braunschweig (1970)
5.359 H. Ewe, E.W. Justi, K. Stephan, Energy Conversion **13**, 109 (1973)
5.360 G. Bronoël, S. Sarradin, A. Percheron, J.C. Achard, M. Bonnemay, J. Loriers: French Patents No. 7 516 160 (1975) and No. 7 723 812 (1977)
5.361 G. Bronoël Sarradin, M. Bonnemay, A. Percheron, J.C. Achard, L. Schlapbach: Communication at ISE. Meeting, Marcoussis, France 1975 and Int. J. Hydrogen Energy **1**, 251 (1976); G. Bronoël, J. Sarradin, A. Percheron-Guegan, J.C. Achard: Mat. Res. Bull. **13**, 1265 (1978)
5.362 J.J.G. Willems: Philips J. Res. **39** (Suppl. 1), 1 (1984)
5.363 J.J.G. Willems, J.R. Beek, K.H. Buschow: European Patent Appl. 84 201 493.8, Oct. 1984
5.364 J.-P. Schultze, P. Antoine: "Application of Hydrides to the Silver–Hydrogen System", Rep. SAS-224/82 JPS/MD, 1982, SAFT Aerospace Dept, F-93230 Romain ville, France
5.365 H.F. Bittner, C.C. Badcock: J. Electrochem. Soc. **130**, 193C (1983)
5.366 H.F. Bittner, M.V. Quinzio, C.C. Badcock: "Metal Hydrides for Hydrogen Storage in Nickel Hydrogen Batteries", Proc. 5th World Hydrogen Energy Conf. (Pergamon, Oxford 1984) pp. 1371–1381
5.367 T. Sakai, A. Takagi, T. Hazama, H. Miyamura, N. Kuriyama, H. Ishikawa, C. Iwakura: Proc. 3rd Int. Conf. Batteries for Utility Energy Storage, Kobe, Japan Mach 18–22, 1991, in press
5.368 H Ogawa, M. Ikoma, H. Kawano, I. Matsumoto: In Power Sources **12**, ed. by T. Keily, B.W. Baxter, Int. Power Sources Symp. Committee (Taylor and Francis, London 1989) p. 393
5.369 T. Sakai, H. Miyamura, N. Kuriyama, A. Kato, K. Oguro, H. Ishikawa, C. Iwakura: J. Less-Common Met. **159**, 127 (1990)
5.370 T. Sakai, K. Oguro, H. Miyamura, N. Kuriyama, A. Kato, H. Ishikawa, K. Iwakura: J. Less-Common Met. **161**, 193 (1990)
5.371 T. Sakai, H. Miyamura, N. Kuriyama, A. Kato, K. Oguro, H. Ishikawa: J. Electrochem. Soc. **137**, 795 (1990)
5.372 A. Boonstra, T.M. Bernards: J. Less-Common Met. **161**, 245, 355 (1990)
5.373 F. Meli, L. Schlapbach: J. Less-Common Met. **172–174**, 1252 (1991)
5.374 T. Sakai, A. Yuasa, H. Miyamura, N. Kuriyama, K. Kato, H. Ishikawa: J. Less-Common Met. **172–174**, 1194 (1991)
5.375 C. Iwakura, Y. Kajiya, H. Yoneyama, T. Sakai, K. Oguro, H. Ishikawa: J. Electrochem. Soc. **136**, 1351 (1989)
5.376 T. Sakai, A. Takagi, K. Kinoshita, N. Kuriyama, H. Miyamura, H. Ishikawa: J. Less-Common Met. **172–174**, 1185 (1991)
5.377 Y. Moriwaki, T. Gamo, H. Seri, T. Iwaki: J. Less-Common Met. **172–174**, 1211 (1991)
5.378 Y. Moriwaki, T. Gamo, A. Shintani, T. Iwaki: Proc. 28th Battery Symp., Japan, 111 (1987); Proc. 29th Battery Symp., Japan, 117 (1988)
5.379 M.A. Fetcenko, S. Venkatesan, K.C. Hong, B. Reichman: "Hydrogen Storage Materials for Use in Rechargeable Ni-Metal Hydride Batteries" in Power Sources **12**, ed. by T. Keily, B.W. Baxter, Int. Power. Sources Symp. Committee (Taylor and Francis, London 1989) p. 411
5.380 B. Reichman, S. Venkatesan, M. Fetcenko, K. Jeffries, S. Stahl, C. Bennett: US Patent 4 716 088 (1987); S. Venkatesan, B. Reichman, M. Fetcenko: US Patent 4 728 586 (1988)

5.381 H. Sawa, K. Ohzeki, M. Ohta, H. Nakano, S. Wakao: Z. Phys. Chemie NF **164**, 1521 (1989); H. Wawa, M. Ohta, H. Nakano, S. Wakao: Z. Phys. Chemie NF **164**, 1527 (1989)
5.381a P.H.L. Notten, P. Hokkeling: J. Electrochem. Soc. **138**, 1877 (1991)
5.382 D.H. Ryan, F. Dumais, B. Patel, J. Kycia, J.O. Ström–Olsen: J. Less-Common Met. **172–174**, 1246 (1991)
5.383 Matsushita News, January 12 (1989)
5.384 Ovonic Battery Company, Troy, MI, USA
5.385 M. Kanda, T. Satoh, K. Sasaki, S. Tsuruta, S. Kohan, E. Tagasaki, K. Kurigara: Proc. 27th Battery Symp., Japan, 9 (1986); Proc. 28th Battery Symp., Japan, 109 (1987)
5.386 M. Mohri, T. Tajima, H. Tanaka, T. Yomeda, M. Kasahara: Sharp Technical Reports **34**, 97 (1986) and **38**, 55 (1987)
5.387 S. Furukawa, K. Inoue, T. Matsumoto, S. Kameoka: Proc. 28th Battery Symp. Japan, 107 (1987); Proc. 28th Battery Symp. Japan, 119 (1988)
5.388 L.R. Dillworth, W.J. Wunderlin: "Fuel Cell and Fuel Cell Electrode Containing Nickel-Rare Earth Intermetallic Catalyst", U.S. Pat. 3 405 008, Oct. 5, 1968
5.389 C. Folonari, G. Iemmi, F. Manfredi, A. Rolli: J. Less-Common Met. **74**, 371 (1980)
5.390 T. Kitamura, C. Iwakura, H. Tamura: Chem. Lett. **965** (1981)
5.391 D.E. Hall, V.R. Shepard, Jr.: Int. J. Hydrogen Energy **9**, 1005 (1984)
5.392 I.R. Harris: J. Less-Common Met. **131**, 245 (1987) and references therein; I.R. Harris, P.J. McGuiness: Proc. 11th Int. Workshop on Rare Earth Magnets and Their Applications, Pittsburgh, 21–14 Oct. 1990, in press
5.393 P. Dalmas de Reotier, D. Fruchart, P. Wolfers, R. Guillen, P. Vulliet, A. Yaouanc, R. Fruchart, Ph. L'Heritier: J. Phys. Colloq (Paris) **46**, 6 (1985); R. Fruchart, R. Madar, A. Rouault, P. L'Heritier, P. Taunier, D. Boursier, D. Fruchart, P. Chaudouët: French Patent 84.10387 (1984); D. Fruchart, S. Miraglia: Proc. MMM Conf., San Diego CA, Nov. 90, J. Appl. Phys. in press (1991)
5.394 E. Rozendaal, J. Ormerod, P.J. McGuiness, I.R. Harris: "A Study of Nd–Fe–B Magnets Produced Using a Combination of Hydrogen Decrepitation and Jet Milling" in *Concentrated European Action on Magnets*, ed. by I.V. Mitchell et al. (Elsevier, London 1989) p. 510
5.395 J. Ormerod: J. Less-Common Met. **111**, 49 (1985)
5.396 L. Schlapbach: J. Less-Common Met. **111**, 291 (1985)
5.397 R.E. Salomon: Solar Energy **23**, 91 (1979)
5.398 P. de Lamberterie, M. Forestier, J. Guitton, A. Rouault, R. Fruchart: J. Less-Common Met. **131**, 427 (1987)
5.399 E.L. Johnson: "The Texas Instruments Solar Energy System Development", Proc. 16th Intersociety Energy Conversion Engineering Conf. (Am. Soc. Mech. Eng, New York 1981) pp. 798–804
5.400 M.S. Bawa, E.A. Ziem: Int. J. Hydrogen Energy **7**, 775 (1982)
5.401 W.D.K. Clark, M.N. Hull, J.T. Arms: "Photoelectrochemical Cell with In-Situ Storage using Hydrogen Storage Electrodes", Canadian Patent 1 142 131, March 1, 1983
5.402 "High-Tech Hydrides on the Way", Chem. Week (Feb. 26, 1986) pp. 49–50
5.403 "KHI Commission to Develop Overall Regenerator System for Research on Wild Energy Storage Technology by Metal Hydride", Kawasaki Topics, No. 143 (Kawasaki Heavy Industries Ltd., Japan 1983)

6. Experimental Techniques I: The Perturbed Angular Correlation Method and Its Application to Hydrogen in Metals

Alois Weidinger

With 16 Figures and 2 Tables

The perturbed angular correlation (PAC) technique uses radioactive nuclei as local probes of the solid. The nuclei interact via their magnetic dipole moments with the local magnetic field, and via their electric quadrupole moments with the local electric field gradient (e.f.g.). They can therefore be used to measure these fields, the information on the interaction being obtained by the observer from the angular correlation of the emitted γ-rays.

The PAC method is similar to Mössbauer spectroscopy, in which radioactive nuclei are also employed. However, there are important differences: the PAC method is sensitive only to magnetic dipole and electric quadrupole interactions, but is completely unaffected by the isomer shift or Debye–Waller factor. This has the advantage that superpositions of different interactions (e.g. the isomer shift and the quadrupole interaction) can be avoided, and that experiments are not restricted to certain temperature ranges by the Debye–Waller factor, but has the disadvantage, of course, that information contained in the isomer shift and Debye–Waller factor is inaccessible.

With the PAC method, only source experiments can be performed, i.e. the radioactive nuclei have to be incorporated into the sample which is to be investigated. This is sometimes rather tedious and may have prevented extensive use of the method in the past. On the other hand, the PAC technique is a very powerful tool and in some cases provides information which cannot be obtained by any other means. The method has been applied extensively in other fields, e.g. for studies of radiation damage [6.1, 2].

This chapter presents a brief outline of the PAC technique, and reviews the application of the method to metal–hydrogen systems. The method is particularly suited to the study of local aspects such as defects in hydride structures or hydrogen trapping at impurities. A special but important application is the investigation of hydrogen diffusion at low temperatures.

6.1 The PAC Method

The essential part of a perturbed angular correlation experiment is a γ–γ cascade in a radioactive nucleus with an intermediate state which has a certain (not too short) lifetime τ_N. The two γ-rays are recorded by two fixed detectors (Fig. 6.1), and the difference between their arrival times is measured. The emission of the first γ-ray (γ_1) in a certain direction leads to an alignment of the nuclear spin

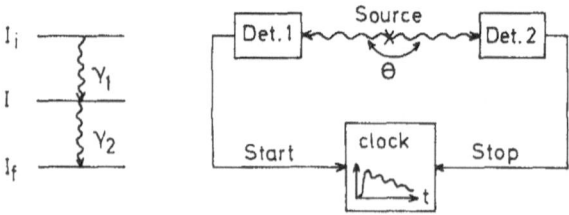

Fig. 6.1. γ–γ cascade of a nucleus and schematic representation of a perturbed angular correlation set-up

in the intermediate state (i.e. unequal population of the M-substates), with the consequence that the second γ-ray (γ_2) is emitted anisotropically. If during the lifetime of the intermediate state the alignment is changed by the nucleus interacting with the local fields in the solid, then the emission characteristics of the second γ-ray are also changed. This leads to an oscillating coincidence count rate which reflects the change of the alignment as a function of time.

The coincidence count rate $N_{12}(\theta, t)$ of detectors 1 and 2 as a function of time t has the following form:

$$N_{12}(\theta, t) = N_0 \, e^{-t/\tau_N} W(\theta, t), \tag{6.1}$$

where τ_N is the mean lifetime of the intermediate state, N_0 a normalizing constant, $W(\theta, t)$ the time-dependent angular correlation, and θ the angle between the two detectors. The general form [6.3–5] of $W(\theta, t)$ is rather complicated and shall not be reproduced here. However, in most practical cases, very simple expressions for $W(\theta, t)$ can be used; these will now be described.

6.1.1 Quadrupole Interaction in a Polycrystalline Sample

All applications of the PAC method to hydrogen in metals have so far used the quadrupole interaction in a polycrystalline sample. In such samples, which have a random orientation of the electric field gradients, $W(\theta, t)$ can be written to first order as

$$W(\theta, t) = 1 + A_2 G_2(t) P_2(\cos \theta), \tag{6.2}$$

where A_2 is the anisotropy of the angular correlation and $P_2(\cos \theta)$ is the Legendre polynomial. The higher order terms are usually neglected. Their contributions are partially included in an effective A_2 term. Only in experiments with very high statistics can the influence of the higher order terms be detected. The physically interesting quantity $G_2(t)$ is given by

$$G_2(t) = s_{20} + \sum_{n=1}^{n_{max}} s_{2n} \cos \omega_n t. \tag{6.3}$$

The ω_n are the transition frequencies in the quadrupole-split intermediate state.

They are uniquely related to the quadrupole coupling constant

$$v_Q = \frac{eQV_{zz}}{h},$$

(6.4)

and to the asymmetry parameter

$$\eta = \frac{V_{xx} - V_{yy}}{V_{zz}},$$

(6.5)

where Q is the quadrupole moment of the nuclear state, and V_{ii} are the components of the electric field gradient tensor in the principal axes frame. The energy-splitting of a nuclear level can be calculated by a diagonalization of the quadrupole interaction hamiltonian

$$H_Q = \frac{h v_Q}{4I(2I-1)} \left[3I_z^2 - I(I+1) + \frac{\eta}{2}(I_+^2 + I_-^2) \right].$$

(6.6)

Here $I_{\pm}(=I_x \pm iI_y)$ and I_z denote spin operators, and I is the spin quantum number of the state. For $\eta = 0$ the following simple expression for the energy splitting is obtained

$$E_Q(M) = \frac{h v_Q}{4I(2I-1)}[3M^2 - I(I+1)],$$

(6.7)

where M is the magnetic quantum number of the spin. The frequencies ω_n in (6.3) are given by

$$\omega_n = \frac{E_i - E_j}{h},$$

(6.8)

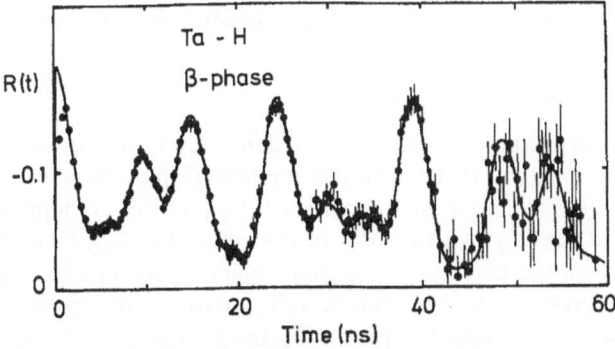

Fig. 6.2. PAC spectrum of ^{181}Hf-^{181}Ta in tantalum hydride at 100 K for [H]/[Ta] = 0.43 [$(\alpha + \beta)$-phase]. The solid line is a fit with $v_Q = 433$ MHz, $\eta = 0.45$ [6.6]

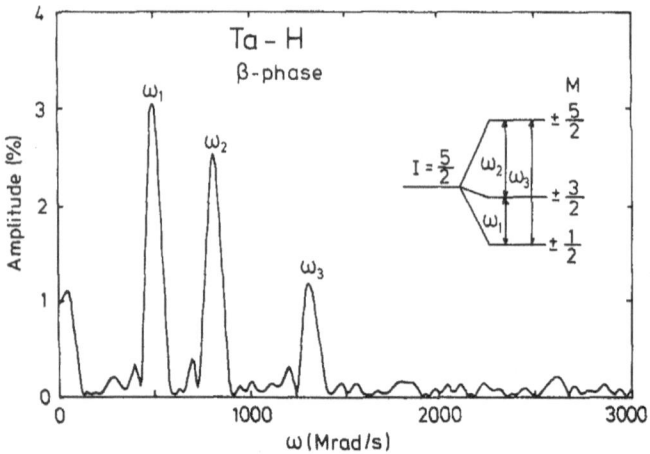

Fig. 6.3. Fourier transform of the PAC spectrum of Fig. 6.2. The insert shows the quadrupole splitting of the $I = 5/2$ state

where E_i and E_j are the eigenvalues of H_Q, see (6.6). The s_{2n} values in (6.3) can be calculated [6.3–5]. For $I = 5/2$ and $\eta = 0$ they have the following values:

$$s_{20} = 0.2 \quad s_{21} = 0.37 \quad s_{22} = 0.29 \quad s_{23} = 0.14. \tag{6.9}$$

As an example, the PAC spectrum [6.6] of ^{181}Ta in tantalum hydride (β-phase) is shown in Fig. 6.2. Here, only the reduced count rate $A_2 G_2(t)$ is displayed. The solid line is a fit to the data with $\nu_Q = 433$ MHz and $\eta = 0.45$, from which $V_{zz} = 7.64 \times 10^{17}$ V/cm^2 is derived.

The concept behind the PAC method can be seen more easily in the Fourier transform of the PAC spectrum (Fig. 6.3). The three lines expected for an $I = 5/2$ intermediate state (see insert of Fig. 6.3) are clearly seen. The unequal spacing of these lines indicates directly that η cannot equal zero. The intensity ratio of the three lines is given by the s_{2n} values in (6.3) and is predicted by theory.

6.1.2 Quadrupole Interaction in a Single Crystal

PAC experiments on single crystalline samples allow a determination of the axes of the electric field gradient (e.f.g.) in relation to the crystallographic directions. The measurement relies on the fact that the intensities of frequencies ω_n depend strongly on the direction of the e.f.g. with respect to the emitted γ-rays, i.e with respect to the arrangement of the detectors. The frequencies ω_n themselves are not changed and are again given by the quadrupole splitting of the intermediate state. The e.f.g directions are determined by comparing the measured intensities with the calculated values for different assumed directions of the e.f.g. Since here the precise relative intensities matter, the higher order anisotropy terms must be considered. Explicit formulae for the calculation of the intensities have been published, e.g. in [6.4].

6.1.3 Magnetic Interactions

In ferro- or antiferromagnetic substances, the PAC probe measures the local magnetic field at the probe site. The behaviour of hydrogen in a magnetic sample can then be studied by its influence on this local field.

The magnetic hyperfine interaction leads to a splitting of the intermediate state according to

$$E(M) = -M\hbar\omega_L, \tag{6.10}$$

where M is the magnetic quantum number and ω_L the Larmor frequency. In turn, ω_L is related to the local magnetic field B_{loc} by

$$\omega_L = \gamma \cdot B_{\mathrm{loc}} \tag{6.11}$$

where γ is the gyromagnetic ratio of the nuclear state. In a PAC experiment, ω_L and its harmonics are observed as precession frequencies.

6.2 Experimental Details

6.2.1 PAC Sources

Only a few nuclei are appropriate for PAC experiments. The most convenient ones will now be briefly described:

^{181}Hf–^{181}Ta (Fig. 6.4 and Table 6.1). ^{181}Hf can be easily produced from ^{180}Hf at a nuclear reactor by the (n, γ) reaction. Unfortunately, no chemical separation

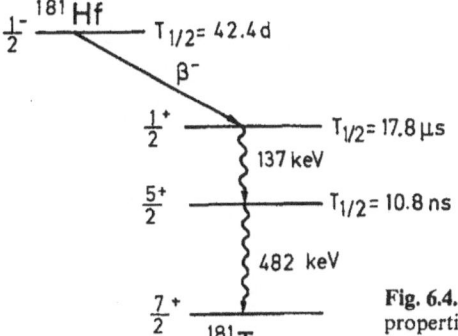

Fig. 6.4. Partial decay scheme of ^{181}Hf. The nuclear properties relevant for the PAC experiment are summarized in Table 6.1

Table 6.1. Some properties of nuclear probes which are often used in perturbed angular correlation experiments. $T_{1/2}$ is the halflife of the parent nucleus or the intermediate state. I^π designates the spin and parity, μ the magnetic moment in nuclear magnetons and Q the quadrupole moment in barns ($1b = 10^{-28}\,m^3$). The γ-γ cascade is characterized by the energies of the two γ-rays and by the anisotropy coefficients A_2 and A_4. The values of Q are from [6.7, 8] and the values of A_2 and A_4 from [6.9–12]. All other data are taken from the table of isotopes [6.13]

Probe	Parent		Intermediate State				γ-γ cascade			
	nucleus	$T_{1/2}$ [days]	$T_{1/2}$ [ns]	I^π	μ[nm]	Q[barns]	$E_{\gamma 1}$ [keV]	$E_{\gamma 2}$ [keV]	A_2	A_4
181Ta	181Hf	42.4	10.8	5/2+	3.24	2.36	137	482	-0.29	-0.076
111Cd	111In	2.8	85	5/2+	-0.77	0.83	171	245	-0.18	0.002
100Rh	100Pd	3.6	214	2+	4.3	0.076	84	75	0.17	0
99Ru	99Rh	15.0	20.5	3/2+	-0.28	0.23	353	90	-0.13	—
							528	90	-0.28	—

of the active from the inactive part is possible. Thus, either mass separation is required or a rather high Hf contamination (minimum ca. 200 ppm, mostly inactive Hf) in the sample must be tolerated. The sample must undergo high temperature treatment in UHV ($p < 10^{-7}$ Pa) in order to avoid oxygen contamination.

^{100}**Pd–**^{100}**Rh** (Table 6.1). ^{100}Pd can be produced by the nuclear reaction ^{103}Rh $(d, 5n)$ ^{100}Pd with a deuteron energy of $E_d > 50$ MeV. A chemical separation of ^{100}Pd from Rh is in principle possible [6.14], but, because of the inertness of Rh, in practice this is rather complicated. Therefore, as with Hf, mass separation has to be performed, or else Rh contamination must be accepted.

^{111}**In–**^{111}**Cd** (Table 6.1). ^{111}In can be produced by ^{110}Cd(d, n) ^{111}In or ^{109}Ag $(\alpha, 2n)$ ^{111}In and is commercially available. ^{111}In is the most convenient PAC probe provided the metallurgical properties fit those of the sample.

^{99}**Rh–**^{99}**Ru** (Table 6.1). The source ^{99}Rh can be produced by the nuclear reaction ^{100}Ru $(d, 3n)$ ^{99}Rh or ^{99}Ru $(d, 2n)$ ^{99}Rh. The separation of the radioactive from the inactive part is again complicated and usually not performed. Because $I = 3/2$ for the intermediate state, the asymmetry parameter, η, cannot be determined.

6.2.2 PAC Setup

A conventional PAC set-up consists of four detectors in a configuration such as the one shown in Fig. 6.5. From each detector one obtains a fast and a slow signal, which are used for timing and energy selection, respectively. After the fast-slow coincidence, each detector provides a start and a stop signal. By using appropriate combinations, a maximum of 12 coincidence spectra can be obtained.

From the coincidence spectra, $N_{ij}(\theta, t)$, of the 4 detectors, (where θ is the angle between the detectors i and j), the following ratio is formed (with similar expressions for the other combinations):

$$R(t) = \frac{2}{3}\left[\left(\frac{N_{13}(180, t) \cdot N_{24}(180, t)}{N_{14}(90, t) \cdot N_{23}(90, t)} \right)^{1/2} - 1 \right]. \qquad (6.12)$$

This procedure eliminates the exponential decay of the coincidence spectra, and even more importantly the influence of detector efficiency, since the efficiency of each detector occurs in both the numerator and the denominator. Taking the experimental ratio $R(t)$, and using the approximation $|A_2G_2(t)| \ll 1$, one obtains the physically interesting quantity $A_2G_2(t)$, see (6.2, 3), from

$$R(t) \approx A_2G_2(t). \qquad (6.13)$$

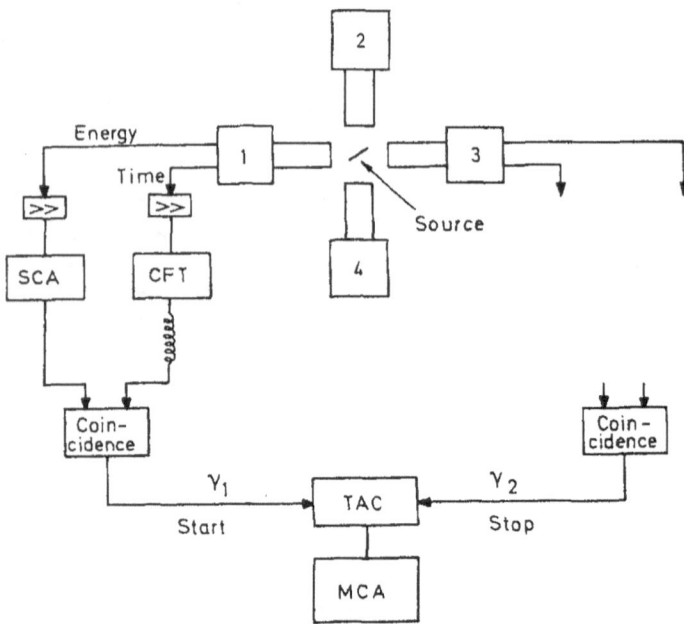

Fig. 6.5. Four-detector PAC set-up and electronics. The following abbreviations are used: SCA: single channel analyser, CFT: constant fraction discriminator, TAC: time to amplitude converter and MCA: multichannel analyser

The function $R(t)$ is usually denoted as the experimental PAC spectrum, and is fitted to the theoretical function $A_2 G_2(t)$. Since the disturbance can be caused by different interactions, a more general form of $G_2(t)$ than in (6.3) has to be used. We set

$$S(t) = \sum_{n=0}^{n_{max}} e^{-\delta\omega_n t} s_{2n} \cos \omega_n t \qquad (6.14)$$

and

$$G_2(t) = \sum_i f_i e^{-\lambda_i t} S_i(t). \qquad (6.15)$$

Equation (6.14) has the same form as (6.3) ($n = 0$ and $\omega_{n=0} = 0$ are included in the summation), but in addition a distribution of the interactions with relative width δ is included (in (6.14) a Lorentzian distribution has been assumed). Each interaction affects a fraction f_i of probe atoms ($\Sigma f_i = 1$) and may cause a spin lattice relaxation λ_i. The $S_i(t)$ in (6.15) have the form $S(t)$ as given in (6.14), but with individual parameters for each interaction. The distribution of interactions causes a dephasing of the precession, with a similar effect on both the $R(t)$ spectrum and the relaxation rate. In principle these two processes can be distinguished, since λ influences the s_{20}-term whereas δ does not, but in practice a clear separation is difficult.

6.3 Applications to Metal–Hydrogen Phase Diagrams

Hydride phases can be investigated by the PAC method via the electric field gradient (e.f.g.) at the probe site. Changes in the hydride structure are in general connected with large changes in the local e.f.g. However, this method does not permit a direct determination of the structure, since the e.f.g. cannot be calculated with sufficient precision to be used for structure assignments. Therefore, model calculations can serve only as a rough guide for the discussion.

The data are often compared with predictions of the point charge model in which one assumes that point charges are located at the hydrogen and lattice atom sites and that the electric field gradient can be represented by

$$V_{zz} = (1 - \gamma_\infty) \frac{e}{4\pi\varepsilon_0} \sum_{i \neq 0} \frac{Z_i}{r_i^3} \left(\frac{3z_i^2}{r_i^2} - 1 \right), \tag{6.16}$$

with equivalent expressions for the other components $V_{\alpha\alpha}$ of the e.f.g. tensor. The Sternheimer factor [6.15] describes the polarization of the electronic shell of the probe atom by the external e.f.g. The summation in (6.16) includes all ions around the probe except the probe itself. Sometimes the screening of the charges by conduction electrons is taken into account by a pseudopotential ansatz. However, since even this does not permit reliable predictions of the e.f.g. to be made, the value of these more complicated calculations is questionable.

A few specific hydride phases will now be discussed.

6.3.1 Palladium Hydride

PAC experiments with the nuclear probes ^{99}Rh–^{99}Ru and ^{100}Pd–^{100}Rh were performed on polycrystalline Pd–H samples [6.16] with hydrogen concentrations between [H]/[Pd] = 0.65 and [H]/[Pd] = 0.96. The samples were prepared by alloying the irradiated target material with Pd metal to yield sources containing approximately 0.2 at.% Ru or Rh.

Pure palladium has *fcc* structure and therefore the electric field gradient at a substitutional site is zero. This is also true for a Pd–H system with [H]/[Pd] = 1, since in this case all the octahedral sites are occupied by hydrogen and the environment of a substitutional probe atom is again cubic. Experimentally, indeed, zero or almost zero electric field gradients were observed in these two extreme cases. However, for intermediate concentrations (0 < [H]/[Pd] < 1) the cubic environment is disturbed, since some of the nearest neighbour octahedral sites are unoccupied, and a non-zero e.f.g. is expected.

Figure 6.6 shows two representative PAC spectra for [H]/[Pd] = 0.72 and 0.65, respectively. From a statistical point of view, approximately four of the

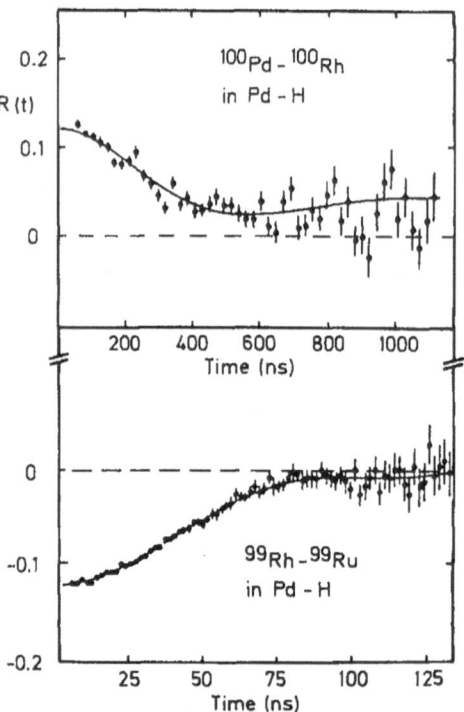

Fig. 6.6. PAC spectra of ^{100}Pd–^{100}Rh and ^{99}Rh–^{99}Ru in Pd–H with [H]/[Pd] = 0.72 and 0.65 respectively. The experiment was performed at 77 K [6.16]

six nearest neighbour octahedral sites should be occupied, while on average two should be vacant. For the source ^{100}Pd, the parent isotope is chemically equivalent to a matrix atom and therefore has the same hydrogen environment as a normal matrix atom. If one assumes that at 77 K the hydrogen environment does not change after the nuclear conversion from ^{100}Pd to ^{100}Rh (because of the low hydrogen mobility at this temperature), then the PAC probe should see distinct e.f.g.s corresponding to hydrogen environments with one, two, three etc. hydrogen vacancies in the first shell. These should have probabilities in agreement with the statistical distribution of hydrogen for [H]/[Pd] = 0.72.

Experimentally, a disturbance of the angular correlation is indeed observed (Fig. 6.6). However, since the e.f.g. is very small, the repetition peak of the oscillation cannot be observed in the experimental time window and therefore the resolution is rather poor. The data were fitted with a single e.f.g. but allowing for a distribution around the mean value. According to this fitting procedure, the experiment gives the average value of the different configurations. The experimental e.f.g. is $V_{zz} = 1.0 \times 10^{17}$ V/cm^2. This value is more than a factor of 10 less than the one calculated using the point-charge model for a single hydrogen hole in the first shell. This strong reduction indicates a very effective shielding of the proton charge by the conduction electrons.

The situation is similar for the ^{99}Rh–^{99}Ru probe (Fig. 6.6). Again, very small e.f.g.s were observed. Here, the authors [6.16] extracted two different e.f.g.s, which they attributed (i) to a single hydrogen hole in the first neighbour shell, and (ii) to more distant holes. The two experimental values were $V_{zz} = 2.8 \times 10^{17}$ V/cm^2 and $V_{zz} = 1.1 \times 10^{17}$ V/cm^2 respectively. It is interesting to note that the V_{zz} produced by a single hole is considerably larger at the ^{99}Ru probe than at the ^{100}Rh probe (if the average value in case of ^{100}Rh is assumed to be representative for a single hole). This difference is not understood at present.

6.3.2 Tantalum and Niobium Hydrides

The Ta–H system was investigated [6.6, 17] with the ^{181}Hf–^{181}Ta probe for several hydrogen concentrations below [H]/[Ta] = 0.43. For these concentrations, the Ta–H system is in the $(\alpha + \beta)$-phase [6.18] below about 280 K, when it changes to the $(\alpha + \varepsilon)$-phase, which is stable around room temperature, and then finally the α-phase is reached. The β- and ε-phases possess ordered hydrogen configurations [6.18]; the structures were determined by neutron scattering for the deuterides [6.19, 20].

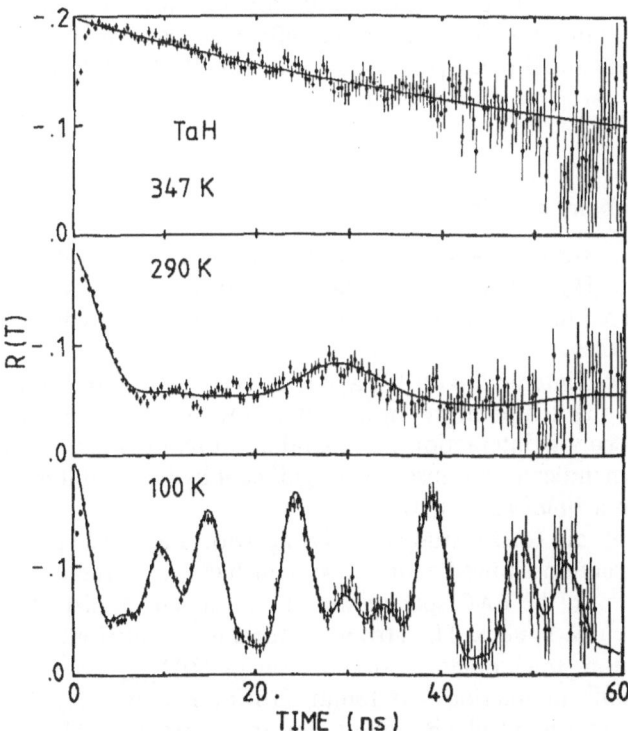

Fig. 6.7. PAC spectra of ^{181}Hf–^{181}Ta in Ta–H with [H]/[Ta] = 0.43. The structure in the spectrum at 100 K is attributed to the β-phase and that in the spectrum at 295 K to the ε-phase [6.6]

The PAC spectra (Fig. 6.7) at representative temperatures for the different phases clearly demonstrate the sensitivity of the PAC method to phase changes. The β-phase quadrupole interactions is experimentally $v_Q = 433$ MHz and $\eta = 0.45$, whereas theoretically $v_Q = 1213$ MHz and $\eta = 0.64$ (calculated with V_{zz} from the point-charge model using $Z = 1$ for the hydrogen charges). The rather large experimental quadrupole interaction indicates that in Ta the hydrogen charge is less completely shielded than in Pd. The sign of the hydrogen charge cannot be determined in a PAC experiment. The ε-phase has a much smaller v_Q ($= 228$ MHz) and almost axially symmetric e.f.g. This result is consistent with the proposed structure in [6.20]. The α-phase is cubic, but a small disturbance from diffusing hydrogen atoms is observed (Fig. 6.7).

At low temperatures the phase diagram of Nb–H [6.18] shows several ordered hydrogen phases depending on the H concentration. For [H]/[Nb] < 0.75 and $T < 200$ K, Nb–H is in the $(\alpha + \varepsilon)$-phase. For the ε-structure three non-equivalent Nb environments are expected, with two, three, and four nearest hydrogen neighbours. Experimentally [6.21] only two different e.f.g.s ($v_Q = 546$ MHz and $v_Q = 481$ MHz, both with $\eta = 0.65$) were found. This discrepancy is probably caused by the fact that the probe ^{181}Hf–^{181}Ta is an impurity in Nb and that, therefore, the local structure around the probe atom is different from the pure Nb–H structure. Above 180 K the PAC data can be fitted with a single e.f.g., but with a distribution around the mean value. No difference was found in the PAC spectra between the ζ- and β-phase [6.18] indicating that the local structure around the probe atom is very similar in these two phases.

6.3.3 Zirconium and Hafnium Hydrides

Pure Zr and Hf have h.c.p. structure and therefore the e.f.g. at the substitutional site is a priori non-zero. Hydriding changes these structures: stoichiometric dihydrides are face-centered tetragonal [f.c.t.), but at somewhat lower hydrogen concentrations they are f.c.c.

The electric quadrupole interaction at ^{181}Ta in f.c.t. ZrH$_{1.97}$ was studied by *Rasera* et al. [6.22] in the temperature range 20–588 K. A static, slightly asymmetric electric quadrupole interaction was found at all temperatures. The static nature of the pattern indicates the absence of significant hydrogen diffusion at these temperatures on a time scale of 100 ns.

De O. Damasceno et al. [6.23] studied HfH$_{1.64}$ with the PAC probe ^{181}Hf–^{181}Ta. X-ray diffraction showed that the system has f.c.c. structure at room temperature. However, the PAC spectra were by no means undisturbed as expected for such a cubic system. The strong disturbance is attributed to hydrogen holes in the non-stoichiometric composition of HfH$_{1.64}$. At room temperature a purely static interaction was found. However, above 373 K a dynamical behaviour was registered which showed motional narrowing at higher temperatures. From the temperature dependence of the relaxation rate an activation energy for hydrogen diffusion was deduced.

6.3.4 Hydrides of Intermetallic Compounds

Heidinger et al. [6.24, 25] studied the intermetallic compounds HfV_2, ZrV_2, TaV_2, and $Hf_{0.5}Zr_{0.5}V_2$ which, without hydrogen, have a cubic Laves phase (C15) structure. These compounds, except TaV_2, undergo a martensitic transition at lower temperatures. In the cubic phase, no electric field gradient is expected if the probe ^{181}Hf–^{181}Ta substitutes an A atom in the AB_2 structure. Experimentally (Fig. 6.8a), a weak disturbance is found which can be attributed to small inhomogeneities of the sample. Hydriding does not change the e.f.g.

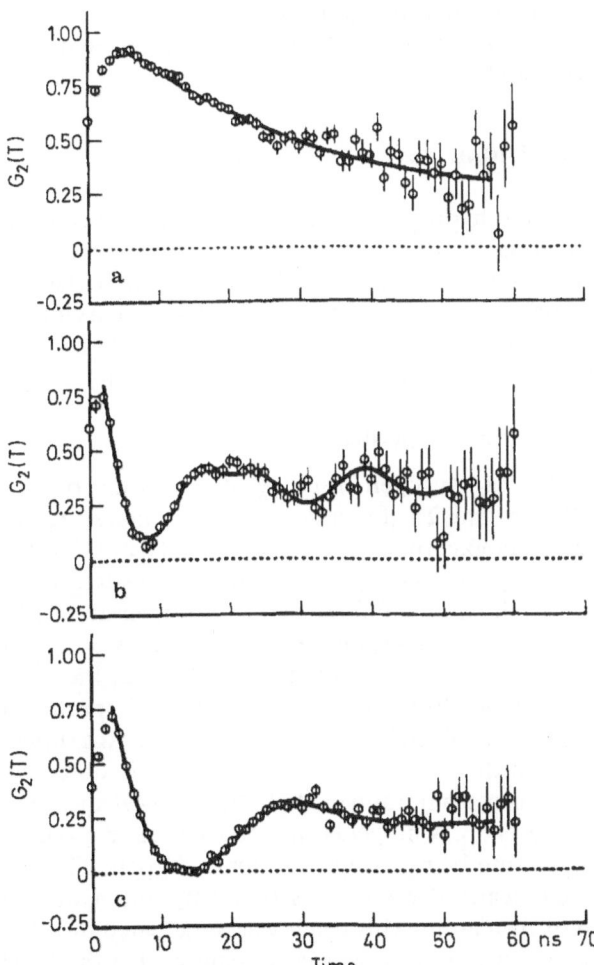

Fig. 6.8. PAC spectra of ^{181}Hf–^{181}Ta in HfV_2 at different temperatures and different hydrogen concentrations: (**a**) hydrogen free and $T = 300\,K$ (cubic AB_2 structure); (**b**) hydrogen free but at $T = 4.3\,K$ (after the martensitic transition) and (**c**) $[H]/[HfV_2] = 3.9$ and $T = 4.3\,K$ [6.24]

much, although the cubic environment should be destroyed for nonstoichiometric hydrogen concentrations. This means that, as in the case of Pd, the proton charge is almost completely screened by the conduction electrons.

After the martensitic transition, a well-defined interaction is observed (Fig. 6.8b) in the hydrogen-free sample. Hydriding changes the interaction frequency considerably (Fig. 6.8c). It is also found that the martensitic transition is shifted to higher temperatures if hydrogen is added.

Vulliet et al. [6.26] studied the system Hf_2Fe for different hydrogen concentrations in the range 0–4.4 H atoms per unit formula. In the non-hridided sample, 3/4 of the probe atoms are in a cubic environment and 1/4 in a non-cubic environment, as expected for the Hf_2Fe structure. Hydriding changes first the 3/4 component, and only at higher H concentrations the 1/4 component as well. From this behaviour the authors conclude that at low concentrations hydrogen is localised near only one type of Hf.

6.4 Hydrogen Trapping at Impurities

In this section, systems with very low hydrogen concentrations will be considered. The purpose of these experiments is to study the behaviour of a single hydrogen atom in a metal and its interaction with impurities. We believe, that the real strength of the PAC method lies in this field, since the local sensitivity of the porbe methods is ideal for these problems. However, in order to obtain high sensitivity it is necessary that hydrogen is trapped at the probe atom.

6.4.1 Hydrogen Trapping at Interstitial Impurities

The classic example of this effect is hydrogen trapping at interstitial oxygen and nitrogen impurities in V, Nb and Ta (for a survey of the literature see [6.27]. As a result of many different experiments, it is proposed [6.27] that the trapped hydrogen is in a tunnelling state between two tetrahedral sites, which are fourth nearest neighbours to the interstitial oxygen or nitrogen atom (the N and O are assumed to be at octahedral sites in *bcc* Nb, Ta and V).

The O–H system in Ta was studied by *Peichl* et al. [6.28] with the PAC probe $^{181}Hf–^{181}Ta$. Figure 6.9 shows Fourier transforms of PAC spectra at low temperatures for $TaO_{0.001}$ and $TaO_{0.001}H_{0.008}$ samples. In the upper spectrum a single e.f.g. (3 lines) with $v_Q = 588\,MHz$ and $\eta = 0.35$ is observed, which is attributed to a single oxygen atom trapped at Hf. In the lower spectrum a clearly different interaction shows up. A fit yields $v_Q = 580\,MHz$ and $\eta = 0.73$, the interaction in the lower spectrum being assigned to an O–H pair. The fact that a unique e.f.g. is observed clearly demonstrates that only one hydrogen atom can be trapped by each oxygen atom. This result was anticipated by other authors [6.29]. However, the uniqueness of the e.f.g. also requires that the O–H pair has a unique position with respect to the probe atom. Since the parent nucleus ^{181}Hf attracts hydrogen (see below), it seems likely that the orientation of the O–H pair is such that hydrogen is nearest neighbour to the probe.

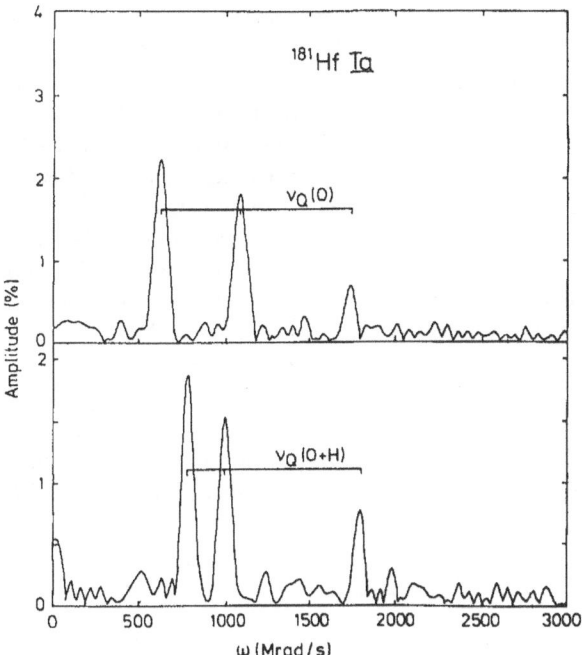

Fig. 6.9. Fourier spectra of ^{181}Hf–^{181}Ta in Ta. *Top:* Oxygen atom trapped at ^{181}Hf *Bottom:* Oxygen–hydrogen pair trapped at ^{181}Hf

The lower spectrum in Fig. 6.9 was obtained by quenching the sample rapidly to low temperatures. If the sample was cooled slowly, only a fraction of the oxygens (approximately 1/2) captured a hydrogen. This indicates a competition between hydrogen precipitation and hydrogen trapping at oxygen. In the case of deuterium, almost all the oxygen is decorated.

Figure 6.10 shows the fraction of probe atoms seeing an oxygen or O–H pair. At 55 K, the fraction of the O–H interaction disappears, and that of the

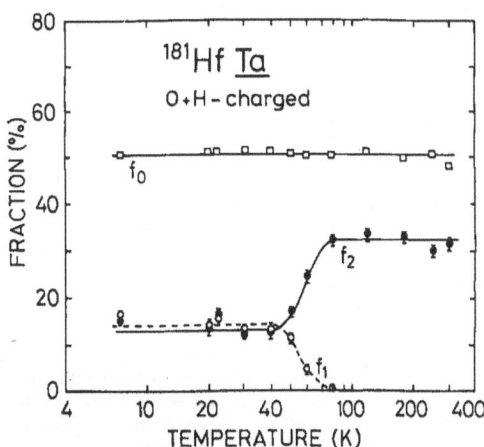

Fig. 6.10. Fraction of probe atoms in different environments. f_0: undisturbed, f_1: e.f.g. ν_{Q1} (O–H complex), f_2: e.f.g. ν_{Q2} (O atom). The concentrations are: [O]/[Ta] ≈ 0.1 at.%, [H]/[Ta] ≈ 1 at.% [6.28]

O interaction increases by approximately the corresponding amount, indicating a conversion of O–H to O at this temperature. Possible explanations are 1) a break-up of the O–H bond or 2) a rearrangement of the pair in such a way that only the oxygen interaction is seen. However, in this latter case, the presence of hydrogen in the neighbourhood of the probe should give rise to at least a small change in the interaction frequency with respect to oxygen alone. Precise experiments should be performed in order to decide this question.

Similar experiments to those described above were performed on the O–H and O–D pairs in Nb [6.30]. The results are comparable with those of Ta. The conversion of O–H(D) to O occurs in Nb at around 65 K. An isotope effect, as reported in [6.28] for Ta, was not found for O–H and O–D in Nb.

6.4.2 Hydrogen Trapping at Substitutional Impurities

Until recently, very little was known about hydrogen trapping at substitutional impurities. Lately, however, an increasing amount of data has become available

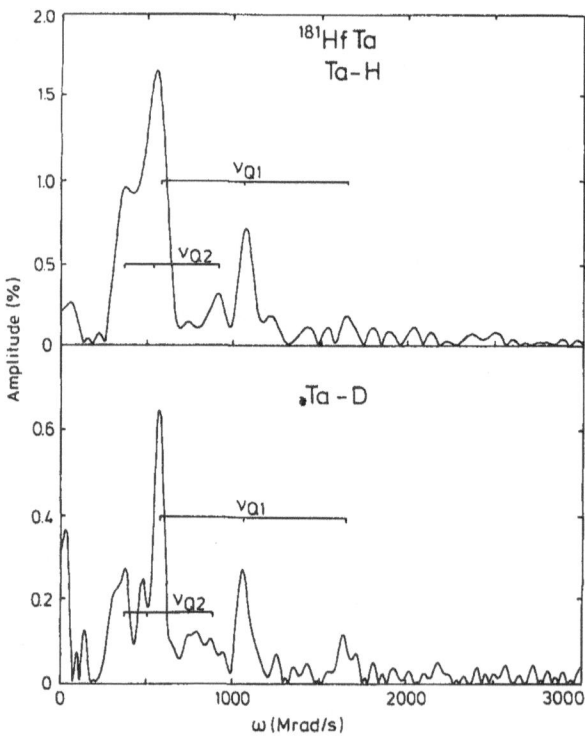

Fig. 6.11. Fourier transforms of PAC spectra measured at 4.2 K for hydrogen- and deuterium-charged ^{181}Hf*Ta* samples which were rapidly quenched in liquid helium from room temperature. The deuterium-charged sample was annealed at 40 K. The concentrations were [H]/[Ta] = 0.006 and [D]/[Ta] = 0.06 respectively. The bars connect the three frequencies belonging to a given interaction

[6.31–33]. The PAC method is particularly suited to studying this problems if the probe atom is the impurity at which hydrogen is trapped. Since the electric field gradient is drastically changed by the presence of a hydrogen atom close to the probe, trapping and detrapping can easily be recognized. In addition, different configurations can be distinguished via the different electric field gradients. As an example, hydrogen trapping at ^{181}Hf in Ta [6.34] will be discussed here.

Tantalum samples containing approximately 300 ppm Hf were charged electrolytically with hydrogen ([H]/[Ta] = 0.006) and deuterium ([D]/[Ta] = 0.06). If the samples were slowly cooled to 4 K, no disturbances of the PAC spectra were found. However, if the samples were rapidly immersed in liquid helium, a strong modulation appeared. Figure 6.11 shows the Fourier spectra of this experiment. The data can be decomposed into two interactions with

$$\nu_{Q1} = 580 \, \text{MHz}, \quad \eta = 0.23$$

and

$$\nu_{Q2} \approx 300 \, \text{MHz}, \quad \eta \approx 0.8 \quad \text{and} \quad \delta \approx 10\%.$$

The first interaction is assigned [6.34, 35] to a single trapped hydrogen atom, and the second to two or more trapped $H(D)$ atoms in different configurations (because of the distribution of ν_{Q2}). The frequencies are very similar for the hydrogen- and deuterium-charged samples. (Figure 6.11).

The fact, that rapid cooling is required in order to obtain trapping, indicates that the Hf–H binding is rather weak, since it cannot prevent precipitation. This behaviour was studied more explicitly in an annealing experiment [6.34] in which the sample was rapidly cooled, and then warmed up to successively higher temperatures. It was found that the fraction of disturbed probe atoms goes to zero after annealing at around 60 K (with a slight difference for the H- and D-charged samples). The PAC measurements were always performed at 4 K. This experiment shows that the Hf–H and Hf–D pairs are stable up to approximately 60 K, when they dissolve, the H(D) atoms then migrating to precipitates.

The positive binding energy of H(D) to Hf in Ta is consistent with the rule [6.33] that elements to the left of the matrix atom in the periodic table attract hydrogen. The Hf–Ta system is similar to Ti–Nb (both group IV impurities in a group V matrix), for which a positive binding energy for H–Ti was also found [6.32].

6.5 Hydrogen Diffusion

6.5.1 Quantum Diffusion at Low Temperatures

The low-temperature diffusion of hydrogen is particularly interesting since large quantum mechanical effects are predicted. It is expected that the classical barrier hopping mechanism ceases at low temperatures, and that tunnelling processes take over. At moderately low temperatures, incoherent tunnelling should be dominant, whereas at very low temperatures coherent tunnelling or band propagation might show up. (For some review articles on this subject see [6.36–39]).

Experimentally, very little is known about this subject. The most direct indication of the onset of quantum diffusion is the bend in the Arrhenius function [6.40] for hydrogen diffusion in Nb and Ta. An extension of these experiments to much lower temperatures is not possible, since hydrogen precipitates and forms hydride phases in which it is immobile at low temperatures. However, precipitation can be avoided by trapping hydrogen at impurities. Usually, this also prevents a measurement of hydrogen diffusion in the intrinsic system. However, if the impurity is radioactive and converts to a matrix atom, then the ideal case of a single hydrogen atom in a pure matrix is realized and the diffusion can be studied.

This idea was applied [6.35] to [181]Hf in Ta. As can be seen from Fig. 6.4, [181]Hf decays first to an excited state of [181]Ta. A hydrogen atom trapped at [181]Hf is released after the nuclear conversion, and may diffuse away during the $25.7\,\mu s$ lifetime of this state or remain at [181]Ta, depending on its intrinsic mobility. The subsequent $\gamma-\gamma$ cascade will record the presence or absence of hydrogen near the probe atom. Therefore, the fraction of disturbed atoms is a direct measure of the jump frequency of hydrogen in pure Ta. In an experiment at various temperatures (Fig. 6.12), it was found that up to about 15 K the fraction of disturbed probe atoms is constant, but then drops to zero between 15 and 30 K. These measurements were completely reversible, i.e. the non-converted [181]Hf kept their hydrogens whereas the converted ones released them.

In the analysis it was assumed that hydrogen resides at tetrahedral sites [6.41] and that the probability of return can be neglected. In this case the

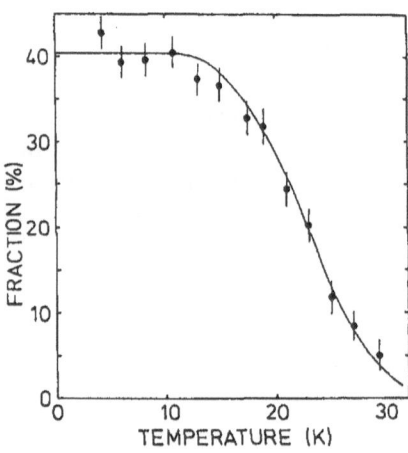

Fig. 6.12. Fraction of probe atoms which are disturbed by a hydrogen atom as a function of temperature. The hydrogen concentration was 0.1 at.% [6.35]

diffusion coefficient D and the mean residence time $\bar{\tau}$ can be determined from the following formulae:

$$\frac{f}{f_0} = \frac{\tau_c}{\tau_c + \tau_N},$$ (6.17)

$$\tau_c = 4\bar{\tau},$$ (6.18)

$$D = \frac{a_0^2}{48\bar{\tau}},$$ (6.19)

where f/f_0 is the relative fraction of disturbed atoms normalized to the constant value f_0 at low temperatures. The factor 4 in (6.18) between the correlation time τ_c and the mean residence time $\bar{\tau}$ accounts for the fact that only one out of four jumps leads away from the probe if tetrahedral sites are considered. Equation (6.19) gives the relation between D and $\bar{\tau}$ for tetrahedral–tetrahedral jumps, where a_0 is the lattice constant of Ta.

The results of this analysis are shown in Fig. 6.13. It can be seen that the diffusion coefficients are much larger than expected from an extrapolation of the published data [6.40] using an Arrhenius function. On the other hand, the data are well represented by a T^7 temperature dependence as predicted by the small polaron hopping theory [6.42] for low temperatures.

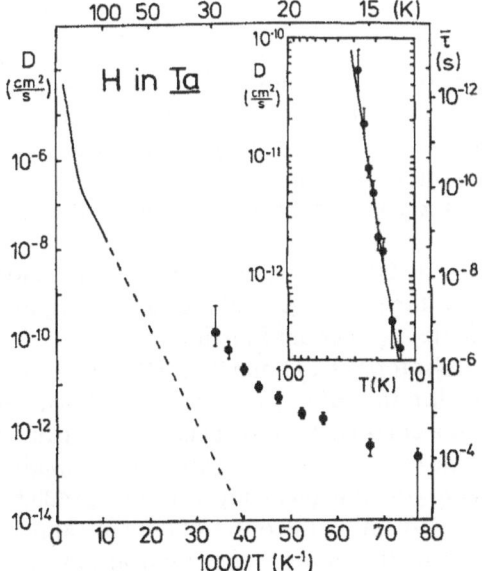

Fig. 6.13. Diffusion coefficient D and mean residence time $\bar{\tau}$ for H in Ta in the α-phase. The thick line represents the data of [6.40], the dashed line is an extrapolation assuming an Arrhenius law. In the insert the data are presented on a log–log scale. The solid line shows a T^7 law [6.35]

6.5.2 Diffusion in Metal Hydrides

In this section, diffusion studies based on measurements of the relaxation rate λ will be discussed. The method relies on the fact that diffusing hydrogen atoms produce a fluctuating electric field gradient (e.f.g.) which gives rise to a relaxation of the PAC spectrum. Since appreciable changes of the e.f.g. at the probe site are expected only in concentrated metal–hydrogen systems, this method is restricted to hydride phases.

Dynamical effects become visible if the fluctuation rate of the e.f.g. is comparable to the precession frequency of the angular correlation. As in NMR and other similar methods, one observes first an increase of the relaxation rate with increasing fluctuation rate and then a decrease due to motional narrowing, the maximum occurring when the fluctuation rate and the precession angular velocity are equal.

A theory of this effect was developed by *Blume* et al. [6.43–45] within the framework of a stochastic model. The application of the general theory to data analysis is fairly complicated and requires large computation times; approximations are thus of great practical interest. In most cases it is appropriate to assume that the dynamical effects can be described by a single exponential function which multiplies the static perturbation function $G_2^{\text{stat}}(t)$, i.e.

$$G_2(t) = e^{-\lambda t} G_2^{\text{stat}}(t). \tag{6.20}$$

In the limit of slow and fast fluctuations further simplifications can be made

$$\lambda \sim v \qquad \text{for } v \ll v_Q \tag{6.21}$$

$$\lambda \sim v_Q^2 \frac{1}{v} \qquad \text{for } v \gg v_Q, \tag{6.22}$$

where v is the jump rate of the diffusing hydrogen atoms and v_Q is the coupling constant, see (6.4), of the fluctuating e.f.g. The proportionality constants in (6.21) and (6.22) depend on details of the interactions and of the jump processes and can be evaluated only with specific assumptions for these quantities. For an order of magnitude estimate, $\lambda \approx v$ and $\lambda \approx v_Q^2/v$ may be assumed in the two asymptotic limits.

In the fast fluctuation limit, $G_2^{\text{stat}}(t)$ in (6.20) is the static perturbation factor for the time-averaged interaction. If the time average of the e.f.g. is zero, one has $G_2^{\text{stat}}(t) = 1$ and $G(t)$ becomes a simple exponential function.

The diffusion of the three hydrogen isotopes H, D and T in Hf was studied by *Forker* et al. [6.46–48]; their data for the $HfT_{1.90}$ system are shown in Fig. 6.14. Up to 350 K the spectra reveal a completely static behaviour as can be seen from the fact that the anisotropy at long times (hard core value) remains at a finite value (20% of the anisotropy at $t = 0$). At higher temperature ($T \gtrsim 400\,\text{K}$) this part of the spectrum relaxes to zero indicating the onset of dynamical processes which in this case are attributed to the motion of tritium atoms. The

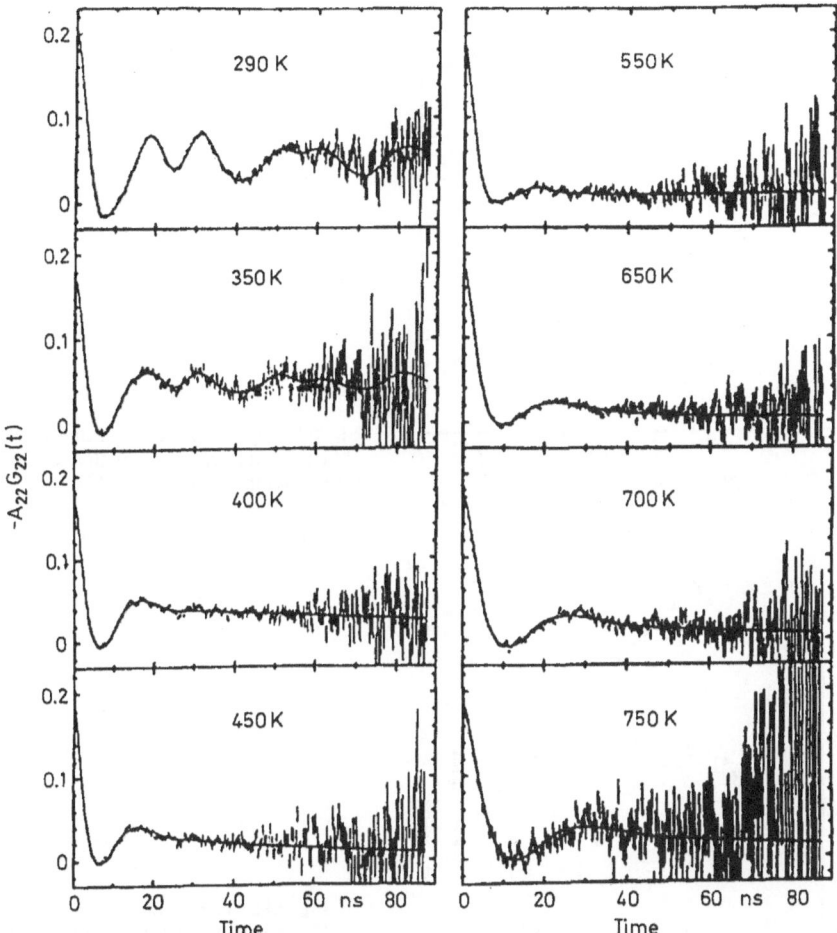

Fig. 6.14. PAC spectra of ^{181}Ta in HfT$_{1.90}$ at different temperatures [6.48]

relaxation rate reaches a maximum around 550 K and then decreased due to motional narrowing.

The relaxation rates extracted from the data are shown in Fig. 6.15. Assuming an Arrhenius behaviour for the jump rates, i.e.

$$\nu = \nu_0 e^{-E_a/kT}, \tag{6.23}$$

and applying (6.21) and (6.22), a plot of $\log \lambda$ versus $1/T$ should result in straight lines with the same slopes but opposite signs in the slow and fast fluctuation regimes. This is indeed fulfilled as can be seen in Fig. 6.15. The authors derive an activation energy for tritium jumps in HfT$_{1.90}$ of $E_a = 0.36(6)\,$eV. Similar results were obtained for the other two isotopes H and D in Hf.

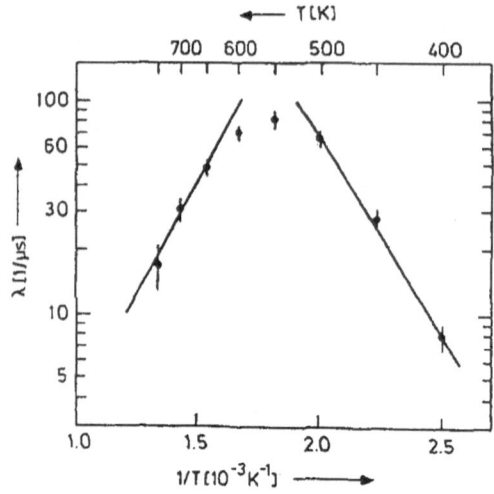

Fig. 6.15. The relaxation rate λ of ^{181}Ta in HfT$_{1.90}$ as a function of the inverse temperature $1/T$ [6.48]

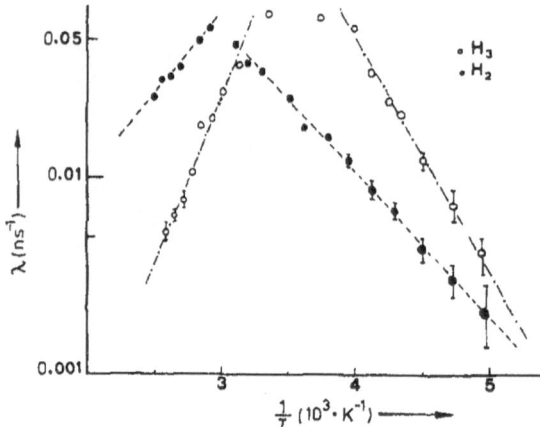

Fig. 6.16. The relaxation rate λ of ^{181}Ta versus $1/T$ for Zr_2NiH_2 (full circles) and Zr_2NiH_3 (open circles) [6.51]

Hydrogen diffussion in Zr_2Ni was studied extensively by the Grenoble group [6.49–52]. The results obtained for the relaxation rate λ are displayed in Fig. 6.16. It can be seen that the slopes of $\log \lambda$ versus $1/T$ are very different for Zr_2NiH_2 and Zr_2NiH_3. The activation energies extracted from the data are $E_a = 0.15(1)\,\mathrm{eV}$ for the H$_2$ and $E_a = 0.27(1)\,\mathrm{eV}$ for the H$_3$ system. A further increase of the activation energy was observed for the $Zr_2NiH_{4.8}$ system ($E_a = 0.38(2)\,\mathrm{eV}$), but there a second process with a much lower activation energy ($E_a = 0.11(1)\,\mathrm{eV}$) was found. The different activation energies are assigned to different pathways in the Zr_2Ni hydrides.

Hydrogen diffusion in amorphous Zr_2Ni has also been investigated [6.51]. The result is that the activation energy in the amorphous system is smaller than in the crystalline sample with the same hydrogen concentration. This is expected since easier pathways are offered by the amorphous lattice for hydrogen diffusion. The authors note that their experiments do not reveal the existence of a broad distribution of activation energies in amorphous $Zr_2NiH_{2.5}$. As an explanation of this rather surprising result they suggest that H–H interactions play a role in smoothing the variation of barrier heights present in the bare amorphous Zr_2Ni system.

Table 6.2. Summary of published PAC experiments on metal–hydrogen systems

Host	Hydrogen Isotope	Probe nucleus	Concentration [H]/[Host]	Temperature	References
Ti	H	^{111}Cd	1.83–1.92	4.2–300	[6.53]
V	H	^{99}Ru	0.50–0.83	77–450	[6.54]
	D	^{99}Ru	0.59–1.65	77–450	[6.54]
Ni	H	^{111}Cd	?^{a}	295	[6.55]
Zr	H	^{181}Ta	1.97	20–588	[6.22]
Nb	H	^{99}Ru	0.78–0.90	77–420	[6.56]
	H	^{100}Rh	0.72–0.83	90–450	[6.57]
	H	^{181}Ta	0–1.0		[6.21, 30, 58–61]
	D	^{181}Ta	≈0.01	4.3–300	[6.30]
Pd	H	^{99}Ru	0.65–0.96	77–300	[6.16]
	H	^{100}Rh	0.72–0.86	77	[6.16]
	H	^{181}Ta	≦0.90	77–300	[6.56]
Dy	H	^{181}Ta	2–3	RT^b	[6.63]
Hf	H	^{181}Ta	1.64–1.99	25–550	[6.23, 46]
	D	^{181}Ta	>1.87	19–725	[6.47]
	T	^{181}Ta	1.90	33–750	[6.48]
Ta	H	^{181}Ta	≦0.43	0.03–380	[6.6, 17, 28, 34, 35, 64–66]
	D	^{181}Ta	≦0.03	4.3–300	[6.34, 64, 66]
	T	^{181}Ta	≈0.01	4.3–140	[6.67]
W	H	^{111}Cd	?^{a}	RT^b	[6.68]
Pt	H	^{111}Cd	?^{a}	RT^b	[6.69]
ZrV_2	H	^{181}Ta	4.3	77–300	[6.25]
HfV_2	H	^{111}Cd	4	28–900	[6.70]
	H	^{181}Ta	0–4	4.2–400	[6.24, 25]
TaV_2	H	^{181}Ta	1.2–2.2	77–300	[6.25]
$(Y, Hf)Fe_2$	H	^{181}Ta	4.08	292	[6.71]
Hf_2Fe	H	^{181}Ta	0–4.4		[6.26]
Zr_2Ni	H	^{181}Ta	2.1–4.8	170–470	[6.49, 51]
	D	^{181}Ta	5	210–470	[6.52]

^{a} ? = concentration unknown
^{b} RT = room temperature

6.6 List of PAC Experiments on Metal–Hydrogen Systems

PAC experiments on metal–hydrogen systems are summarized in Table 6.2. They are ordered according to the atomic number of the host materials with the elemental systems before the compounds. The references were collected in a literature search in December 1989.

6.7 Conclusion

The perturbed angular correlation (PAC) method has been used relatively little in hydrogen physics so far. This is probably related with the fact that the method is still fairly complicated, although simple and standardized set-ups have been developed recently; it also requires the handling of radioactive materials. On the other hand, quite a number of groups are familiar with the technique, but most have not yet applied it to hydrogen problems.

Particular promising are experiments in which the local structure around the probe atom is investigated. For example, hydrogen trapping at substitutional impurities (probe atoms), can be investigated with very high precision and many details which are inaccessible with other methods can be studied. The advantage of PAC lies in the fact that different hydrogen configurations can be distinguished and that changes of structure, e.g. as a function of temperature, can be detected. The diffusion studies described in Sect. 6.5.1 are virtually impossible with classical methods and can only be performed in the way described there. Although the number of systems to which this method can be applied is rather limited, it is nevertheless important in that it enables one to deduce the principal features of hydrogen diffusion at low temperatures.

For structural studies, the PAC method is inferior to diffraction measurements, as are all local probe methods. However, local defects in ordered structures, in particular missing hydrogen atoms in hydrides, can be sensitively investigated. One can also obtain information on the electronic structure of hydrides via the measured electric field gradient. In principle, the method is also surface sensitive if the probe atoms are deposited at the surface, but as yet no such experiments have been performed on hydrogen systems.

Acknowledgement. This work was supported by the Bundesministerium für Forschung und Technologie of the Federal Republic of Germany. We would like to thank Professor *E. Recknagel* for his continuous interest in and support of this work.

References

6.1   E. Recknagel, G. Schatz, Th. Wichert: In *Hyperfine Interactions of Radioactive Nuclei*, ed. by J. Christiansen, Topics Curr. Phys. Vol. 31. (Springer, Berlin, Heidelberg 1983) p. 133
6.2   L. Niesen: Hyperfine Interact. **10**, 619 (1981)
6.3   H. Frauenfelder, R.M. Steffen: In *Alpha-, Beta-, and Gamma-Ray Spectroscopy*, Vol. 2, ed. by K. Siegbahn (North-Holland, Amsterdam 1965) p. 997

6.4   K. Alder, H. Albers-Schönberg, E. Heer, T.B. Novey: Helv. Phys. Acta **26**, 761 (1953)
6.5   H.H. Rinneberg: Atomic Energy Review **17**, 477 (1979)
6.6   A. Weidinger: J. Less-Common Met. **103**, 285 (1984)
6.7   T. Butz, A. Lerf: Phys. Lett. **97 A**, 217 (1983)
6.8   R. Vianden: Hyperfine Interact. **15/16**, 1081 (1983)
6.9   Nuclear Data Sheets **9**, 337 (1973)
6.10 Nuclear Data Sheets **11**, 312 (1974)
6.11 Nuclear Data Sheets **6**, 65 (1971)
6.12 Nuclear Data Sheets **12**, 468 (1974)
6.13 C.M. Lederer, V.S. Shirley (eds.): *Table of Isotopes*, 7th edition, (Wiley, New York 1978)
6.14 J.S. Evans, R.A. Naumann: Phys. Rev. B **138**, 1017 (1965)
6.15 F.D. Feiock, W.R. Johnson: Phys. Rev. **187**, 39 (1969)
6.16 J. Trager, M. Karger, T. Butz, F.E. Wagner: Hyperfine Interact. **15/16**, 795 (1983)
6.17 P.J. Mendes, J.M. Gil, N. Ayres de Campos, R. Peichl, A. Weidinger: Hyperfine Interact. **15/16**, 791 (1983)
6.18 T. Schober, H. Wenzel: In *Hydrogen in Metals II*, Topics Appl. Phys. Vol. 29 ed. by G. Alefeld, J. Völkl (Springer, Berlin, Heidelberg 1978) p. 11
6.19 V.F. Petrunin, V.A. Somenkov, S.Sh. Shil'shtein, A.A. Chertkov: Sov. Phys.-Crystallogr. **15**, 137 (1970)
6.20 H. Kaneko, T. Kajitani, M. Hirabayashi, N. Niimura, A.J. Schultz, P. Leung: J. Less-Common Met. **103**, 45 (1984)
6.21 J.M. Gil, P.J. Mendes, C. Gil, A.P. de Lima, A. Weidinger, N. Ayres de Campos: J. Less-Common Met. **103**, 227 (1984)
6.22 R.L. Rasera, G.K. Shenoy, B.D. Dunlap, D.G. Westlake: J. Phys. Chem. Solids **40**, 75 (1979)
6.23 O. de O. Damasceno, A.L. de Oliveira, J. de Oliveira, A. Baudry, P. Boyer: Solid State Commun. **53**, 363 (1985)
6.24 R. Heidinger, P. Peretto, S. Choulet: Solid State Commun. **47**, 283 (1983)
6.25 R. Heidinger, P. Peretto, S. Choulet: Hyperfine Interact. **15/16**, 787 (1983)
6.26 P. Vulliet, G. Teisseron, J.L. Oddou, C. Jeandey, A. Yaouanc: J. Less-Common Met. **104**, 13 (1984)
6.27 A. Magerl, J.J. Rush, J.M. Rowe, D. Richter, H. Wipf: Phys. Rev. B **27**, 927 (1983)
6.28 R. Peichl, A. Weidinger, E. Recknagel, J.M. Gil, P.J. Mendes, Ayres de Campos: Hyperfine Interact. **15/16**, 463 (1983)
6.29 G. Pfeiffer, H. Wipf: J. Phys. F **6**, 167 (1976)
6.30 P.J. Mendes, J.M. Gil, N. Ayres de Campos, R. Peichl, A. Weidinger: Z. Phys. Chem. NF **145**, 141 (1985)
6.31 H. Kronmüller, P. Vargas: Philos. Mag. A **51**, 59 (1985)
6.32 G. Cannelli, R. Cantelli, G. Vertachi: Appl. Phys. Lett. **39**, 832 (1981)
6.33 A.I. Shirley, C.K. Hall: Acta Metall. **32**, 49 (1984)
6.34 R. Peichl, A. Weidinger, P. Ziegler: Z. Phys. Chem. NF **143**, 197 (1985)
6.35 A. Weidinger, R. Peichl: Phys. Rev. Lett. **54**, 1683 (1985)
6.36 K.W. Kehr: In *Hydrogen in Metals I*, Topics Appl. Phys., Vol. 28, ed. by G. Alefeld, J. Völkl (Springer, Berlin, Heidelberg 1978) p. 197
6.37 K.W. Kehr: Hyperfine Interact. **17–19**, 63 (1984)
6.38 D. Richter: In *Neutron Scattering and Muon Spin Rotation*, Springer Tracts in Modern Physics, Vol. 101, ed. by G. Höhler (Springer, Berlin, Heidelberg 1983) p. 85
6.39 A. Seeger: Hyperfine Interact. **17–19**, 75 (1984)
6.40 Zh Qi, J. Völkl, R. Lässer, H. Wenzel: J. Phys. F. **13**, 2053 (1983)
6.41 E. Yagi, T. Kobayashi, S. Nakamura, Y. Fukai, K. Watanabe: J. Phys. Soc. Japan **52**, 3441 (1983)
6.42 C.P. Flynn, A.M. Stoneham: Phys. Rev. B **1**, 3966 (1970)
6.43 M. Blume: Phys. Rev. **174**, 351 (1968)
6.44 H. Winkler, E. Gerdau: Z. Phys. **262**, 363 (1973)
6.45 S. Dattagupta: Hyperfine Interact. **11**, 77 (1981)
6.46 M. Forker, L. Freise, D. Simon, H. Saitovitch, P.R.J. Silva: Hyperfine Interact. **35**, 829 (1987)
6.47 M. Forker, L. Freise, D. Simon, H. Saitovitch, P.R.J. Silva: Z. Naturforsch. **41a**, 403 (1986)
6.48 M. Forker, W. Herz, D. Simon, R. Lässer: Z. Phys. Chem. NF **164**, 889 (1989)
6.49 P. Boyer, A. Baudry: J. Less-Common Met. **129**, 213 (1987)
6.50 A. Chikdene, A. Baudry, P. Boyer: J. Phys. F **18**, L187 (1988)
6.51 A. Chikdene, A. Baudry, P. Boyer: Z. Phys. Chem. NF **163**, 443 (1989)
6.52 A Baudry, A. Chikdene, P. Boyer: J. Less-Common Met. **143**, 143 (1988)

6.53 H. Foettinger, D. Forkel, H. Plank, W. Witthuhn: Hyperfine Interact. **35**, 765 (1987)
6.54 L. Iannarella, M. Baier, M. Zelger, F.E. Wagner: J. Less-Common Met. **130**, 173 (1987)
6.55 G.S. Collins, R.B. Schuhmann: Phys. Rev. B **34**, 502 (1986)
6.56 M. Berneis, J. Trager, R. Wordel, M. Zelger, F.E. Wagner, T. Butz: Z. Phys. Chem. NF **145**, 129 (1985)
6.57 P. Lidbjörk, B. Lindgren: Hyperfine Interact. **35**, 769 (1987)
6.58 J.M. Gil, P.J. Mendes, A.P. de Lima, N. Ayres de Campos, Sheng Yuqin, R. Peichl, A. Weidinger: J. Less-Common Met. **129**, 145 (1987)
6.59 J.M. Gil, P.J. Mendes, A.P. de Lima, N. Ayres de Campos, A. Weidinger: In *Nuclear Physics Application on Materials Science*, ed. by E. Recknagel, J.C. Soares, Nato ASI Series E Vol. **144**, 413 (1988)
6.60 J.M. Gil, P.J. Mendes, A. Weidinger, N. Ayres de Campos: Z. Phys. Chem. NF **163**, 193 (1989)
6.61 R.-D. Roitzheim, H.-J. Rudolph, R. Schumacher, U. Wrede, R. Vianden: Z. Phys. Chem. NF **164**, 999 (1989)
6.62 O. Boebel, A. Weidinger, P. Ziegler, W. Lehn, Y.Q. Sheng: Z. Phys. Chem. NF **164**, 1005 (1989)
6.63 L.P. Fereira, G. Teisseron, P. Vulliet, J.M. Gil, P.J. Mendes, A.P. de Lima, N.A. de Campos: J. Less-Common Met. **130**, 155 (1986)
6.64 R. Peichl, P. Ziegler, A. Weidinger: J. Less-Common Met. **129**, 243 (1987)
6.65 A. Weidinger, R. Peichl, E. Hagn, E. Zech, K.H. Ebeling, R. Eder, T. Butz, S. Saibene: Hyperfine Interact. **35**, 773 (1987)
6.66 A. Weidinger: In *Nuclear Physics Application on Materials Science*, ed. by E. Recknagel, J.C. Nato ASI Series E Vol. **144**, 275 (1988)
6.67 O. Boebel, A. Weidinger, P. Ziegler, W. Lehn, Y.Q. Sheng, R. Lässer: Z. Phys. Chem. NF **164**, 1005 (1989)
6.68 K. Post, F. Pleiter: Hyperfine Interact. **35**, 615 (1987)
6.69 G.S. Collins, H.-J. Yang, S. Shropshire: In *Nuclear Physics Application on Materials Science*, ed. by E. Recknagel, J.C. Soares, Nato ASI Series E. Vol. **144**, 415 (1988)
6.70 M. Forker, W. Herz, D. Simon: Z. Phys. Chem. NF **163**, 91 (1989)
6.71 M. Budzynski, O.I. Kochetov, M. Subotovicz, H. Niezgoda, H. Spustek, W. Tanska-Krupa: Phys. Status Solidi B **140**, 589 (1987)

7. Experimental Techniques II: Adaptation of New Techniques to Study Surface and Bulk Properties of H–Metal Systems

Moshe H. Mintz, Isaac Jacob, and David Shaltiel

With 11 Figures and 3 Tables

The increased interest in metal–hydrogen interactions has led to the application of a large variety of experimental techniques for the study of these systems. Numerous data have been acquired during the last decade on the bulk and surface properties of metal hydrides. Yet, stimulating problems related to this research field still remain.

In this chapter three newly developed experimental techniques adapted to the research of hydrogen–metal systems are presented. One of these techniques (time-of-flight analysis of direct surface recoils) deals with hydrogen–surface interactions. The second technique (nuclear resonant scattering of gamma rays) is related to bulk properties of binary and intermetallic hydrides. The third experimental method (thermal desorption spectroscopy) involves both bulk and surface properties of these compounds. In the following sections, each of these methods is described. The basic physical principles underlying the technique and the experimental setup and data interpretation are discussed. Recent developments achieved by the application of the method to hydrogen–metals research are then reviewed. Some possible future applications and prospects conclude each of the sections.

7.1 Time-of-Flight Analysis of Direct Recoils

7.1.1 Detection of Surface Hydrogen

The study of the interactions between gaseous hydrogen and the outermost surfaces of hydride-forming metals [7.1, 2] was the focus of intense activity in the last decade. This research field is important in many applications (e.g., catalysis, hydride formation and hydrogen embrittlement). In these studies, information about the surface concentration of hydrogen, the structure of adsorption sites, formation of surface compounds, etc., is required. One of the most intriguing questions addressed by the kinetic studies of hydrogen–metal reactions is the understanding of the initial stages of the process, which involves a complex set of parameters reflected in a variety of phenomena such as initial incubation periods, poisoning of the reaction by the presence of gas phase impurities, hydriding activation or passivation mechanisms, etc. [7.3–5]. Other examples of the significance of such surface studies are given by the catalysis of gas phase hydrogenation reactions by metals, including hydride-forming intermetallic compounds [7.6–8] and by hydrogen isotope exchange and separation [7.9].

Unfortunately, the direct detection of surface hydrogen is not possible by most of the conventional experimental techniques. *Schlapbach* describes the currently employed methods of surface analysis in Chap. 2 of this volume. We thus limit the present discussion to a comparison of some of these methods with the newly developed TOF-DR technique in regard to the ability to obtain compositional and chemical information related to hydrogen–surface interactions. Also, some structural determinations accessible by the method are briefly discussed.

Among the diverse variety of surface analysis techniques we shall refer to two main groups of methods: electron spectroscopy (e.g., AES, XPS, UPS, EELS) and ion beam analysis (e.g., SIMS, LEIS) [7.10]. The former group of methods indirectly probe surface hydrides by monitoring the changes induced by the hydrogen–metal electronic interactions reflected in the corresponding variations in the electron emission peaks. Although they provide valuable information on the electronic structures of hydrides, such indirect detection methods are limited by several shortcomings:

(i) Surface impurities (e.g., oxygen, carbon) result in changes in the electronic structure similar to those produced by hydrogen. This interference of impurities makes the interpretation of the data obtained by these methods (for hydrogen absorption) meaningful only for a completely clean hydrogen–metal system. Obviously, interactions between hydrogen and contaminated surfaces (which are relevant to most practical cases) cannot be studied by these techniques.

(ii) Due to the probing depths of electron spectroscopy techniques, which usually extend beyond the topmost surface layer, accommodated subsurface hydrogen may contribute to the above-mentioned changes induced in the electron-emission spectra. The quantitative interpretation of such data is thus complicated by the convolution of the subsurface concentration profiles, which usually are unknown. Consequently, it is in most cases not possible to determine the absolute hydrogen surface coverages.

The application of ion beam techniques to the analysis of surface and subsurface hydrogen has been described by several researchers (for some recent reviews see [7.11, 12]). In addition to high-energy methods utilizing incident ions accelerated to the MeV energy range (e.g., Rutherford back scattering, RBS), there are two well-known low-energy techniques: secondary-ion mass spectrometry (SIMS) [7.13] and low energy ion scattering (LEIS) (or ion scattering spectroscopy (ISS)) [7.14]. Both methods apply low-energy (1–10 keV) noble gas ions, which are irradiated on the studied surface. In SIMS measurements of the secondary ions sputtered from the surface are detected, whereas in LEIS measurements the energy spectra of the scattered noble gas primaries are monitored.

A direct qualitative detection of hydrogen and its isotopes is possible by SIMS. However, quantitative interpretation of SIMS peak intensities is hampered by two factors: i) The destructive nature of the measurement, resulting from the high irradiation doses (10^{15}–10^{16} ions cm^{-2}) applied to obtain reasonable spectra. ii) The lack of knowledge of secondary ion yields. This latter factor also impedes the quantitative analysis of other ion-sensitive techniques such as LEIS or ion desorption spectroscopy (IDS) [7.15].

The present TOF-DR method may be considered a modification of the LEIS method. However, unlike LEIS, which is incapable of directly probing hydrogen [7.16], this modification is suitable for this purpose. Moreover, it obviates the above-mentioned limitations of SIMS, thus providing a quantitative method for the determination of surface hydrogen [7.17]. It is important, however, to point out the sensitivity of that method to the structural details of the surface [7.17–19], which will be discussed later on.

7.1.2 The TOF-DR Method: General Description

The TOF-DR technique is based on the bombardment of the studied surface with pulsed beams of low energy (1–10 keV) rare gas ions incident at grazing angles. Directly recoiled surface atoms ejected at low forward angles (as well as forward scattered primaries) are then sufficiently energetic to be directly detected either as neutrals or as ions in a conventional electron multiplier positioned in this direction (Fig. 7.1). TOF measurements of these direct recoil (DR) particles and scattered primaries allow the energy analysis of both ions and neutrals. The ability to detect both ions and neutrals obviates the problem of taking into account the neutralization phenomena that cause difficulties in the interpretation of ion-sensitive measurements. Also, due to the high detection sensitivity of this method, very low irradiation doses (below 10^{12} ions cm^{-2}) are sufficient and the analysis of the surface is carried out in an almost non-destructive way. Metals and adsorbed surface atoms including light atoms such as hydrogen and its isotopes can be detected as DRs displayed as distinct peaks in the TOF spectra [7.17, 19].

Assuming binary elastic collisions between the incident rare gas primary ions M_i[amu] and the recoiled target atom M_j of type j and utilizing the corresponding energy relations for single collision events [7.14, 17] the TOFs (in microseconds) of the recoiled surface particles are given by

$$t_{ji}^{DR} = 1.139 \times 10^{-2} \frac{L(M_i + M_j)}{\sqrt{E_0 M_i} \cos \theta_r}, \tag{7.1}$$

where L is the flight length [cm], E_0 is the primary energy [keV] of the incident ions and θ_r is the recoil angle ($0 \leq \theta_r < \pi/2$), which for the experimental arrangement illustrated in Fig. 7.1 is the angle between the ion beam line and the detector (denoted as θ in the inset of Fig. 7.1). For a typical flight length of 1 m and incident 3 keV Ar$^+$ ions the TOFs obtained for some common light adsorbates recoiled at low angles ($\theta_r \cong 10°$–$30°$) are of the order of a few microseconds (e.g., 4.3–4.9 μs for H, 5.5–6.2 μs for C, 5.9–6.7 μs for O). Utilizing (7.1), the mass

288   M.H. *Mintz* et al.

resolution attributed to single collision events is then given [µs/amu] by

$$\frac{dt_{ji}^{DR}}{dM_j} = \frac{1.139 \times 10^{-2} L}{\sqrt{E_0 M_i} \cos \theta_r}. \tag{7.2}$$

In the above example we obtain a mass resolution of about 0.1 µs/amu.

The distinction between different DR peaks in the TOF spectra is determined by the relative magnitude of the mass separation and the intrinsic line widths of the respective peaks. These line widths are affected by a variety of multiple collision events, which cause broadening of the TOF (or energy) DR peaks (as well as of the scattering peak associated with the incident primaries). This broadening is strongly dependent on the recoil (or scattering) angles, causing smearing of the DR lines at high recoil angles. Consequently, even though a better mass resolution is anticipated at high recoil angles, see (7.2), it is not possible to distinguish the different peaks associated with different surface species. Only at low recoil angles ($\leq 30°$) are well-resolved TOF spectra obtained (with typical line widths of a few tenths of a microsecond).

In addition to the DR peaks, scattering peaks associated with the primary ions appear in the TOF patterns. Assuming single elastic collisions between a primary ion i and a surface atom j, the corresponding scattering peaks are given by

$$t_{ij}^s = \frac{2.277L(M_i/E_0)^{1/2}(M_i + M_j)}{[M_i \cos \theta_s + (M_j^2 - M_i^2 \sin^2 \theta_s)^{1/2}]}, \tag{7.3}$$

where θ_s is the scattering angle, which for $M_i < M_j$ lies in the range $0 - \pi$, while for $M_i > M_j$ it is limited to a maximum value of $\sin^{-1}(M_j/M_i)$. With the experimental arrangement illustrated in Fig. 7.1, θ_s is equal to θ.

The instrumental setup of a TOF-DR spectrometer is illustrated in Fig. 7.1. Different modifications and improvements can be added to this basic design

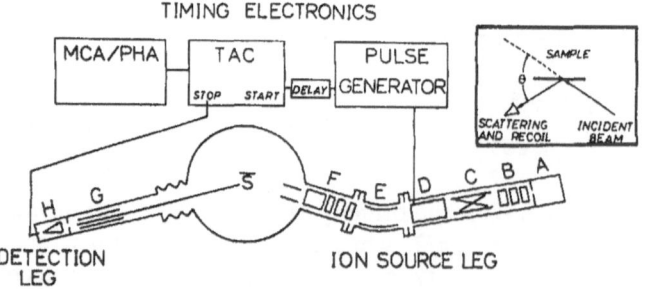

Fig. 7.1. Schematic block diagram of a TOF–DR spectrometer: (A) ion source; (B) lenses; (C) Wien filter (velocity selection); (D) pulsing plates; (E) electrostatic sector (energy selection); (F) focusing lenses and X–Y deflectors; (S) sample; (G) Electrostatic parallel plates deflector; (H) Electron-multiplier. *Inset*: Angular arrangement of incident pulsed beam and detector. θ: scattering and recoil angle. (From [7.17])

[7.20–23]. However, for the purpose of the present demonstration this simplified scheme is adequate. The two main components are:

(i) an ion source leg which produces a pulsed beam of rare gas ions with a well defined energy (in the range 1–10 keV), and
(ii) a detection leg with an electron-multiplier probing the forward scattered and direct recoil particles.

Both legs are positioned at grazing angles relative to the sample's surface (see Fig. 7.1 inset). The ion source leg consists of an ion beam which is velocity selected, pulsed, energy analyzed (mainly to eliminate unpulsed neutral contaminants), and directed onto the sample at low incidence angles. The trigger output of the pulse generator used to form the ion pulse goes to the input of a time-to-amplitude converter (TAC) after being delayed by slightly less than the time necessary for the ion pulse to travel from the pulsing region to the target. The TAC is stopped when a single scattered primary or direct recoil particle arrives at the detector. A voltage pulse of height proportional to the time between start and stop signals is produced by the TAC and fed into a multichannel analyzer (MCA) operated in the pulse height analysis (PHA) mode. The channel number of the MCA is proportional to the particle time of flight, hence by repeating this sequence TOF spectra are generated as a histogram of the distribution of particles' flight times.

It should be noticed that with a single TAC detection the mean number of particles in the detection leg per bunch (i.e., pulse) of ions has to be less than one; otherwise the particles with the highest velocity will be measured preferentially. This requirement limits the allowed counting rate to about 10%–20% of the applied pulsing rate.

An additional capability is provided by a simple (parallel plates) deflector positioned in front of the electron-multiplier detector (G in Fig. 7.1). By periodically applying deflection voltages on these plates, intermittent TOF countings of ions plus neutrals (deflection voltage-off) and neutrals only (deflection voltage-on) are obtained and stored in two different parts of the MCA memory. The TOF ion spectrum (and hence the ion fractions in each of the TOF peaks) is determined by subtracting the neutrals from the total spectrum.

The experimental data acquired by the TOF-DR measurements provide the following information on the studied surface:

(i) Surface composition. Assuming that for a given detection angle and a fixed irradiation flux the intensities (or integrated areas) of the direct-recoils peaks in the total TOF spectra (i.e., neutrals and ions) are proportional to the surface concentrations of the corresponding species, the relative variations in the concentrations of these atoms can be determined. Surface concentrations may be obtained versus gas exposure doses or as a function of thermal treatments. The assumption of a linear proprotionality between TOF-DR signal intensities and surface concentrations is justified only if multiple collisions [7.14, 24–28] as well as shadowing and blocking [7.29–34] or focusing [7.35] effects have a small contribution. Evidently, more theoretical and experimental work is still needed in order to generally quantify the TOF-DR method.

(ii) Surface chemistry and electronic structure. By measuring the ion fractions of both scattered and direct recoil particles, some insight may be gained into the chemical nature of the originating surface compounds and on the electronic structure of the solid substrate, as demonstrated in Sect. 7.2.3. However, the interpretation of the measured ion fractions data is not straightforward, since different neutralization mechanisms may participate in this process [7.36].

(iii) Surface geometrical structure. Multiple encounters and shadowing and blocking effects strongly depend on the structure of the atomic array on the surface (first and/or second layers) [5.24–35]. The analysis of the TOF peak intensities as a function of incidence and/or azimuthal angles for well-defined single crystal faces may provide quantitative surface structure information as well as characterization of adsorbed layers, as demonstrated for the Pt–H_2 system [7.18, 37]. A very interesting structural determination performed recently on the H/W(211) surface (utilizing the TOF-DR method) [7.38] provided direct experimental evidence for the existence of delocalized hydrogen positions in agreement with previous theoretical calculations [7.39, 40]. In the present chapter however, we concentrate mainly on the application of the TOF-DR method for studying the reactions between gaseous hydrogen and polycrystalline metal surfaces (the application that is more frequently met in the hydride research field). The exploitation of this technique for surface crystallography will not be discussed further. Nevertheless, it is worth mentioning that the sensitivity of the TOF-DR method to the structural arrangement of the surface atomic arrays may, in certain cases, permit one to distinguish between adsorption processes occurring simultaneously on different types of planes composing a poly-crystalline surface [7.41a]. This possibility was demonstrated recently for oxygen adsorption on polycrystalline copper [7.41a, b] and on a copper–lithium alloy [7.42]. Combined measurements with both TOF-DR and Auger electron spectroscopy (AES) or X-ray photoelectron spectroscopy, (XPS) enabled the evaluation of apparent sticking probabilities on different types of structural arrays and pointed to significant contributions of interplane spill-over processes. Such a detection ability is very important in quantitative kinetic studies of adsorption on complex surfaces and may provide a useful tool for both basic and applied surface studies (e.g., in catalysis and corrosion).

7.1.3 Adaptation of TOF-DR Measurements to the Study of Some Metal–Hydrogen Systems

The interactions of gaseous hydrogen and clean or oxidized polycrystalline magnesium surfaces were studied using TOF-DR measurements performed with a pulsed 3 keV Ar^+ beam [7.43]. The total TOF spectra (of ions and neutrals)

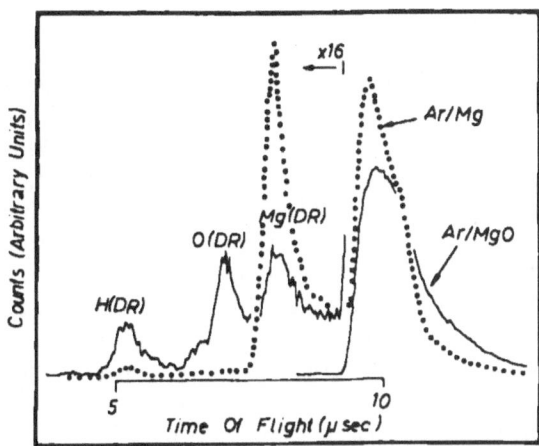

Fig. 7.2. The time-of-flight spectra obtained for a polycrystalline magnesium surface with a pulsed 3 keV Ar$^+$ ion beam incident at an angle of about 15° relative to the surface (detection of scattered primaries and direct surface recoils, θ, and θ_r, of about 25°). *Dotted line*; sputter-cleaned surface. *Solid line*: oxidized surface (exposed to 20 langmuirs of oxygen). Each of the direct recoil peaks corresponding to the surface species S_i is denoted by S_i(DR). The Ar scattering peaks off the clean and the oxidized surfaces are denoted by Ar/Mg and Ar/MgO, respectively. (From [7.44])

obtained for the sputter-cleaned and for the oxidized magnesium surfaces are illustrated in Fig. 7.2. The hydrogen direct recoil peak, H(DR), is well resolved from all other peaks, providing a convenient way to follow its surface concentration. It has been found that the accumulation rates of surface hydrogen on clean polycrystalline magnesium exposed to gascous H_2 in the pressure range $< 5 \times 10^{-6}$ torr (which is below the room temperature dissociation pressure of MgH_2) are negligible [7.43]. Up to doses of about 2000 langmuirs of H_2 no H(DR) peaks were detected in the TOF-DR spectra. On the o her hand, oxidized magnesium surfaces (i.e., pre-exposed to 20 langmuirs of oxy en, which causes the saturation of the O(DR) peak in the TOF-DR spectrum [7.22]) accumulate substantial amounts of surface hydrogen under similar exposure conditions. Some possible adsorption models were fitted to the H(DR) peak intensity versus H_2 exposure curve [7.43]. A one-site adsorption model seemed to yield the best fit to the experimental results obtained for the oxidized Mg surface.

The absence of surface hydrogen on the H_2-exposed clean magnesium seems to contradict theoretical model calculations made on the hexagonal Mg(0001) plane [7.44]. Several possible explanations have been proposed to account for this disagreement [7.17].

An interesting feature displayed by Fig. 7.2 is the accumulation of surface *hydrogen* concomitant with the *oxygen* exposure of the magnesium sample (compare the solid line obtained after 20 langmuirs of O_2 exposure with the dotted spectrum of the clean surface). Actually, the H(DR) peak intensity versus oxygen exposure curve displays a monotonic saturation-type behavior [7.45]. It has been shown [7.45] that in this case the contribution of possible contaminant gas reactions (e.g., H_2, H_2O constituting the residual pressure in the UHV chamber) is less than 2% of the total accumulated surface hydrogen. This effect has been attributed [7.45] to oxygen-induced surface segregation of bulk hydrogen, which is known to be present in many metals. The migration of this hydrogen to the surface is driven by the stabilization occurring by its reaction with the adsorbed surface oxygen. Such adsorption-enhanced segregation of bulk impurities [7.46] (e.g., carbon and sulfur) and of reactive metal constituents in alloys and intermetallic compounds [7.2, 5] has been well documented in the literature. However, so far the lack of a quantitative technique for probing surface hydrogen concentrations has impeded such observations regarding hydrogen. It is likely that similar H-segregation effects occur in many other hydrogen-containing metals, e.g., uranium [7.47] and nickel [7.48]. In the latter case a TOF-DR study of the interactions between gaseous H_2, O_2, H_2O and a Ni(111) surface [7.48] also indicated an oxygen-induced segregation of bulk hydrogen. However, the H(DR) peak intensity versus oxygen exposure curve displayed in this case a more complex behavior than in the magnesium case [7.45]. A maximum in the exposure curve occurred at about 1.5 langmuirs of O_2. The exact nature of the hydrogen–oxygen–surface interactions (or reactions) which result in these segregation effects is still not satisfactorily understood. The detailed study of more metallic systems corroborated by theoretical calculations may elucidate these interesting phenomena.

Conversion of the TOF-DR peak intensities into actual surface concentrations requires an appropriate calibration procedure. Such a procedure has been demonstrated for the $MgO–H_2$ system [7.49]. It is based on the controlled exposure of the clean Mg surface to gaseous methanol, which forms a surface methoxide located above the uppermost Mg layer. Consequently, for each carbon atom, three H atoms are present. The amount of carbon on the surface can then be measured by a complementary reference technique (e.g., XPS). The intensity is then related to the absolute surface hydrogen coverage through the reference carbon measurement and the known chemistry of the surface methoxide. Assuming a structural model for the H_2 adsorption process such a procedure provides values of the sticking coefficients. A comparison between the model predicted and actual surface coverages may point to the location of the adsorbed hydrogen atoms (e.g., on top or beneath the uppermost surface layer).

The chemisorption of H_2, O_2 and H_2O on polycrystalline lanthanum has been studied by means of combined surface recoiling, XPS and ultraviolet photoelectrom spectroscopy (UPS) measurements [7.50], which allowed the distinction between the topmost surface reactions and subsurface processes. As mentioned in Sect. 7.1.1, such a distinction is not possible by electron spectroscopy measurements alone.

Correlations between measured ion fractions in the TOF scattering peak (of the rare gas primaries) and the electronic structures of the clean and of the adsorbate-covered metal substrates have been reported for polycrystalline lanthanum and ytterbium [7.51]. The variations in these ion yields as a function of H_2, O_2 and H_2O exposure have been accounted for by corresponding changes in the valence band structures induced by the formation of the respective surface compounds.

The variations in the ion fractions of H, C and O direct recoil peaks, induced by matrix effects have been determined for different substrates [7.52].

Oxygen–hydrogen inter-relations on a niobium surface have recently been studied by combined measurements using TOF-DR, AES and XPS [7.53]. It was found that oxygen surface segregation occurred upon heating, with hydrogen "sticking" beneath the segregated oxygen topmost layer.

A generalized scheme for the distinction between different possible gas–surface reaction mechanisms has been proposed recently [7.47], based on the comparison between exposure curves obtained by TOF-DR and electron spectroscopy techniques. This approach is applicable to single crystal and polycrystalline surfaces. In the latter case, such a method provides a unique technique for basic studies of complex surfaces encountered in "real", technologically oriented systems. An interesting application of such a comparative TOF-DR and electron spectroscopy study has been demonstrated for the hydrogen–oxygen accumulation kinetics on polycrystalline uranium surfaces [7.54]. It has been shown that predosing the uranium surface with oxygen then exposing it to hydrogen induced changes in the accumulation kinetics of the latter (changing from island growth on the clean metallic surface to a random dissociative chemisorption on the oxygen predosed surface). Reversing the exposure sequence, i.e., predosing with hydrogen then exposing to oxygen, modified the mechanism of oxygen incorporation and the structure of the product over layer.

7.1.4 Future Prospects

The interactions of gaseous H_2 with a large variety of metals, alloys and inter-metallic compounds may be studied by the TOF-DR technique. Clean and oxidized surfaces may be compared, providing valuable information on the initial stages preceding the hydriding reactions. For metals which form very stable hydrides (e.g., the lanthanides) it is possible to produce the corresponding hydride phases in situ and study their surface properties (e.g., H/M composition ratios at the surface, location of hydrogen at the uppermost layers, effects of formation of surface hydrides on measured ion fractions). These studies may be performed on either polycrystalline materials or well-defined single crystal faces. In the latter case, a more accurate compositional analysis combined with structure models is expected. Heterogeneous hydrogen isotope exchange processes may be detected directly by TOF-DR measurements. The exchange rates on a variety of clean and oxidized metallic surfaces may thus be investigated.

Surface segregation and adsorption-induced segregation of bulk hydrogen (and its isotopes) may be studied. Diffusion rates and effects of different parameters (e.g., temperature and concentration) may be evaluated.

7.2 Nuclear Resonance Scattering of Gamma Rays

7.2.1 Bonding Strengths in Metal Hydrides

The bonding strength of a hydride is directly related to its stability, i.e., its heat of formation, as demonstrated in [7.55, 56]. In this relation, the Debye temperature θ_D is regarded as an integrated quantitative measure of the binding forces. θ_D may be derived by various techniques, such as specific heat, ultrasonic measurements, X-ray scattering (see for example [7.57, 58]), as well as nuclear resonant scattering of gamma rays [7.59]. The latter technique can supply important additional information that is not directly accessible by other methods. This information, which is associated with the mean kinetic energy (including the zero-point energy) of the atomic lattice vibrations of particular elements, may have a significant impact on the analysis of neutron inelastic scattering data of compounds in general and hydrides in particular. The mean kinetic energy is a model-independent quantity, while θ_D is both model dependent (the Debye approximation) and method dependent (the various experimental techniques are sensitive to different parts of the phonon spectra). A survey of experiments in solid-state physics using resonant scattering of gamma rays is given in [7.60].

In a detailed microscopic picture one should consider the characteristic lattice vibrations and interatomic forces, which present an important solid-state problem. Today the most powerful experimental tool for investigating lattice vibrations (phonons) is undoubtedly inelastic neutron scattering. This method gives a close view of the vibrating atoms and provides knowledge of phonon dispersion curves and phonon densities of state. A large amount of experimental and analytical data on this subject may be found in [7.61–64], including data on hydrides [7.63–64]. As shown below, the nuclear resonance scattering method may provide information complementary to neutron scattering experiments.

7.2.2 The Nuclear Resonance Scattering Method: Principles

The most significant physical parameter measured by the gamma resonance scattering technique in the present context is the mean kinetic energy of certain elements in a compound. This parameter is related to the phonon frequency spectrum by

$$\varepsilon \equiv kT_{eff} = \int_0^\infty n(v, T)hvg(v)dv, \tag{7.4}$$

$$n(v, T) = [\exp(hv/kT) - 1]^{-1} + \tfrac{1}{2},$$

where ε is the mean energy per mode of vibration in the harmonic approximation (e.g. [7.65]) and $g(v)$ is the normalized phonon frequency distribution, i.e., $g(v)dv$ is the number of phonon states per degree of freedom in the frequency range

$(v, v + dv)$. The mean energy ε also includes the zero-point energy. The notion of the effective temperature T_{eff} was first introduced by *Lamb* [7.66]. He treated the thermal motion of atoms in a solid in the same manner as atoms in a gas, but used an effective temperature which was higher than (or equal to in a limiting case) the thermodynamic temperature T of the material. In the harmonic approximation the mean kinetic (or potential) energy per vibrational mode is equal to exactly half of the corresponding total energy ε.

The nuclear resonance scattering method utilizes an accidental overlap between an incident gamma line from some (n, γ) source and an excited nuclear level in a certain target nucleus. This overlap, represented schematically in Fig. 7.3, causes resonant scattering to occur. So far, the occurrence of an overlap that can be utilized for deriving mean kinetic energies has been found in about 10% of the elements (Table 7.1).

Excited nuclear levels at energies $E_\gamma \approx 7$ MeV usually display natural radiation widths Γ_0 between 10 meV (which is approximately the lowest detectable limit of the technique) and 1 eV. However, the nuclear levels are also Doppler broadened due to the thermal and zero-point vibrational motion of the scattering atoms. The Doppler width Δ is given by

$$\Delta = E_\gamma \left(\frac{2kT_{\text{eff}}}{Mc^2} \right)^{1/2}, \tag{7.5}$$

where M is the mass of the resonance nucleus and c is the velocity of light. The Doppler widths are of the order of 5–10 eV at room temperature and therefore the dominant factor determining the energy width of the resonant levels. The intensity $C(T, \psi)$ of the scattered radiation at temperature T and angle ψ is determined by the overlap between the incident gamma line and the excited nuclear level, as illustrated in Fig. 7.3. $C(T, \psi)$ depends on the nuclear parameters $\Gamma_0, \Gamma, \delta, g$ of the level, on the Doppler width Δ and therefore on the effective

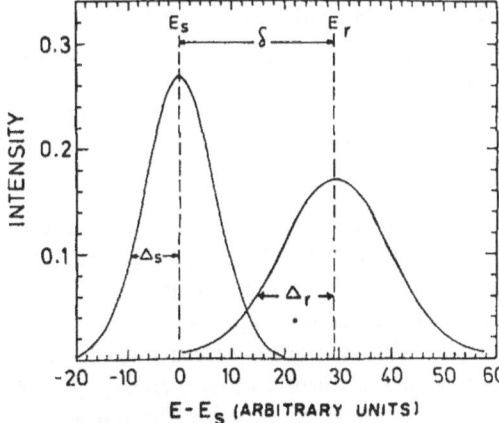

Fig. 7.3. The condition for resonant scattering of incident monoenergetic photons of peak energy E_r by an excited level of peak energy E_s. The two lines are Doppler broadened (Δ_r and Δ_s) and are separated by an energy δ. (From [7.60])

Table 7.1. Some resonant scatterers with significant temperature effect, $R(T_1, T_2, 135°)$, (7.6). (From [7.67, 68])

(n, γ) source	Resonant scatterer	Percentage abundance	Resonance energy [MeV]	Temperature effect[a]
Fe	^{62}Ni	3.66	7.646	0.90 ± 0.006
Fe	^{208}Pb	52.3	7.279	0.937 ± 0.003
Fe	^{50}Cr	4.3	8.888	0.84 ± 0.02
Fe	^{141}Pr	100	7.632	0.95 ± 0.01
Fe	^{139}La	99.9	6.018	0.969 ± 0.003
Ni	^{86}Sr	9.9	7.82	0.96 ± 0.01
Ni	^{150}Sm	7.4	8.998	0.95 ± 0.007
V	^{48}Ti	74	6.600	0.80 ± 0.02
V	^{146}Nd	17.2	7.163	0.93 ± 0.01
Cr	^{15}N$_2$	0.366	6.324	0.72 ± 0.03
Ti	^{65}Cu	30.9	6.556	0.94 ± 0.02

[a] The temperature effect R was measured at $T_1 = 115$ K, $T_2 = 293$ K for N_2^{15} and at $T_1 = 77$ K, $T_2 = 300$ K for the rest of the resonant scatterers in the table

temperature $T_{\rm eff}$ of the resonant nucleus and on the temperature T_s of the (n, γ) source:

$$C(T, \psi) = f(\Gamma, \Gamma_0, \delta, g, T_{\rm eff}, T_s).$$

Γ and Γ_0 are the total and the partial ground state radiative widths, respectively, δ is the energy separation between the incident γ line and the resonance level, g is a statistical factor depending on the spins of the resonance and ground states. $C(T, \psi)$ is proportional to the resonance scattering cross section and its explicit form is given elsewhere [7.69]. Obviously, a change in the effective temperature affects the Doppler width (7.5). Two ways of changing $T_{\rm eff}$ are as follows:

(i) Stiffening of the phonon spectra increases the mean total energy ε as indicated by (7.4) and therefore increases the effective temperature and the Doppler width, and vice versa—softening of the phonon spectrum decreases ε, $T_{\rm eff}$ and Δ.

(ii) Another way to change $T_{\rm eff}$ is to vary the temperature T of the scatterer, thus changing the mean occupation number $n(v, T)$—an increase of T increases ε, $T_{\rm eff}$ and Δ, and vice versa.

The change of the Doppler width (Δ_s in Fig. 7.3) may or may not cause a variation in the intensity of the resonantly scattered radiation. Such a variation depends on the specific relative position between the incident gamma line and the resonance level (Fig. 7.3). A temperature variation experiment may serve as a convenient probe for the sensitivity of the scattered intensity (in a certain source–target combination) to a change of the atomic kinetic energy (i.e., $T_{\rm eff}$ or Δ). More specifically, it is convenient to utilize as a probe the temperature effect R defined as the ratio

$$R(T_1, T_2, \psi) = \frac{C(T_1, \psi)}{C(T_2, \psi)}; \qquad (T_1 < T_2). \tag{7.6}$$

A suitable (n, γ) source–scatterer combination for the present study should provide a relatively high scattered intensity and a significant temperature effect $(R \lesssim 0.95)$. Table 7.1 lists some of the intensive scatterers together with the corresponding resonance energies and measured R values at $T_1 = 77\,\mathrm{K}$ and $T_2 = 300\,\mathrm{K}$. From the measured R values it is possible to calculate T_{eff} at one temperature, provided T_{eff} at the second temperature is known. Usually the determination of $T_{\mathrm{eff}}(T)$ requires a careful measurement of $C(T, \psi)$ over a wide T range, including sufficiently high temperatures. At such (high) temperatures one can assume that $T_{\mathrm{eff}} \approx T$ and consequently calculate $T_{\mathrm{eff}}(T)$ at lower temperatures. This can be seen from (7.4): For $kT > h\nu$ one can approximate the mean occupation number $n(\nu, T)$ by $kT/h\nu$; T_{eff} then equals T.

A check of the γ-ray scattering method may be performed for the pure elements by comparing T_{eff} obtained by this technique with the corresponding T_{eff} derived by inelastic neutron data [by averaging over the whole phonon frequency distribution $g(\nu)$ according to (7.4)]. In this case the two T_{eff} values should be equal, since a single element causes the scattering events in the two experiments. Such a comparison between the two methods has been done recently [7.70] for Ni and Pb utilizing the wide temperature range γ-ray scattering data presented in [7.59] and the inelastic neutron scattering data compiled in [7.63]. A very good agreement (1%–2% difference) was obtained in this comparison [7.70], *indicating the consistency of the two techniques.*

The γ-ray measurements may be of great importance in multicomponent compounds (e.g., hydrides), where the information on T_{eff} obtained by this technique is additional to that provided by neutron inelastic scattering. In such compounds the mean energy per degree of freedom ε_i of the ith component in a compound of c components is given by

$$\varepsilon_i \equiv kT_{i,\mathrm{eff}} = \frac{\int\limits_0^\infty S_i(\nu)n(T, \nu)h\nu g(\nu)d\nu}{\int\limits_0^\infty S_i(\nu)g(\nu)d\nu};$$

$S_i(\nu)$ is the fraction of the phonon energy carried by the ith atom at frequency ν. Evidently, the equality $\sum\limits_{i=1}^{c} S_i(\nu) = 1$ holds for every ν. The derivation of $S_i(\nu)$ for a certain component in a compound is a complicated problem and is associated with the determination of the eigenvectors for a certain set of eigenfrequencies. In other words, the quantity $T_{i,\mathrm{eff}}$ is not directly accessible from the neutron inelastic scattering experiment, and it is model dependent. On the other hand, the nuclear resonance scattering of gamma rays can provide $T_{i,\text{eff}}$ directly through (7.5), independently of any model, and therefore may assist in testing the validity of various interpretations of the phonon spectra.

7.2.3 Derivation of Debye Temperatures from the γ Resonance Scattering and Comparison with Other Techniques

The effective temperature T_{eff} gives a measure of the binding forces at a certain temperature T. The higher T_{eff} the stronger the binding forces. It is sometimes more convenient to extract the Debye temperature θ_D from T_{eff}, (7.4), by utilizing the Debye approximation $g(v) = 3v^2/v_D^3$, where v_D is the Debye cutoff frequency:

$$\frac{T_{\text{eff}}}{T} = 3\left(\frac{T}{\theta_D}\right)^3 \int_0^{\theta_D/T} x^3 \left(\frac{1}{e^x - 1} + \frac{1}{2}\right) dx. \tag{7.7}$$

θ_D of (7.7) is sensitive to the zero-point atomic oscillations in the lattice and therefore high-energy phonons will make an important contribution at all temperatures. Consequently, the Debye temperature derived from the nuclear resonant scattering experiment may be expected not to be too strongly dependent on the temperature. This was experimentally demonstrated for the case of Ni [7.59]. Debye temperatures are also derived from other experimental methods, of which the best-known is the specific heat. In this case θ_D^s (s stands for specific heat) is insensitive to the zero-point energy as the latter term vanishes upon temperature differentiation of the total vibrational energy. It is worth stressing the difference between θ_D^{γ} and θ_D^x (from X-ray diffraction) as in the two cases a Debye temperature is derived from photon scattering experiments and confusion of similarity may arise. In the X-ray experiment, as in any other diffraction scattering, there is a decrease in intensity, but not in the sharpness of the Bragg reflections with an increase in temperature. The "lost" intensity, which reappears as a diffuse "background", is associated with the mean square vibrational displacement of a lattice point. This means that a diffraction scattering experiment may provide the average $\langle \varepsilon(v)/v^2 \rangle$ over the whole frequency spectrum while one can get $\langle \varepsilon(v) \rangle$ from the nuclear resonance scattering. In terms of moments, $\theta_D^{\gamma}(T=0)$, i.e., the Debye temperature derived from the zero-point energy, is associated with the first moment of the frequency distribution function; $\theta_D^{\gamma}(T=\infty)$, $\theta_D^x(T=0)$, $\theta_D^x(T=\infty)$, $\theta_D^s(T=0)$ and $\theta_D^s(T=\infty)$ are associated with the second, the first negative, the second negative, the third negative and the second moments, respectively [7.71, 72].

7.2.4 Experimental Setup

A detailed description of the measurement facilities used for the γ-ray resonance scattering performed at the Nuclear Research Centre—Negev (NRCN) may be found in [7.69, 73]. The schematic experimental arrangement is illustrated in Fig. 7.4.

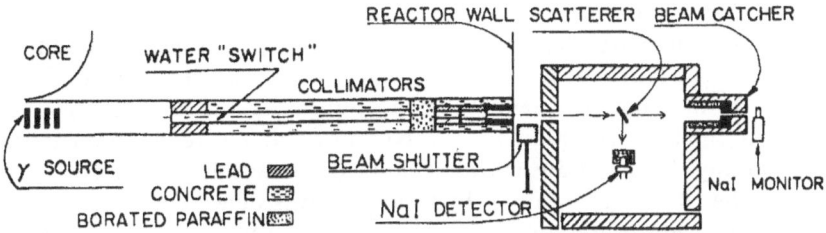

Fig. 7.4. Schematic representation of the experimental arrangement for γ-scattering

The incident γ source consists of several metal discs inserted along a horizontal tube near the core of the IRR-2 reactor (Fig. 7.4). Thermal neutron capture by the metal nuclei is immediately ($\sim 10^{-16}$ s) followed by the emission of a discrete γ spectrum up to energies equal to the binding energy of the last neutron in the compound nucleus. The most intense lines are concentrated in the range 6–9 MeV. The incident γ beam is collimated along a distance of several meters and then impinges upon a target situated in an experimental lead chamber. The scattered radiation is detected by a 12.7×12.7 cm NaI(Tl). The neutron flux was 2×10^{13} neutrons cm^{-2} s^{-1} yielding an intensity on the target of about 10^6 photons cm^{-2} s^{-1} per strong single gamma line. The temperature dependence of the scattered radiation was determined by measuring the scattered intensity of the target at room temperature and at liquid nitrogen temperature.

7.2.5 Recent Applications of the Technique to Metal Hydrides

It should be pointed out that from a single temperature effect measurement (which gives the ratio R of the scattered intensities at two different temperatures) only one parameter may be derived, provided all the relevant nuclear factors (e.g., $\Gamma, \Gamma_0, \delta, g$) are known. Such a parameter may be either T_{eff} at one of the

Table 7.2. Experimental values of the temperature effect $R(77\,\text{K}, 300\,\text{K}, 135°)$, (7.6), for ^{62}Ni and ^{48}Ti in several compounds and deduced values of θ_D. (From [7.74, 75])

Compound	$R(77\,\text{K}, 300\,\text{K}, 135°)$	θ_D [K]
Ni (metallic)	0.900 ± 0.006	430 ± 34
Mg$_2$Ni	0.898 ± 0.005	420 ± 30
Mg$_2$NiH$_4$	0.915 ± 0.006	520 ± 40
Mg$_2$NiD$_4$	0.924 ± 0.004	580 ± 30
LaNi$_5$	0.900 ± 0.006	430 ± 34
LaNi$_5$H$_6$	0.900 ± 0.006	430 ± 34
Ti (metallic)	0.79 ± 0.01	420 ± 30^a
TiH$_{1.8}$	0.79 ± 0.01	420 ± 30^a

[a] The present value for θ_D of Ti is taken from [7.65] and is used as a "calibration" point for deducing the θ_D values for TiH$_{1.8}$

utilized temperatures or the Debye temperature, if the Debye model (7.7) is assumed. Single temperature effect measurements were performed on a number of hydrides [7.74, 75] and the respective θ_D^γ values were derived. They are listed in Table 7.2. In these measurements the contributions of the Ni component to θ_D in some intermetallic compounds (Mg_2Ni and $LaNi_5$) were determined and the corresponding changes induced by hydriding (or deuteriding) were evaluated [7.74]. Also, the change of θ_D upon hydriding Ti was determined [7.75]. The qualitative conclusion of these results is that while the binding forces of Ni in Mg_2Ni increase significantly upon hydriding no observable changes occur in the binding forces of either Ni or Ti upon hydrogenation of $LaNi_5$ or metallic Ti, respectively. The absence of a detectable effect of the Ni in $LaNi_5H_x$ is in good agreement with the reported behavior of this compound as a whole, where a small decrease in θ_D resulting from hydrogenation was found by specific heat measurements [7.76]. Actually, this small effect is within the experimental accuracy of the γ-ray scattering method. The results obtained for Ni in Mg_2Ni [7.74] fit fairly well an empirical relation derived for θ_D changes induced by hydriding metallic elements and compounds [7.55, 56]. For Ti, on the other hand, the empirical relation [7.55, 56] predicts an increase in θ_D upon hydride formation. However, as mentioned above, no changes were detected by the γ-ray scattering experiment. Moreover, a decrease in θ_D was claimed by some researchers using low-temperature specific heat measurements [7.77, 78]. It is worth pointing to the possibility that the tetragonal distortion of titanium hydride (Jahn–Teller effect) [7.79–83] may be responsible for the non-increased θ_D values obtained experimentally. At the moment we cannot propose a clear correlation between the results obtained by the γ-ray scattering method for Ti in $TiH_{1.8}$ [7.75] and the corresponding results obtained from specific heat measurements [7.77, 78]. We believe that a better understanding of these results requires further experiments combining γ-ray scattering and neutron inelastic scattering measurements. The procedure for evaluating the Debye temperatures is illustrated in Fig. 7.5 for Mg_2NiH_x and Mg_2NiD_x ($x = 0,4$) [7.74].

The sloping line in Fig. 7.5 shows the calculated temperature effect $R(77\,K, 300\,K, 135°)$ of Ni as a function of θ_D using (7.6) and the explicit function $C(T, \psi)$

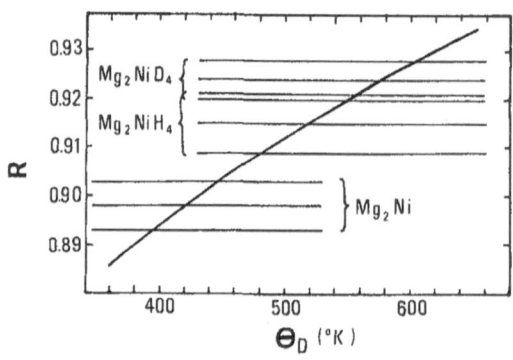

Fig. 7.5. Calculated values (*sloping line*) of the temperature effect $R(77\,K, 300\,K, 135°)$ (defined in (7.6)) of Ni versus the Debye temperature θ_D. The horizontal lines indicate the experimental values of R for Mg_2Ni, Mg_2NiH_4 and Mg_2NiD_4 and the respective erors. The intersections between the horizontal and sloping lines yield the corresponding θ_D values

given in [7.69]. The horizontal lines indicate the experimental values of R and the respective errors. The intersection of the horizontal and calculated lines determines the θ_D values of Ni in the compounds.

A good agreement between nuclear resonant photon scattering (NRPS) and inelastic neutron scattering has been obtained for the bonding strength of some relatively simple cases, namely Ni and Pb [7.70], and Ti in TiC [7.75]. Consequently, we used NRPS to investigate the Ni cohesion variation in $LaNi_{5-x}Al_x$ intermetallics as a function of x and some of their hydrides [7.84]. These compounds ($0 \leq x \leq 1.5$) have been extensively investigated since the discovery of their interesting hydriding properties in 1977 [7.85, 86]. The variation of the hydrogen-sorption behavior of this system as a function of x, however, is not comprehensively understood. The Al substitution for Ni drastically and monotonically increases the hydride stability, but concomitantly the hydrogen absorption ability is gradually inhibited toward high Al concentrations ($x \approx 1.5$). We believe we have provided an explanation for this puzzling hydriding behavior, based on an apparent correlation between the Ni cohesion and the hydrogen uptake. The Ni bonding strength was determined in the present case by using the experimentally measured ratio R:

$$R(Ni^c, Ni^p, T_1, T_2, \psi) = C(Ni^c, T_1, \psi)/C(Ni^p, T_2, \psi); \qquad (7.8)$$

Ni^c and Ni^p denote Ni in compound c and pure Ni, respectively. It is convenient to set $T_1 = T_2 = 298$ K (room temperature). Equation (7.8) can be used to derive the effective temperature $T_{eff}(Ni^c, T_1)$. The intensity ratios together with the deduced values of $T_{eff,Ni}$ are given in Table 7.3. The present results of the $LaNi_{5-x}Al_x$ ($0 \leq x \leq 1.5$) system and of some of the corresponding hydrides exhibit the following features: (1) A sharp decrease in the resonantly scattered intensity by Ni in $LaNi_{5-x}Al_x$ occurs at $x = 0.25$. (2) A significant increase in R is observed for $0.25 < x \leq 1.5$, the value of 1.035 being reached for $x = 1.5$.

Table 7.3. Measured scattered intensity ratios R, see (7.8), with an accuracy of about 2%, normalized to pure nickel (after background and self-absorption corrections), and deduced effective temperature $T_{eff,Ni}$ at $T = 298$ K. The numbers in parantheses indicate errors. (From [7.84])

Compound	R	$T_{eff,Ni}$ [K]
Ni (metal)	1	324(3) [7.70]
$LaNi_5$	1.013	349(30)
$LaNi_{4.9}Al_{0.1}$	1.019	360(41)
$LaNi_{4.75}Al_{0.25}$	0.920	202(36)
$LaNi_{4.5}Al_{0.5}$	0.946	239(30)
$LaNi_4Al$	0.966	269(21)
$LaNi_{3.5}Al_{1.5}$	1.035	391(36)
$LaNi_{4.5}Al_{0.5}H_{5.8}$	1.046	413(40)
$LaNi_4AlH_{4.9}$	1.041	403(33)
$LaNi_{3.5}Al_{1.5}H_{3.8}$	1.049	418(43)

(3) The R values measured for the three hydrides $LaNi_{5-x}Al_xH_y$ ($x = 0.5, 1, 1.5$) are in general higher than in the corresponding unhydrided compounds. It is interesting to note that these R values are all around the same value, namely 1.045.

The variations of the R values directly reflect the variation in the bonding strength of the Ni atoms in the different compounds. The very distinct change of the Ni bonding strength in the $LaNi_{5-x}Al_x$ system as a function of x may be qualitatively explained as follows: The substitution of the larger Al atoms for the Ni atoms expands the original $LaNi_5$ hexagonal lattice. At relatively small x values ($x = 0.25$) the Ni bonding strength is drastically decreased because of the increased distance of the Ni atoms to their neighbors. Further increase of the Al content, although accompanied by further expansion, enhances the Ni bonding because of increased d–p hybridization.

In view of our results we suggest that the decreased hydrogen absorption capacity in $LaNi_{5-x}Al_x$ for high values of x is associated with the increased bonding or rigidity of the lattice. This may be formulated as a *rule of reverse capacity* (RRC)—in analogy to the rule of reverse stability [7.87]—in the following way: The hydrogen absorption capacity of isostructural intermetallic compounds *may* decrease with increasing lattice rigidity, and, vice versa, the capacity *may* increase with decreasing lattice rigidity (softening). "May" stresses that the RRC does not represent a sufficient condition. For example, softening of the lattice will not cause an increased hydrogen absorption capacity if: (1) hydrogen absorbing elements are not present in the intermetallic (in such a case there is no hydrogen absorption) or (2) the hydrogen capacity of the unsoftened intermetallic is so high that any further increase in the hydrogen content is limited by other factors such as repulsion between too closely located hydrogen atoms. This is probably the reason for the non-increased hydrogen absorption in $LaNi_{5-x}Al_x$ for small values of x.

The notion of a restrained H capacity of stiff crystal lattices is also supported by the hydrogen-induced changes in the R values of $LaNi_{5-x}Al_xH_y$ ($x = 0.5, 1, 1.5$)—see Table 7.3. The enhanced scattered intensities of these relatively stable hydrides ($\Delta H = -39, -53$ and -61 kJ/mole H_2 [7.85], respectively) with respect to the corresponding unhydrided compounds are in qualitative, but not quantitative, agreement with the previously mentioned empirical rule [7.55, 56]. According to this rule, the change in Debye temperature (and hence the change in bonding strength) upon hydrogenation is proportional to the heat of hydrogen formation. Therefore, we expect for the $LaNi_{5-x}Al_xH_y$ compounds studied in this work (a) larger R values (7.8) with respect to the corresponding intermetallic compounds, and (b) an increase in R with x for the three hydrides. In Table 7.3 we see clearly that the experimental data confirm (a), but do no confirm (b). This last fact is most probably due to a restraint of the hydrogen capacity, induced by the increased rigidity at larger x values. We consider this further evidence for the proposed RRC.

7.2.6 Possible Future Research

The unique possibility of directly determining mean kinetic energies by the nuclear resonant scattering of γ rays may shed more light on the binding forces in metals, hydrides, and other compounds. The information deduced from the present technique may be complementary to that from inelastic neutron scattering. In the latter case, detailed knowledge of the phonon dispersion relations is obtained, but no direct derivation of the kinetic energies of the different atoms in a compound (e.g., a hydride) is possible. We therefore believe that the data provided by the present technique will also be important in the analysis of neutron inelastic scattering results. Some of the suggestions for future research into the bonding strength in metal hydrides utilizing the present technique include:

1) Investigation of the changes induced in the binding forces of pure metals upon hydrogenation (binary hydrides). The results may provide an experimental answer to the ambiguity in the θ_D derivations from low-temperature data. This ambiguity has been pointed out in [7.56] and occurs because of the different treatments of the H-atom contributions. While some researchers include these contributions in deriving θ_D values from heat capacity and elastic measurements, others ignore it. The analysis of the combined results obtained by the present technique and by neutron inelastic scattering is likely to clarify such ambiguities.

2) Determinations of the bonding strengths of one or more metallic components in intermetallics and the respective changes induced by hydrogenation. In our opinion, the determination of the Fe cohesion in $Zr(Al_xFe_{1-x})_2$ compounds would be a good choice for the continuation of this class of experiments. Such an experiment would provide a twofold benefit. First, it would firmly establish the use of NRPS for deriving kinetic vibrational energies of a metal element in a compound crystal matrix. Second, additional confirmation of the RRC may be expected. We suggest that the $Zr(Al_xFe_{1-x})_2$ system [7.88] provides a comprehensive demonstration of the RRC. $ZrFe_2$ does not absorb a significant quantity of hydrogen. Substituting Al for small amounts of Fe probably leads to a softening of the intermetallic lattice, as in the case of $LaNi_{5-x}Al_x$, and to a sharp increase in the H capacity. Further substitution of Al for Fe, which is probably accompanied by a concomitant stiffening of the crystal lattice, first causes a gradual decrease in hydrogen absorption and finally inhibits it completely. On the other hand, the stability of the hydrides monotonically increases with an increase in the Al content, in agreement with the higher affinity of Al to hydrogen than that of Fe. Additional Al-containing systems exhibit similar behavior [7.88–90].

3) Measurements of the bonding strengths in metal nitrides, which being interstitial compounds may resemble metal hydrides in some ways. The isotope ^{15}N is a strong resonant scatterer [7.68] and is suitable for the study of nitrogen binding properties, as demonstrated by the extensive work performed by *Moreh* and *Shahal* [7.68, 91–95]. Hence, the evaluation of the

contributions of both the metal and the interstitial nitrogen may be possible in these compounds.

We propose two experimental methods of deducing $T_{eff}(T)$ of elements which form metal hydrides (Table 7.1). The first one requires a measurement of the resonantly scattered radiation over a wide temperature range extending up to sufficiently high temperatures where $T_{eff} \cong T$ (as done for Ni and Pb [7.59]). In the second way, T_{eff} of a pure element at a given temperature may be obtained from inelastic neutron scattering data via (7.4). Then, T_{eff} of this element in any form (e.g., a compound) may be deduced by comparing the corresponding intensity of the γ resonant scattering with the scattered intensity of the pure element at the given temperature. Relations of the kind (7.8) may then yield the corresponding T_{eff} at any other temperature. This was done for Ni in $LaNi_{5-x}Al_xH_y$ compounds [7.84]. Both proposed procedures present some experimental difficulties (e.g. the decomposition of the hydrides at elevated temperatures), but we believe they can be overcome.

7.3 Thermal Desorption Spectroscopy

7.3.1 Methods of Investigation of Desorption Processes

The desorption or absorption processes of hydrogen in metals involve two main steps, namely permeation through the surface layer and diffusion in the bulk. The kinetics are affected by both steps and, as they are in series, the desorption (absorption) rate is determined by the slowest one. Extensive basic kinetic studies [7.96–98] as well as application-oriented investigations [7.99] have been performed on metal–hydrogen systems. Various methods have been used in such studies. The method of isothermal desorption and absorption monitors the overall kinetics of hydrogenation (or dehydrogenation) [7.100, 101]. Other methods such as microsocopical observations [7.102] directly provide intrinsic kinetic parameters (e.g. nuclei growth rates). Bulk dynamics (i.e., hydrogen diffusion) may be studied by NMR [7.103], the Gorski effect [7.104] and other methods, whereas surface dynamics may be investigated by a variety of surface analysis techniques (e.g., the TOF–DR method discussed in Sect. 7.1 or other methods described in Chap. 2 of this volume).

Thermal desorption spectroscopy (TDS) has been conventionally used as a surface kinetic method yielding valuable information on the energy levels of the surface adsorbates [7.105, 106]. The TDS measurement has been applied to the study of hydrides [7.107–118] where, in addition to the surface states, the bulk hydrogen states are also involved. This technique measures the desorption rate of hydrogen into vacuum as a function of temperature, which, in most cases, is linearly increased with time. It can be compared with the conventional desorption where the desorption rate is measured at constant temperature and pressure [7.117, 118]. Both methods may in principle give the same information, but, as shown below, the TDS has certain advantages in obtaining rapidly a whole temperature scan behavior in a single experiment. Repeating such temperature scans for various initial concentrations yields a temperature–concentration grid of data which describes the thermal behavior of the system.

7.3.2 General Description of the Experimental TDS Method

The experiments are performed by first loading the metal to an initial hydrogen concentration x_0, followed by cooling to a temperature where desorption into vacuum is negligible. The tempeature is then increased linearly with time under vacuum conditions and the desorption rate is measured. The experiment is repeated for different x_0 values and, if necessary, for various heating rates.

The experimental setup for the TDS is schematically shown in Fig. 7.6. It consists of a chamber, an oven that increases the sample temperature at a constant rate (which can be varied), a thermocouple and an evacuating system. The desorption rate is measured either under dynamic pumping conditions or with a static pre-evacuated volume. In the first case the desorption rate is measured directly, while in the second case the time-dependent integrated desorption is obtained. The desorption is measured via a proper gauge (e.g., a vacuum gauge [7.108], a mass spectrometer [7.119], or a resistivity measuring system [7.118]). The information from the thermocouple and the gauge may be fed into a data acquisition system. Alternatively, data can be registered manually or via a recorder. However, as the data analysis is very elaborate, the use of a computerized data acquisition system is preferable. For the analysis of the results the time behavior of both the desorption rate and the total desorbed amount are needed. There is no distinct preference for one of the two detection schemes shown in Fig. 7.6. When desorption into a pre-evacuated volume is used, the build-up of pressure imposes the consideration of an additional reabsorption term, in particular towards the end of the desorption cycle, where the equilibrium pressure of the hydride is approached.

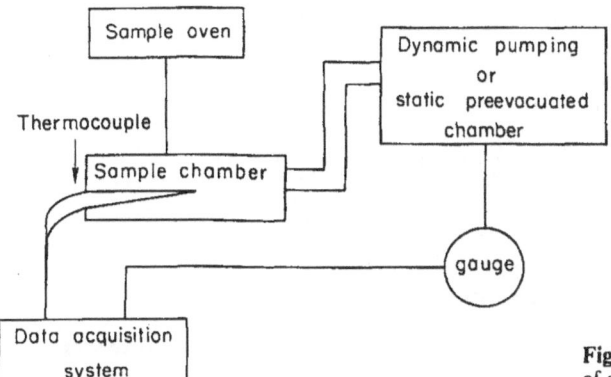

Fig. 7.6. Schematic block diagram of a TDS experiment

In some cases oxidation of the surface during the experiment may affect the experimental results [7.117]. In such cases ultra high vacuum systems have to be used. Usually the hydrogen loading is performed in the sample chamber. However, when ultra high vacuum systems are utilized an introduction chamber has to be used if high equilibrium pressures are involved [7.119].

The metal prior to hydrogenation is in the form of a powder or of very thin plates and is always covered with an oxide layer that inhibits hydrogenation. A large number of hydrogen desorption–absorption cycles are thus needed in order to enhance the hydrogen permeation into the metal. A criterion for the number of cycles to be performed is the achievement of reproducible desorption spectra.

7.3.3 Theoretical Considerations of the TDS Method

The main basic assumption that is discussed here is that the bulk is in thermal equilibrium with the surface [7.108, 112]. This occurs when the rate-limiting step is on the surface. In principle one may discuss the alternative conditions that bulk diffusion [7.116, 120] or a phase transformation is the rate-limiting step [7.121]. However, the solution of such a problem is very complicated, depending on shape and size with additional parameters to be introduced, and therefore it is difficult to interpret the results in a straightforward way [7.122]. We will briefly discuss these possibilities further below.

First, we discuss the case where the rate-limiting step is determined by the recombination of a hydrogen atom with a molecule at the hydride surface. Figure 7.7 shows a schematic potential energy diagram of hydrogen atoms in the bulk and on the surface. The bulk is presented by one or more potential

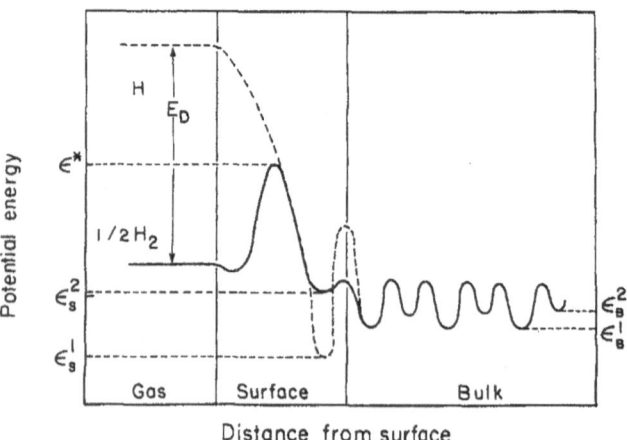

Distance from surface

Fig. 7.7. Schematic potential energy diagram of hydrogen atoms in bulk and surface sites

wells with ground state energies ε_B^i, $i = 1, 2, \ldots$, and the surface with surface sites ε_S^i. A simple model restricted to the α phase of the metal–hydrogen system, where the interactions between the hydrogen atoms are neglected, has been developed in detail by *Davenport* et al. [7.112]. The energies of hydrogen atoms in the bulk and on the surface sites are presented by a single ε_B and a single ε_S respectively, with the surface sites being lower in energy. Using the lattice gas model and assuming a second-order process, these authors solved the kinetic equations, considering the fluxes from the surface sites into and out of the bulk and into and out of the gas phase. They obtained the second-order desorption rate equation [7.112]

$$-\frac{dx}{dT} = \left(\frac{K_0'}{\alpha N_l}\right) x^2 \exp\left(-\frac{2E_B}{kT}\right), \tag{7.9}$$

where the constant K_0'/N_l is determined by the experimental system defined in [7.112], α is the heating rate, T the temperature, and E_B the bulk solution energy.

A different approach to the problem has been used by *Stern* et al. to investigate TDS at high hydrogen concentrations [7.108, 115]. Here the bulk is presented by its chemical potential $\mu_b(x, T)$ in quasi equilibrium with the surface: $\mu_s = \mu_b = \mu$. Using the transition state theory, the second-order rate equation is given by [7.115]

$$-\frac{dx}{dT} = \frac{M^* v_0}{\alpha} \theta_p^2 \exp\left(-\frac{2(f^* - f_p)}{kT}\right), \tag{7.10}$$

where θ_p is the average occupancy of the surface sites, which is given by

$$\theta_p = \frac{\exp\left[(\mu - f_p)/kT\right]}{1 + \exp\left[(\mu - f_p)/kT\right]}, \tag{7.11}$$

M^* is the number of active sites on the surface, v_0 is the jump frequency, $2f^*$ is the free energy of a H_2 molecule in the transition state (with one degree of freedom omitted) and f_p is the free energy related to the precursor site given by $f_p = -kT \ln \sum_v \exp(-E_p^v/kT)$, E_p^v being the energy levels of the precursor well. Substituting (7.11) into (7.10) and considering the limiting dilute precursor case when $\theta_p \ll 1$ (which is obtained when $f_p - \mu \gtrsim 4kT$) yields

$$-\frac{dx}{dT} = \left(\frac{M^* v_0}{\alpha}\right) \exp\left(-\frac{2(f^* - \mu)}{kT}\right). \tag{7.12}$$

The desorption rate depends then on the hydrogen concentration via $\mu(x, T)$. In the other extreme case, when $\mu - f_p \gtrsim 4kT$, the saturated precursor site condition $\theta \approx 1$ occurs and hence

$$-\frac{dx}{dT} = \left(\frac{M^* v_0}{\alpha}\right) \exp\left(-\frac{2(f^* - f_p)}{kT}\right). \qquad (7.13)$$

The desorption rate is then independent of μ and therefore of x. For noninteracting hydrogen atoms (7.12) leads to a second-order expression similar to that in (7.9). *Malinowski* [7.114] has obtained an explicit formula from the principle of detailed balance of the hydride in equilibrium with the gas phase where the desorption term is equal to the absorption term. In TDS the external pressure is zero and therefore only the desorption rate term is present and can be calculated from the corresponding absorption term:

$$-\frac{dx}{dT} = \frac{1}{4}\left(\frac{8N_A}{\pi M}\right)^{1/2} \sigma(T, x, P) T^{1/2} P_{eq}(T, x). \qquad (7.14)$$

$P_{eq}(T, x)$ is the hydride equilibrium pressure at temperature T and concentration x, σ the sticking coefficient of hydrogen, M the molecular weight of the gas molecule and N_A Avogadro's number. $P_{eq}(T, x)$ may be obtained from the phase diagram, however, $\sigma(T, x, P)$ is usually not known and its values are derived from the TDS experiment and may be compared to theoretical models. By presenting $\sigma(T, x, P)$ as a variable, this approach obviates the model-dependent presentation of σ utilized implicitly in the treatments discussed previously.

7.3.4 Experimental Results and Analysis

TDS experiments have been reported for some elemental metal hydrides [7.111, 114, 115] and for some intermetallic hydrides [7.108, 113, 115]. As discussed above, the interpretation of TDS data requires that the rate-limiting step is on the surface. Hence, at any given time it is assumed that the maximum variation in hydrogen concentration in the bulk across the reacting particle, Δx, is small compared to $x(T)$. An expression for Δx is [7.115]

$$\Delta x(T) = \alpha \frac{dx}{dT} \tau_D(T), \qquad (7.15)$$

where $\tau_D = AR^2/D(T)$ and $D(T)$ is the diffusion constant, R being the smallest dimension of the particle and A a constant depending on the particle shape (sphere, a plate, etc.).

Figure 7.8 illustrates the effect of the sample size and shape on the TDS of Pd samples in the form of a wire, a thin plate and fine powder [7.109]. In the wire the condition $\Delta x/x \ll 1$ is not satisfied, thus the rate-limiting step is not on the surface. On the other hand, in the plate and powder the assumption of the surface rate-limiting step is valid. Yet, the two latter spectra are different,

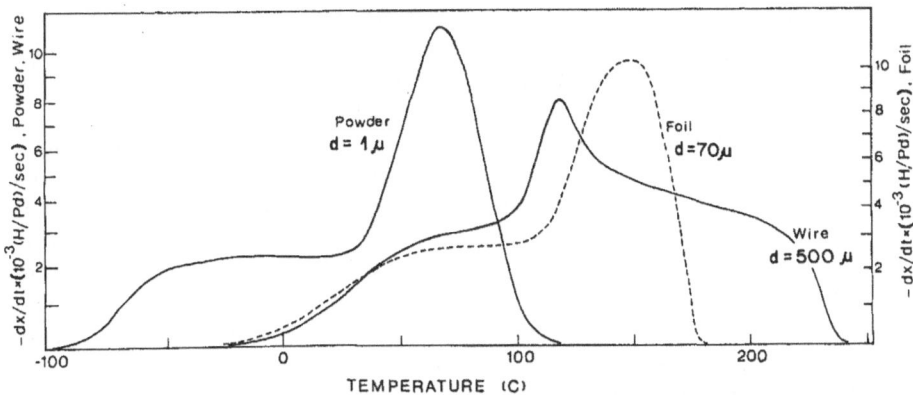

Fig. 7.8. TDS for Pd in the form of a wire, plate or powder. The numbers indicate the smallest dimension in micrometers. (From [7.115])

due to the different number of surface sites on the two samples (being much smaller for the plate [7.115]). In fact, when one of these spectra is shifted along the T axis it approximately coincides with the other. The shape of the surface-controlled TDS spectra may be different for different systems as demonstrated for Pd [7.115] and for ZrV_2 [7.113].

Figure 7.9a shows the TDS spectra of Pd for various initial hydrogen concentrations x_0 [7.115]. Their evolution as x_0 increases reflects the corresponding phase diagram characteristics. At $x_0 < 0.6$ only one peak is observed with a small shoulder at the high temperature side. This peak corresponds to the $(\alpha + \beta)$ region, while the small shoulder corresponds to the

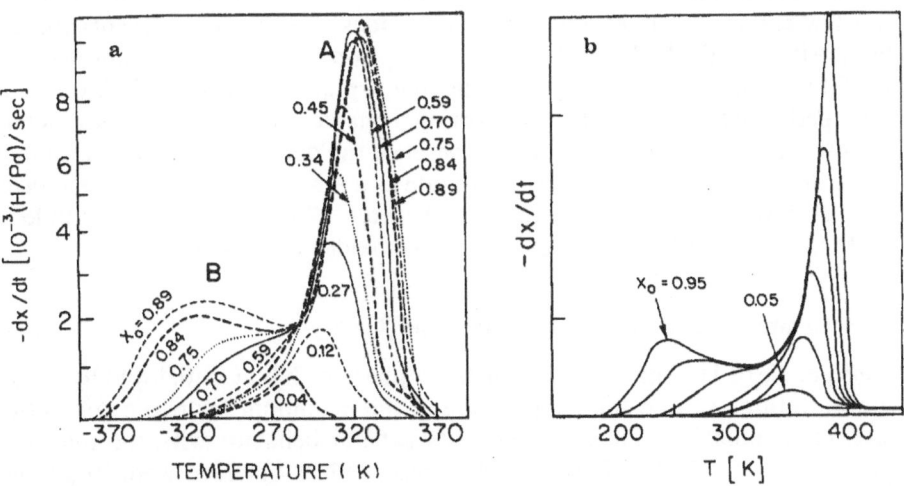

Fig. 7.9. (a) Experimental TDS spectra of Pd for various initial concentrations x_0 [7.115]. **(b)** Calculated TDS [7.92]

α phase. The β phase is reflected for $x_0 > 0.6$ by the shoulder in the lower temperature region that develops into another peak for increasing x_0 values. Similar spectra for the Ti system at $x_0 = 0.6$ corresponding to the desorption in the γ and $\alpha + \gamma$ phase regions are shown in Fig. 7.10 [7.114].

As discussed in Sect. 7.3.5, the presence of different peaks in the TDS spectra is not necessarily the result of a multiphase structure. This is illustrated in Fig. 7.11a for ZrV_2. The phase diagram of ZrV_2 exhibits a single phase for $T > 300\,K$ up to hydrogen concentrations > 5.6 hydrogen atoms per formula unit; nevertheless, three peaks are observed for high hydrogen concentrations.

The effect of increasing the heating rate is to shift the maxima in the spectra towards higher temperatures (Fig. 7.10) (see also [7.115]). However, it does not affect the temperature where the onset of the desorption occurs.

In the region close to the onset of desorption, the desorption curves are of Arrhenius type [7.115]

$$I = I_0 \exp(-E/kT). \tag{7.16}$$

Hence a plot of $\ln I$ versus $1/T$ gives a straight line that enables the derivation of the corresponding energies and the pre-exponential factors.

Recently a different type of thermal desorption experiment has been introduced. In the experiments described above the basic assumption is that the number of surface sites does not vary with temperature. This can be justified for Pd, which does not oxidize readily. For some intermetallic compounds it was shown [7.123] that usually one of the elements oxidizes much faster than the other, and therefore oxidation does not influencce the desorption of hydrogen. This is not the case for elemental hydrides such as Ti, V, Nb, and Ta, whose surfaces oxidizes readily. In fact, it is possible to store the hydrogen in such metals by exposing the corresponding hydrides to the atmosphere. The formation of a few surface layers of closely packed oxides [7.120] then prevents the decomposition of the hydrides. Thermal desorption experiments on thin plates of Ti [7.117] and Ta [7.124] hydrides having a predetermined oxide layer have been performed recently. The desorption behavior in these cases is different from those of the clean surfaces. The onset of the desorption occurs at a much higher temperature and the desorption rate as a function of temperature reflects the dissolution of the oxide layer into the bulk [7.124]. It is thus possible to study these dissolution processes of surface oxides using TDS.

7.3.5 Comparison Between Experiment and Theory

Ghosh et al. [7.111] have shown for the α phase of Nb at initial low hydrogen concentrations x_0 that the temperature of the maximum hydrogen desorption rate for different x_0 values agrees in its general behavior with the theory of *Davenport* et al. [7.112]. However, the energy values obtained were 50% larger than those obtained experimentally by other methods. The origin of this discrepancy is not understood. Using a model calculated $\mu(x, T)$ reflected by

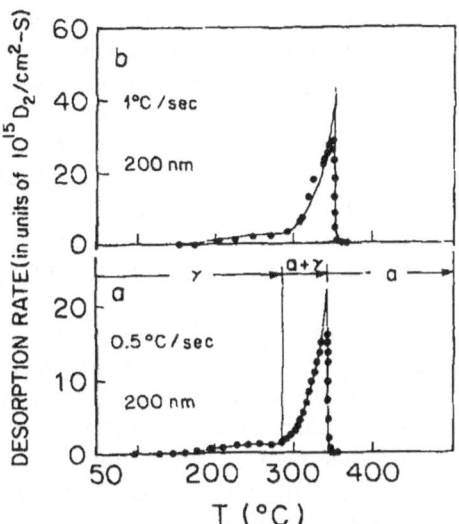

Fig. 7.10. Experimental (*dots*) and calculated (*full lines*) TDS spectra of Ti for an initial hydrogen concentration $x_0 = 1.6$ [7.114]

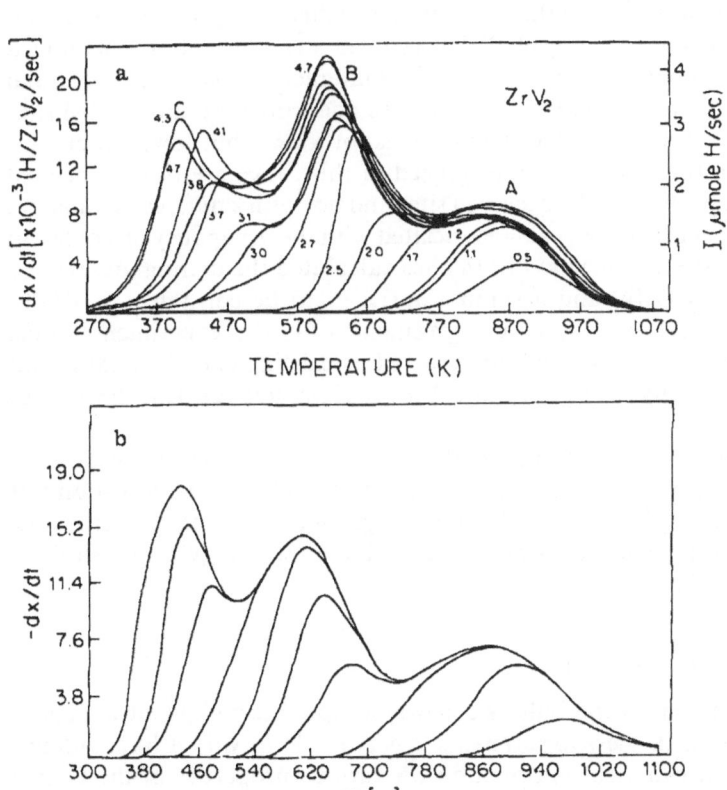

Fig. 7.11. (a) Experimental TDS spectra of ZrV_2 for various initial concentrations [7.113]. (b) Calculated TDS [7.103]

experimental pressure–composition isotherms of the Pd–H system, *Kreitzman* et al. [7.110] have solved (7.10) and have simulated the spectra shown in Fig. 7.9b. The calculated results given in Fig. 7.9b are in good qualitative agreement with the experimental results given in Fig. 7.9a: in the β phase region (line B) μ is strongly dependent on x and T, resulting in the onset of the desorption depending strongly on the initial concentration x_0. However, in the $\alpha + \beta$ phase region, where μ does not depend on x and is a much weaker function of T, the desorption curve does not depend on x_0 and the curves associated with the $\alpha + \beta$ region (line A) coincide on the rising side of the desorption for all x_0 values.

Kreitzman et al. have shown that calculated three-peak TDS patterns can be obtained by assuming that in the bulk three types of interstial site are occupied, with different potential energies (Fig. 7.11b). The results agree qualitatively with those obtained for ZrV_2 (Fig. 7.11a). The appearance of two peak patterns in Fig. 7.11a for $1 \leqq x \leqq 2.8$ seems to contradict neutron diffraction results [7.125] indicating that at this concentration range only a single interstitial site is occupied. However, it should be emphasized that different TDS peaks are no neccessarily associated with the occupation of different crystallographic sites but rather with different energy levels of the occupied sites. In the above-mentioned example it can be shown that above $x \approx 1$, H–H interactions reduce the potential energy of the site [7.126], thereby producing an additional set of occupied sites having the same crystallographic symmetry as for $x_0 < 1$ but lower in energy. The two peaks in the TDS spectra produced at this range are thus associated with the two energy levels of the occupied sites and not with a different symmetry. The third, lower, temperature peak is associated with the occupancy of a different crystallographic site. *Malinowski* [7.114] has calculated the thermal desorption of the $TiH_{1.6}$ using (7.14) and assuming $\sigma(T, x, P)$ to be constant ($\sigma = 0.009$). The calculated results are in good agreement with the experimental ones (Fig. 7.10). This agreement is quite unexpected as existing experimental results indicate that σ actually changes from 0.2 for a clean unhydrided film to 0.02 for a hydrided film.

As follows from (7.9) or (7.12), at the temperature region of the onset of desorption, where x is practically temperature independent, the desorption rate has an exponential temperature behavior in agreement with the experimental form given in (7.16). Hence it is possible to evaluate the energy E in a straightforward way [7.115].

5.3.6 Possible Future Research

The TDS of hydrogen in the bulk is a newly adopted technique at a stage of initial development. In these experiments it is possible to vary a large number of parameters that can yield information on various properties of the system, such as the equilibrium thermodynamics, the energetics of hydrogen in the bulk and on the surface, and kinetic behavior. So far, only a few elemental hydrides and intermetallic compounds have been investigated. The study of a larger number of hydrides by the TDS technique may provide a systematic way for evaluating important physical parameters by this method.

Surface treatments may change the TDS spectra, therefore it can be used to study surface effects (in particular surface poisoning by gas phase impurities). This is a very important problem related to the use of hydrides for some technological applications.

The case where the kinetics is determined at the hydride's surface is the simplest one. When the kinetics is either determined by diffusion of hydrogen atoms in the bulk, or by a phase transformation, the processes are more complicated. We refer only briefly to these two cases.

TDS studies of $LaNi_5H_x$ have shown us that the hydrogen release from the β phase (i.e., for $x > 6$) is controlled by diffusion of hydrogen atoms through the bulk [7.98]. A convenient way to calculate the activation energies of desorption is by applying a method published by *Farrell* and *Carter* [7.127]. They found that under certain conditions the activation energy of desorption can be calculated without knowing the hydrogen distribution inside the metal. We have found that the case of the $LaNi_5$ hydride [7.116], at the very beginning of the desorption process, and at low temperatures (~ 150 K), the conditions of *Farrell* and *Carter* are fulfilled. This enabled us to calculate the activation energies at temperature ranges which usually hinder other forms of diffusion measurements.

The case where a phase transformation is the rate-controlling step of the desorption was recently discussed by *Han* et al. [7.121]. By applying the TDS method it has been found that this rate-controlling step may have a role in determining the desorption kinetics of hydrogen from $LaNi_5H_x$, for $x < 6$ [7.116], as well as in the $ZrMn_2$ hydride.

Finally, this approach may be used in thermal absorption experiments, where the absorption rate and not the desorption rate is measured. Such a technique has an inherent difficulty as, being an exothermic process, the linear increase of temperature may get out of control due to heat released during absorption. Nevertheless, temperature-programmed absorption has already been used to investigate the absorption of hydrogen at low pressures [7.128].

Acknowledgements. We are indebted to all the authors who allowed us to use the previously published results and figures that are presented in this chapter. In particular, we would like to thank the following colleaguess for collaboration: MHM is grateful to J.W. Rabalais and J.A. Schultz from the University of Houston, Texas, who constructed the experimental setup described in Sect. 7.1, for their collaboration and kind hospitality during a sabbatical leave in their laboratory; IJ would like to thank R. Moreh, O. Shahal and A. Wolf who used the γ-ray scattering method extensively and participated with some of the authors in the application of that technique to hydride research; DS is indebted to A. Stern, A. Resnik and S.R. Kreitzman for collaboration in the TDS work.

References

7.1   S.A. Steward, R.M. Alire: "Applicability of Surface Compositional Analysis Techniques for the Study of the Kinetics of Hydride Formation", in *Transition Metal Hydrides*, ed. by R. Bau (American Chemical Society, Washington DC 1978), Chap. 26, pp. 382–397
7.2   L. Schlapbach, A. Seiler, F. Stucki, H. Siegman: J. Less-Common Met. **73**, 145 (1980)
7.3   E.H. Van Deventer, T.A. Renner, R.H. Pelto, V.A. Maroni: J. Nucl. Mater. **64**, 241 (1977)
7.4   H.D. Rohrig, R. Hecker: Thin Solid Films **45**, 247 (1977)
7.5   L. Schlapbach, T. Riesterer: Appl. Phys. A **32**, 169 (1983)
7.6   G.C. Bond: *Catalysis by Metals* (Academic, New York 1962)
7.7   K. Soga, H. Imamura, S. Ikeda: J. Phys. Chem. **81**, 1762 (1977)
7.8   K. Soga, K. Otsuka, M. Sato, T. Sano, S. Ikeda: J. Less-Common Met. **71**, 259 (1980)
7.9   G. Sicking, P. Albers, E. Magomedbekov: J. Less-Common Met. **89**, 373 (1983)
7.10   For a review see: H.W. Werner, R.P.H. Garten: Rep. Prog. Phys. **47**, 221 (1984)
7.11   C.W. Magee: Nucl. Instrum. Methods **191**, 297 (1981)
7.12   R.A. Zuhr, J.B. Roberto, B.R. Appleton: Nucl. Sci. Appl. **1**, 617 (1984)
7.13   See, for example: H.W. Werner: Surf. Sci. **47**, 301 (1975)
7.14   J.A. Van den Berg, D.G. Armour: Vacuum **31**, 259 (1981)
7.15   P.J. Schneider, W. Eckstein, H. Verbeek: Nucl. Instrum. Methods Phys. Res. **218**, 713 (1983)
7.16   H.H. Brongersma, T.M. Buck: Nucl. Instrum. Methods **149**, 569 (1978)
7.17   M.H. Mintz, J.A. Schultz: J. Less-Common Met. **103**, 349 (1984)
7.18   B.J.J. Koeleman, S.T. de Zwart, A.L. Boers, B. Poelsema, L.K. Verheij: Phys. Rev. Lett. **56**, 1152 (1986)
7.19   J.W. Rabalais: CRC Crit. Rev. Solid State Mater. Sci. **14**, 319 (1988)
7.20   J.A. Schultz, R. Kumer, J.W. Rabalais: Chem. Phys. Lett. **100**, 214 (1983)
7.21   J.W. Rabalias, J.A. Schultz, R. Kumar, P.T Murray: J. Chem. Phys. **78**, 5250 (1983)
7.22   J.A. Schultz, M.H. Mintz, T.R. Schuler, J.W. Rabalais: Surf. Sci. **146**, 438 (1984)
7.23   O. Grizzi, M. Shi, H. Bu, J.W. Rabalais: Rev. Sci. Instrum. **61**, 740 (1990)
7.24   W. Heiland, W. Englert, E. Taglauer: J. Vac. Sci. Technol. **15**, 419 (1978)
7.25   D.G. Armour, J.A. Van den Berg, L.K. Verheij: J. Radioanal. Chem. **48**, 359 (1979)
7.26   A.J. Algra, S.B. Luitjens, H. Borggreve, E.P. Th. M. Suurmeijer, A.L. Boers: Radiat Eff. **62**, 7 (1982)
7.27   A.J. Algra, S.B. Luitjens, E.P. Th. M. Suurmeijer, A.L. Boers: Nucl. Instrum. Methods **203**, 515 (1982)
7.28   R.S. Williams: J. Vac. Sci. Technol. **20**, 770 (1982)
7.29   W. Heiland: Appl. Surf. Sci. **13**, 282 (1982); Vacuum **32**, 539 (1982)
7.30   M. Aono, Y. Hou, R. Souda, C. Oshima, S. Otani, Y. Ishizawa, K. Matzuda, R. Shimizu: J. Appl. Phys. **21**, L670 (1982)
7.31   R.P.N. Bronckers, A.G.J. DeWit: Surf. Sci. **104**, 384 (1981)
7.32   T.M. Buck, G.H. Wheatley, D.P. Jackson: Nucl. Instrum. Methods Phys. Res. **218**, 257 (1983)
7.33   O.S. Oen: Surf. Sci. **131**, L407 (1983)
7.34   H. Niehus: Nucl. Instrum. Methods Phys. Res. **218**, 230 (1983)
7.35   Y. Yamamura, W. Takeuchi: Phys. Lett. A **94**, 109 (1983);
       N. Winograd, B.J. Garrison, D.E. Harrison, Jr.: Phys. Rev. Lett. **41**, 1120 (1978)
7.36   See, for example: D.P. Woodruff: Nucl. Instrum. Methods **194**, 639 (1982)
7.37   B.J.J. Koeleman, S.T. deZwart, A.L. Boers, B. Poelsema, L.K. Verhey: Nucl. Instrum. Methods Phys. Res. **218**, 225 (1983)
7.38   O. Grizzi, M. Shi, H. Bu, J.W. Rabalais, R.R. Rye, P. Nordlander: Phys. Rev. Lett. **63**, 1408 (1989); M. Shi, O. Grizzi, H. Bu, J.W. Rabalais, R.R. Rye, P. Nordlander: Phys. Rev. B **40**, 10163 (1989)
7.39   M.J. Puska, R.M. Nieminen, M. Manninen, B. Chakraborty, S. Holloway, J.K. Nørskov: Phys. Rev. Lett. **51**, 1081 (1983)
7.40   K.W. Jacobsen, J.K. Nørskov: Phys. Rev. Lett. **59**, 2764 (1987)
7.41   M.H. Mintz, U. Atzmony, N. Shamir: Phys. Rev. Lett. **59**, 90 (1987); Surf. Sci. **185**, 413 (1987)
7.42   M.H. Mintz, U. Atzmony, J.A. Schultz, N. Shamir: J. Vac. Sci. Technol. A **5**, 1136 (1987)
7.43   M.H. Mintz, J.A. Schultz, J.W. Rabalais: Surf. Sci. **146**, 457 (1984)
7.44   H. Hjelmberg: Surf. Sci. **81**, 539 (1979)

7.45   M.H. Mintz, J.A. Schultz, J.W. Rabalias: Phys. Rev. Lett. **51**, 1676 (1983)
7.46   E.E. Latta, H.P. Bonzel: "Surface Segregation and Gas Adsorption" in *Interfacial Segregation*, ed. by W.C. Johnson, J.M. Blakely (American Society for Metals, Metal Park, OH 1979) pp. 381–404
7.47   E. Swissa, J. Bloch, U. Atzmony, M.H. Mintz: Surf. Sci. **214**, 323 (1989)
7.48   T.R. Schuler, J.A. Schultz, N. Shamir, J.W. Rabalais: Private communication (1984)
7.49   J.A. Schultz, J.W. Rabalais: Chem. Phys. Lett. **108**, 328 (1984); J.A. Schultz, S. Contarini, Yang-Sun Jo, J.W. Rabalais: Surf. Sci. **154**, 315 (1985)
7.50   R. Kumar, M.H. Mintz, J.W. Rabalais: Surf. Sci. **147**, 37 (1984)
7.51   R. Kumar, M.H. Mintz, J.A. Schultz, J.W. Rabalais: Surf. Sci. **130**, L311 (1983); R. Kumar, M.H. Mintz, J.W. Rabalais: Surf. Sci. **147**, 15 (1984)
7.52   J.A. Schultz, Y.S. Jo, J.W. Rabalais: Solid State Commun. **55**, 957 (1985); J.A. Schultz, Y.S. Jo, S. Tachi, J.W. Rabalais: Nucl. Instrum. Methods Phys. Res. B **15**, 135 (1986); J.N. Chen, J.W. Rabalais: Nucl. Instrum. Methods. Phys. Res. B **13**, 597 (1986)
7.53   N. Shamir, U. Atzmony, M.H. Mintz: J. Vac. Sci. Technol. A **5**, 1024 (1987)
7.54   E. Swissa, I. Jacob, U. Atzmony, N. Shamir, M.H. Mintz: Surf. Sci. **223**, 607 (1989)
7.55   I. Jacob, A. Wolf, M.H. Mintz: Solid State Commun. **40**, 877 (1981)
7.56   I. Jacob: J. Less-Common Met. **89**, 309 (1983)
7.57   K.A. Gschneidner Jr.: Solid State Phys. **16**, 348 (1964)
7.58   M. Blackman: "Specific Heat of Solids", in *Encyclopedia of Physics*, ed. by S. Flugge, Vol. 7, Part 1, Crystal Physics I (Springer, Berlin, Heidelberg 1955) pp. 325–382
7.59   R. Moreh, O. Shahal, I. Jacob: Nucl. Phys. A **228**, 77 (1974)
7.60   R. Moreh: Nucl. Instrum. Methods **166**, 45 (1979)
7.61   H. Bilz, W. Kress: *Phonon Dispersion Relations in Insulators*, Springer Ser. Solid-State Sci., Vol. 10 (Springer, Berlin, Heidelberg 1979)
7.62   H. Schober, P.H. Dederichs: "Phonon Dispersion, Frequency Spectra and Related Properties of Metallic Elements" in Landolt-Börnstein, New Series, Group 3, Vol. 13a, *Phonon States of Elements; Electron States and Fermi Surfaces of Alloys*, ed. by K.H. Hellwege, J.L. Olsen (Springer, Berlin, Heidelberg 1981) Chap. 1, pp. 1–191
7.63   W. Kress: "Phonon Dispersion, One-Phonon Density of States and Impurity Vibrations in Metallic Compounds and Disordered Alloys", in Landolt-Börnstein, New Series, Group 3, Vol. 13b, *Phonon States of Alloys; Electron States and Fermi Surfaces of Strained Elements*, ed. by K.H. Hellwege, J.L. Olsen (Springer, Berlin, Heidelberg 1983) Chap. 4, pp. 259–405
7.64   T. Springer: "Investigation of Vibrations in Metal Hydrides by Neutron Spectroscopy" in *Hydrogen in Metals I*, ed. by G. Alefeld, J. Völkl, Topics Appl. Phys., Vol. 28 (Springer, Berlin, Heidelberg 1978) Chap. 4, pp. 75–100
7.65   C. Kittel: *Introduction to Solid State Physics*, 4th ed. (Wiley, New York 1971) p. 205
7.66   W.E. Lamb, Jr.: Phys. Rev. **55**, 190 (1939)
7.67   A. Wolf: "Properties of High and Low Energy Nuclear Levels Studied by the (γ, γ') Reaction"; Ph.D. Thesis, Weizmann Institute of Science, Rehovot (1973)
7.68   R. Moreh, O. Shahal, V. Volterra: Nucl. Phys. A **262**, 221 (1976)
7.69   R. Moreh, S. Shlomo, A. Wolf: Phys. Rev. C **2**, 1144 (1970)
7.70   I. Jacob, A. Wolf, O. Shahal, R. Moreh: Phys. Rev. B **33**, 5042 (1986)
7.71   L.S. Salter: Adv. Phys. **14**, 1 (1965)
7.72   B. Yates, M.J. Overy, O. Pirgon: Philos. Mag. **32**, 847 (1975)
7.73   A. Wolf, R. Moreh, O. Shahal: Nucl. Phys. A **227**, 373 (1974)
7.74   I. Jacob, M.H. Mintz, O. Shahal, A. Wolf: Phys. Lett. A **82**, 145 (1981)
7.75   I. Jacob, R. Moreh, O. Shahal, A. Wolf: Phys. Rev. B **35**, 8 (1987). See also O. Shahal, R. Moreh, A. Wolf, M.H. Mintz, I. Jacob: J. Less-Common Met. **103**, 401 (1984)
7.76   Y. Chung, T. Takeshita, O.D. McMasters, K.A. Gschneidner Jr.: J. Less-Common Met. **74**, 217 (1980)
7.77   F. Ducastelle, R. Caudron, P. Costa: J. de Phys. **31**, 57 (1970)
7.78   K. Bohmhammel, G. Wolf, G. Gross, H. Madge: J. Low Temp. Phys. **43**, 521 (1981)
7.79   A.C. Switendick: J. Less-Common Met. **49**, 283 (1976)
7.80   M. Gupta: Solid State Commun. **29**, 47 (1979)
7.81   N.I. Kulikov, N.V. Borzunov, A.D. Zvonkov: Phys. Status Solidi B **86**, 83 (1978)
7.82   J.H. Weaver, D.J. Peterman, D.T. Peterson, A, Franciosi: Phys. Rev. B **23**, 1692 (1981)
7.83   R.C. Bowman, W.K. Rhim, Phys. Rev. B **24**, 2232 (1981)

316   *M.H. Mintz* et al.

7.84   I. Jacob, R. Moreh, O. Shahal, A. Wolf, G. Zamir: Phys. Rev. B**38**, 7806 (1988)
7.85   M.H. Mendelsohn, D.M. Gruen, A.E. Dwight: Nature **269**, 45 (1977); See also M.H. Mendelsohn, D.M. Gruen, A.E. Dwight: J. Less–Common Metals **63**, 193 (1979)
7.86   M.H. Mendelsohn, D.M. Gruen, A.E. Dwight: Adv. Chem. Ser. **173**, 279 (1979)
7.87   H.H. van Mal, K.H.J. Buschow, A.R. Miedema: J. Less-Common Metals **35**, 65 (1974)
7.88   I. Jacob, D. Shaltiel: Solid State Commun. **27**, 175 (1978);
       see also I. Jacob, D. Shaltiel: In *Hydrogen Energy Systems*, Vol. 4, ed. by T.N. Veziroglu, W. Seifritz (Pergamon, Oxford 1979) p. 1689
7.89   T. Takeshita, W.E. Wallace: J. Less-Common Met. **55**, 61 (1977)
7.90   I. Jacob, D. Shaltiel: Mater. Res. Bull. **13**, 1194 (1978)
7.91   O. Shahal: "Investigation of Atomic and Molecular Binding by Nuclear Resonance Scattering"; Ph.D. Thesis, Weizmann Institute of Science, Rehovot (1977)
7.92   R. Moreh, O. Shahal: Nucl. Phys. A **252**, 429 (1975)
7.93   O. Shahal, R. Moreh: Phys. Rev. Lett. **40**, 1714 (1978)
7.94   R. Moreh, O. Shahal: Phys. Rev. Lett. **43**, 1947 (1979)
7.95   R. Moreh, O. Shahal: Solid State Commun. **43**, 529 (1982)
7.96   J.F. Lynch, Ted B. Flanagan: J. Phys. Chem. **77**, 2628 (1975)
7.97   E. Wicke, H. Brodowsky: "Hydrogen in Palladium and Palladium Alloys" in *Hydrogen in Metals II*, ed. by G. Alefeld, J. Völk, Topics Appl. Phys., Vol. 29 (Springer, Berlin, Heidelberg 1978) pp. 134–151 and references therein
7.98   J. Bloch, M.H. Mintz: J. Less-Common Met. **81**, 301 (1981)
7.99   M. Ron, D. Gruen, M. Mendelsohn, I. Shaft: J. Less-Common Met. **74**, 445 (1980); P.D. Goodel, G.D. Sandrock, E.L. Huston: J. Less-Common Met. **73**, 135 (1980)
7.100 P.S. Rudman: J. Less-Common Met. **89**, 93 (1983)
7.101 F. Pourarian, V.K. Sinha, W.E. Wallace, H. Kevin Smith: J. Less-Common Met. **88**, 441 (1982)
7.102 J. Bloch, Z. Hadari, M.H. Mintz: J. Less-Common Met. **102**, 311 (1984); J. Bloch, F. Simca, M. Kroup, A. Stern, D. Shmariahu, M.H. Mintz, Z. Hadari: J. Less-Common Met. **103**, 163 (1984)
7.103 R.M. Cotts: "Nuclear Magnetic Resonance on Metal–Hydrogen Systems" in *Hydrogen in Metals I*, ed. by G. Alefeld, J. Völkl, Topics Appl. Phys., Vol. 28 (Springer, Berlin, Heidelberg 1978)
7.104 K.W. Kehr: "Theory of the Diffusion of Hydrogen in Metals" in *Hydrogen in Metals I*, ed. by G. Alefeld, J. Völkl, Topics Appl., Vol. 28 (Springer, Berlin, Heidelberg 1978)
7.105 D. Menzel: "Thermal Desorption" in *Chemistry and Physics of Solid Surfaces IV V*, ed. by R. Vanselow, R. Howe, Springer Ser. Chem. Phys., Vol. 20 (Springer, Berlin, Heidelberg 1982)
7.106 R. Burch: Chemical Physics of Solids and their Surfaces **8**, 1 (1980)
7.107 M.H. Mendelsohn, D.M. Gruen: Mater. Res. Bull. **16**, 1027 (1981)
7.108 A. Stern, S.R. Kreitzman, A. Resnik, D. Shaltiel, V. Zevin: Solid State Commun. **40**, 837 (1981)
7.109 A. Stern, A. Resnik, D. Shaltiel, S.R. Kreitzman: "Thermal Desorption Spectra of PdH$_x$ System $0 < x < 0.9$ in Different Samples; A Powder, a Foil and a Wire", in *Electronic Structure and Properties of Metal Hydrides*, ed. by P. Jena, C.B. Satterthwaite (Plenum, London 1983) p. 55
7.110 S.R. Kreitzman, A. Stern, D. Shaltiel: "Fundamental Aspects of Thermodesorption Spectra in Bulk Metal–Hydride Systems" in *Electronic Structure and Properties of Metal Hydrides*, ed. by P. Jena, C.B. Satterthwaite (Plenum, London 1983) p. 61
7.111 V.J. Ghosh, M.A. Pick, D.O. Welch, G.J. Dienes: "Thermal Desorption Spectroscopy (TDS) of Hydrogen in Niobium" in *Electronic Structure and Properties of Metal Hydrides*, ed. by P. Jena, C.B. Satterthwaite (Plenum, London 1983) p. 67
7.112 J.W. Davenport, G.J. Dienes, R.A. Johnson: Phys. Rev. B**25**, 2165 (1982)
7.113 A.Stern, A. Resnik, D. Shaltiel: J. Less-Common Met. **88**, 431 (1982)
7.114 M.E. Malinowski: J. Less-Common Met. **89**, 27 (1983)
7.115 A. Stern, A. Resnik, D. Shaltiel: J. Phys. F **14**, 1625 (1984)
7.116 A. Resnik, M. Stioui, A. Grayevsky, D. Shaltiel: J. Less-Common Met. **131**, 117 (1987)
7.117 M.E. Malinowski: J. Nucl. Mater. **85/86**, 957 (1979)
7.118 M.A. Pick: Phys. Rev. B**24**, 4287 (1981)
7.119 M.E. Malinowski: J. Vac. Sci. Technol. **15**, 398 (1978)
7.120 T. Schober, A. Grayevsky, P.S. Rudman: Scr. Metall. **14**, 101 (1980)
7.121 J.S. Han, M. Pezat, J.Y. Lee: J. Less-Common Met. **130**, 395 (1987)

7.122 S.E. Donnely, D.C. Ingram, R.B. Webb, P.G. Armour: Vacuum **29**, 303 (1979)
7.123 L. Schlapbach, A. Seiler, F. Stucki, H.C. Siegman: J. Less-Common Met. **73**, 145 (1980)
7.124 E. Pörschke, D. Shaltiel, H. Wenzl: J. Phys. Chem. Solids **47**, 1003 (1986)
7.125 J.J. Didisheim, K. Yvon, P. Fischer, D. Shaltiel: J. Less-Common Met. **73**, 355 (1980)
7.126 I. Jacob, D. Shaltiel: J. Less-Common Met. **65**, 117 (1979)
7.127 F. Farrell, G. Carter: Vacuum **17**, 15 (1966)
7.128 A. Resnik, A. Stern, A. Moran, D. Shaltiel: J. Less-Common Met. **103**, 173 (1984)

List of Tables

Subject Index